WIND ENERGY ESSENTIALS

WIND ENERGY ESSENTIALS

Societal, Economic, and Environmental Impacts

RICHARD P. WALKER
ANDREW SWIFT

Published by John Wiley & Sons, Inc., Hoboken, New Jersey
Published simultaneously in Canada

For general information on our other products and services or for technical support, please contact our Customer Care Department within the United States at (800) 762-2974, outside the United States at (317) 572-3993 or fax (317) 572-4002.

Wiley also publishes its books in a variety of electronic formats. Some content that appears in print may not be available in electronic formats. For more information about Wiley products, visit our web site at www.wiley.com.

Library of Congress Cataloging-in-Publication Data:

Walker, Richard P., 1957–
Wind energy essentials : societal, economic, and environmental impacts / Richard P. Walker, Andrew Swift.
pages cm
Includes bibliographical references and index.
ISBN 978-1-118-87789-0 (cloth : alk. paper) 1. Wind power plants. 2. Wind energy.
I. Swift, A. H. P. (Andrew H. P.) II. Title.
TK1541.W346 2015
333.9′2–dc23

2015007886

Printed in the United States of America

10 9 8 7 6 5 4

CONTENTS

FOREWORD

This comprehensive work by Richard Walker and Andrew Swift occupies a unique position in the literature about wind energy. There are many books on wind energy—most, however, are technical in nature and serve as guides for designers. To my knowledge, there are none that span the broad range of environmental, financial, policy, and other topics that define and determine the relationships between wind energy technology and our energy-dependent society. As a consequence of their teaching and industrial experience, both authors are comfortable with the technology of wind energy and other renewables. In addition, the Texas Tech educational program in wind energy always has moved beyond technology to include the topics covered in this book. While the focus is on wind energy, much of the material in this book is applicable to other renewables as well. In that sense, this contribution bridges the technology of renewables with their societal impacts.

As may be seen from the Table of Contents, the authors have included just enough technology description to serve as a foundation for their discussions of environmental and societal issues. The technology descriptions should enable the reader to have a deeper insight and appreciation of the interdependencies between the capabilities of the technology and the impacts that go beyond the provision of electricity. For example, it is well known that the energy and economic productivity of the technology has improved by orders of magnitude since the early 1980's onset of significant installations in Europe and the United States. What is not as well known is that wind energy technology also has evolved significantly in response to societal concerns. These concerns have included impacts on wildlife, land use and permitting, safety and potential public health concerns, and interactions with the electrical grid. The authors deal with these as well as with many other relevant concerns.

The full title of the book gives a preview of the breadth of the subject. In particular, the subtitle *Societal, Economic, and Environmental Impacts* provides a first hint of the breadth of the treatment provided in this book by Walker and Swift. For those who wish to have a grounding in the technology and an objective, comprehensive treatment of the societal impacts, this book will provide the needed background and insight. It is a unique resource.

Time-Variable Systems, LLC
May 26, 2014

DR. JAMIE CHAPMAN

ABOUT THE AUTHORS

Richard P. Walker, PhD, PE

Dr. Richard Walker first joined the wind energy industry in 1994 and since that time has been a pioneer of wind development efforts in Texas beginning with the development of the first utility-scale wind project in Texas. His career has spanned over 30 years including working in the electric utility industry, wind project development, consulting, and teaching at Texas Tech University. He has been involved in the development of over 1600 MW of wind farms currently in operation, including the first wind development in Sweetwater, Texas area (the Trent Wind Farm), and the Roscoe Wind Farm, which was the world's largest wind energy project at 782 MW for several years. Dr. Walker is the President of Sustainable Energy Strategies, Inc., which provides consulting services to utilities, developers of renewable energy projects, and large landowners, and he is the President of RD Energy Group, LLC, an early-stage developer of renewable energy projects. He is a registered professional engineer in the state of Texas and holds BS (Civil Engineering) and MBA (Finance) degrees from Texas A&M University and a PhD in Wind Science and Engineering from Texas Tech University. While at Texas Tech, he developed and taught several of the university's courses in wind energy. In 2003, Dr. Walker was awarded the American Wind Energy Association's award for outstanding contribution to the wind industry. He has served on the board of directors for several organizations including the American Wind Energy Association, the Texas Renewable Energy Industries Association, the Utility Wind Interest Group, the Solar Electric Power Association, and the Central Electric Vehicle Consortium. Dr. Walker and his wife, Connie, have been married for 30 years and have three children.

Andrew Swift, ScD, PE

Dr. Andrew Swift is presently Professor of Civil and Environmental Engineering and Associate Director of the National Wind Institute at Texas Tech University—focused on Wind Energy Education and Workforce Development. His previous academic appointments include Director of the Texas Wind Energy Institute and the Wind Science and Engineering Research Center at Texas Tech and Dean of the College of Engineering at the University of Texas at El Paso. He completed his engineering graduate work obtaining a Doctor of Science degree at Washington University in St. Louis where he began conducting research in wind turbine systems engineering with a focus on the dynamics and aerodynamics of wind turbine rotors. Dr. Swift has worked in wind energy research and education for over 30 years; has over 100 published articles and book chapters in the area of wind turbine engineering and renewable energy; and in 1995 received the American Wind Energy Association's Academic Award for continuing contributions to wind energy technology as a teacher, a researcher, and an author. He is a native of Upstate New York, and his background includes military service as an Air Force pilot and flight instructor. His wife, Linda is a native of West Texas, and they have two daughters and three grandchildren.

DEDICATION AND ACKNOWLEDGMENTS

This book is dedicated to our wives, Linda Swift and Connie Walker, for their support over many years working in the uncertain field of renewable energy and tolerating the countless hours that we worked on this book, and also to our children and their spouses—Luke and Sarah Walker, Amy and Ben Marcotte, Molly Walker, Carolyn and Kevin Kuhnel, and Karen and Jeff Eagleston.

We also wish to acknowledge the contributions of Amy Marcotte, Kyle Jay, Chance Howe, Charles and Dorothy Norland, and Elizabeth Paulk for their help with proofreading, graphics design, photography, and research.

PREFACE

As a nation, the United States uses a lot of energy and in most years, energy use increases. While the efficiency with which the United States uses energy has increased over time, this improvement has not been able to offset overall growth in energy use. Electricity generation accounts for about 40% of the nation's energy use, with the vast majority of this being produced using either fossil fuels or nuclear energy. Renewable resources are used to meet less than 15% of US electricity needs as of the writing of this book, with hydroelectric generation accounting for most of that.

While installed wind generating capacity has grown rapidly in the United States over the last several years, its primary competition in the electricity market, natural gas–fired generation, has benefited from new extraction technologies such as horizontal drilling and fracking, resulting in low natural gas prices, as well as power plant efficiency improvements. Wind energy's growth has been spurred on by the Federal production tax credit (PTC), which has helped make wind very cost-competitive with natural gas generation. However, Congress has allowed the PTC to lapse several times since it was first enacted in 1992, causing severe disruptions in the US wind energy industry. On the other hand, public opposition and increasingly stringent environmental regulations make it very difficult to construct new coal-fired or nuclear power plants in the United States. Thus, most new electric generating capacity in the country has come from natural gas–fired power plants or wind power plants in recent years, although advances in solar energy technology resulting in higher efficiency and lower cost indicate that it will become a major contributor to the nation's power supply in coming years.

Globally, wind power's growth has been more stable than in the United States, with an average annual growth rate of over 25% since 1998. The demand for electricity in highly populated developing nations such as China, India, Indonesia,

Brazil, and Pakistan will place increasing demand on fossil fuels, further compounding environmental issues such as smog and global climate change. However, many of these nations are also looking to renewable energy as a means to meet their rapidly growing demand for electricity with their own domestic resources.

Wind energy will continue to grow rapidly, and although it is one of the cleanest and most environmentally neutral energy sources, wind projects can negatively affect wildlife habitat and individual species unless careful and thoughtful consideration is given to the potential impacts of these wind projects. This book is intended to familiarize the reader with wind technology, the economics of electricity generation, and energy policy. In addition, our hope is that it also helps to enable the wind energy industry to conduct itself in a socially responsible manner consistent with high standards for environmental stewardship, while helping to preserve the Earth's natural resources, including air quality, water, wildlife, and scenic areas.

DR. RICHARD P. WALKER
DR. ANDREW SWIFT

1

IMPACTS OF ENERGY AND ELECTRICITY ON SOCIETY

1.1 WHAT ARE "SOCIAL AND ENVIRONMENTAL IMPACTS"?

1.1.1 Interactions and Effects of Technology on Society and the Environment

We begin with two fundamental characteristics of human nature. First, humans develop and use technology, beginning with stone tools, the use of fire and heat, the plow, and agriculture—to modern times where we have developed electric utilities, computers, and cell phones. Second, humans are social beings and live in groups. Since the earliest times, these two elements of human development have been major contributors to modern civilized society. Technological developments used to the benefit of society usually provide a general improvement in the quality of life (QOL), to include security (such as defense against other people or animals; warfare activities; or natural phenomena such as earthquakes, floods, and windstorms). Other developments, such as politics, economics, philosophy, and education have also been key elements in this development, but our focus in this text is on the interactions of technology, society, and the environment with a particular emphasis on the impacts of wind energy development.

In addition to societal impacts, technology development often impacts the natural environment. The process of generating energy has very significant impacts on the natural environment. This began from the earliest cave dwellers harvesting wood to burn for warmth and light through today where modern society depends on fossil fuels to provide the majority of our energy needs. As will be discussed, the

Wind Energy Essentials: Societal, Economic, and Environmental Impacts, First Edition.
Richard P. Walker and Andrew Swift.

environmental impact of the production, distribution, and use of energy has significant impact on the natural environment, especially as the need for energy has grown with an expanding population.

1.1.2 Sustainable Development

Over the last several decades, the impact of rapid technological progress on the global environment, as well as growing populations, has heightened concerns about negative environmental effects and the growing demand for limited natural resources. These concerns have led to the concept of "sustainable development." The word "sustainability" is derived from the Latin word *sustinere* (*tenere*, to hold; *sus*, up). Dictionaries provide more than 10 meanings for *sustain*, the main ones being to "maintain," "support," or "endure." However, since the 1980s *sustainability* has been used more in the sense of human sustainability on Earth and this has resulted in the most widely quoted definition of sustainability and sustainable development—that of the Brundtland Commission of the United Nations on March 20, 1987:

> Sustainable development is development that meets the needs of the present without compromising the ability of future generations to meet their own needs. [1]

In other words, sustainable development minimizes the impact of resource use so that the needs of the present generation are met without diminishing the ability of future generations to meet their needs.

1.1.3 Wind Power, Technology, and Society

Our study will focus on one of the most basic elements of our planetary environment: the wind. This chapter will examine wind technologies developed over many centuries to harness the power of the wind for an improved QOL and how it has impacted society, both in centuries past and today. We will begin with a historical overview of wind power technological accomplishments, such as the age of discovery using sailing ships, the importance of wind power in providing transportation across the developing United States and settling the central Great Plains, to early electricity production using wind power. This will be followed by an overview of wind science and technology with an in-depth focus on modern global utility-scale wind power development for electrical power production.

1.2 EARLY WIND POWER INNOVATION AND DISCOVERY

1.2.1 Age of Sail Power

Using wind to power sailing vessels has had major impacts on society throughout the history of civilization. Sailing vessels have allowed humans greater mobility for thousands of years and have increased the capacity for fishing, trade, commerce, transport, naval defense, and warfare. The earliest image of a ship under sail was painted on a disk found in the Middle East dating to the fifth millennium BC. In the sixth century, development of the Lateen Rig in Arabia, shown in Figure 1.1, allowed vessels to travel in an upwind direction. Sails used previously could only develop a motive force moving with the wind direction (downwind) and required oarsmen to travel in an upwind direction. This was a major innovation since the vessel could now travel in all directions solely with the power of the wind.

Sailing ships became considerably larger over the centuries, as well as more seaworthy, with improved techniques for harnessing the wind. These advances along with improved navigational techniques allowed sailors to travel the seas worldwide. Sailors learned to use global wind patterns to reduce the time of long trips connecting distant societies in ways that were previously not possible.

One of the periods of most significant change and impact occurred during what is often referred to as the "Age of Discovery" from the fifteenth through the seventeenth century. During this time, such familiar names as Columbus and Magellan set out on famous and historical sailing journeys. Columbus discovered the "New World," while Magellan was the first to lead an expedition that circumnavigated the globe. Leaving Portugal in 1519 with five ships, his fleet returned to Spain in 1522 led by Juan Sebastian, due to the death of Magellan in the Philippines during the 3-year voyage.

The golden age of sail, however, is usually considered to be during the nineteenth century, when sailing vessels had become quite large and the efficiency of long-distance sailing was at its peak. Trade during the golden age was dominated by huge numbers of sailing vessels, following routes defined by the "Trade Winds" and navigating to all parts of the globe providing trade, commerce, and immigration of large numbers of people—changing societies and cultures around the world. This was also a time when the most powerful nations on Earth had large naval fleets of sailing vessels, not only for their own sovereign protection but also to protect shipping lanes and spread power and influence throughout the world. The British Empire, for example, depended heavily on its strong navy of sailing warships to build, to expand, and to protect its empire during this period in the nineteenth century. Figure 1.2 shows the USS Constitution. Named in 1797 by President George Washington, it was one of the ships commissioned for the newly formed U.S. Navy. The ship is best known for her actions in the War of 1812 where she earned the name "Old Ironsides."

In addition to trade, immigration, and national defense, nineteenth century commercial sailing vessels harvested the seas for food and commodities—the most well-known being the whaling fleet. Whaling ships, like the Charles W. Morgan shown in Figure 1.3, would embark on multiyear journeys to hunt whales. In the nineteenth century, whales were abundant and were harvested for the high-quality oil they contained, as it was a valued commodity due to the clean-burning light provided by a whale oil lamp.

Without electricity, candles and lamps provided the only light. Whaling ships could hold in the order of 2000 barrels of oil, valued between $200 and $1500 per barrel (2003 US dollars). Voyages would last until the ship's hold was full, sometimes up to 5 years. Driven by the high value of the whale oil and ever-improving

(a)

(b)

FIGURE 1.1 A Dhow sailing vessel with Lateen rigged sails (a) and one of the most popular recreational sailboats, the Sunfish (b), which use the same ancient sail design. This was the first sail design that allowed sailboats to tack (go back and forth at an angle) allowing travel upwind. Modern wind turbines are driven by similar crosswind (lift) forces. See more about lift forces in Figure 4.5 (Photo Credit—upper photo: Xavier Romero-Frias, http://en.wikipedia.org/wiki/File:Sd2-baggala.JPG; lower photo: Dierde Santos, http://en.wikipedia.org/wiki/File:SunfishRacing.jpg).

FIGURE 1.2 The restored USS Constitution under sail, a warship of the first U.S. Navy (Source: Photo Courtesy of U.S. Navy).

FIGURE 1.3 Charles W. Morgan Whaling Ship, Mystic Seaport, CT (Photo Credit: Mystic Seaport, http://en.wikipedia.org/wiki/File:Charles_W_Morgan.jpg).

sailing vessels, the industry flourished in the nineteenth century, driving the whale population to near extinction. The discovery of petroleum products, in particular kerosene, led to the replacement of whale oil and the decline of the industry. None of these aspects of world history would have been possible without the use of wind-driven ships.

1.2.2 Wind Power and the Transcontinental Railroad

Late in the eighteenth century, the steam locomotive was invented nearly simultaneously in England and in the United States. England, however, was the location of the development of the first railway system, built at the turn of the nineteenth century. The locomotives used steam produced with a water boiler and firebox, usually fueled with wood or coal, to provide the heat needed to create steam used to drive the large steam-pistons that powered the locomotive. For more than 150 years, the railroad dominated freight and passenger land transportation, as sail power dominated sea transportation. Prior to the nineteenth century, most development in the United States was east of the Mississippi river and along the West Coast. Both of these areas had plentiful quantities of wood, coal, and water to provide the fuel and steam to power large-scale railroad networks with steam locomotives.

By 1850, a network of rail lines had connected most parts of the eastern half of the United States and coastal areas along the West Coast, but there was no effective way to connect the coasts and to cross what was then called the "Great American Desert," now known as the Great Plains. The wagon trains of the early 1800s and Pony Express riders carrying the mail were not the solution a growing nation needed. In order to unify the nation, it was important to connect the eastern and western portions of the United States with a means of bulk transportation that was efficient in both time and cost and could move people and goods effectively and rapidly.

Throughout the decades of the 1840s and 1850s, there was significant interest to build a rail line to connect the east and west portions of the nation. The task was substantial and of a magnitude that required government support. Construction was finally authorized by the Pacific Railroad Act of 1862 and 1864—at the same time the American Civil War was being waged. It was funded with 30-year US bonds and extensive grants of government-owned land to the railroad companies to build the line (Fig. 1.4).

The final link of the "Transcontinental Railroad," as it was called, was a route from the twin cities of Council Bluffs, Iowa, and Omaha, Nebraska, in the central part of the United States—via Ogden, Utah, and Sacramento, California, ending at the Pacific Ocean in Oakland, California. The coast-to-coast rail line was popularly known as the Overland Route and continued passenger rail service until 1962—almost 100 years. The Overland Route's final link was built by the Central Pacific Railroad of California from the west and the Union Pacific Railroad from the east between the years of 1863 and 1869 when the last spike was driven at Promontory Summit, Utah on May 10, 1869. That final spike completed the Overland Route, establishing a rail link for transcontinental transportation that not

FIGURE 1.4 Photograph of the driving of the Golden Spike, Promontory, Utah, 1867 (Photo Credit: Andrew J. Russell, http://en.wikipedia.org/wiki/File:1869-Golden_Spike.jpg).

only united the country from coast to coast but also opened the heartland for settlement and development. The Transcontinental Railroad is considered one of the greatest accomplishments of the nineteenth century, surpassing the building of the Erie Canal in the 1820s and crossing the Isthmus of Panama by the Panama Railroad in 1855.

But what is the connection between wind power and the transcontinental railroad? In the continental United States, areas west of the Mississippi river receive much less rain than areas east of the Mississippi river. As mentioned earlier, in nineteenth century America, many people referred to the area as the "Great American Desert" due to the lack of rain and surface water, and since it was mostly grassland and prairie. Steam locomotives, however, required water to operate—large quantities of water. In fact, steam engines at the time required 100–200 gallons of water for each mile that they travelled. As a result, crossing the arid region of the Great Plains was a significant challenge to railroad planners as they looked for large sources of boiler feed water for the steam locomotives. Wind power offered the solution.

Driven by the geography of the Rocky Mountains to the west and large flat expanses across the plains to the Mississippi River, the Great Plains are well known for their almost constant winds that blow across the region. Water-pumping windmills would use the winds of the Great Plains to drive pumps and to lift abundant

FIGURE 1.5 Railroad depot water tower (Photo Credit: Wdiehl, http://en.wikipedia.org/wiki/File:487_at_water.jpg).

underground water to storage tanks providing the needed water for the steam locomotives (Fig. 1.5). Companies such as Eclipse developed large, wooden, multibladed wind pumpers and installed them along the rail lines (Fig. 1.6).

Thus, as the Transcontinental Railroad developed, a wind-powered water-pumping industry developed in the country as well—to meet the needs and large water appetites of the steam engine. It would later turn out that this same industry would play a key role in settlement of the central United States.

Throughout the development of the American west, the lack of water was a major problem for not only the Transcontinental Railroad but the rail feeder-lines developed to support transportation to larger towns throughout the region. Locomotives could only travel approximately 20 miles between water stops, leading to the development of large numbers of small towns along the rail lines. A number of words and phrases of the time survive within our language and society today. For example, the many small towns that were developed as water and fuel stops needed names, and of course nicknames. If the water stop had surface water available but no wind-pumping or gravity feed system, men with buckets would have to take surface water from streams and ponds by tying ropes to the buckets and hauling the water into tanks for the steam locomotives—usually about 2000 gallons for every 20-mile stop. This bucket and rope process was called "jerking." If a town required this method of moving water, it was called a "jerk-water" town. It was, of course, populated by people of the same name (Fig. 1.7).

FIGURE 1.6 Eclipse wind pumper (Courtesy of I.G. Holmes, U.S. National Park Service).

FIGURE 1.7 1923 Burlington Route Steam Engine with fuel/water tender; National Ranching Heritage Center, Lubbock, TX (Charles Norland, Norland Photographic Art, St. Louis, MO. Picture taken at the National Ranching Heritage Center in Lubbock, TX).

1.2.3 Wind Power for Settling the Great Plains

Along with the Transcontinental Railroad connecting the eastern and western parts of the United States came settlement of the Great Plains with cities such as Denver, Colorado; Cheyenne, Wyoming; Albuquerque, New Mexico; and many more. Most of the region, however, was settled by farmers and ranchers who had water needs well in excess of that available with the minimal rains and sparse surface waters that existed throughout the Great Plains. Wind and wind-driven water pumpers, developed and adapted for farm and ranch use, allowed this settlement to occur at a scale that could not have been achieved otherwise (Figs. 1.8 and 1.9).

One of the most successful farm and ranch wind-driven water-pumping units was developed by Daniel Halladay in 1854. This wind-driven pumping unit had steel blades and used a steel tower that replaced older wooden construction and became a common sight throughout the mid-western United States providing water for both farms and livestock.

Before electrical pumps replaced wind-driven water pumpers in the 1930s, there were an estimated 600,000 units in operation with a total rated capacity equivalent to 150 MW of power. With installation numbers of this magnitude, a robust industry was required. The need for designers, engineers, manufacturers, part suppliers, installers, sales, marketing, and repair windsmiths provided a significant number of jobs and opportunities for this industry.

Halladay provided many features that improved the wind-driven water pump making it a reliable and effective way to pump underground water and to provide water services in areas where surface water was not available in sufficient quantities. The mid-western United States could not have been settled and developed without wind power before the electrification of the Great Plains (providing electric-driven pumping) in the 1940s.

Stationary steam power available at the time was just too cumbersome and expensive and, of course, required water in order to drive the units themselves. The impact of these wind-powered water pumps on regional development, not only in the United States but in many other parts of the world, can probably not be overstated (Figs. 1.9, 1.10, and 1.11).

1.2.4 The Dutch Experience

The country of the Netherlands, also called Holland, is so closely associated with the history of windmills that it is often the first fact that people recall when they think about this country on the western edge of the European continent. In addition to windmills, Holland is known for being a world power during the Dutch Golden Age and the commercial dominance of the Dutch West Indies Company as well as in tulips, commerce, trade, and art.

The Netherlands is a low-lying coastal country where 25% of the land is below sea level. The country is extremely prone to flooding due to its geography and in 1287, one of the most destructive floods in recorded history, the Saint Lucia's

FIGURE 1.8 Windmill used for water pumping to a nineteenth century ranch home. Texas Tech University, National Ranching Heritage Center, Lubbock, TX (Charles Norland, Norland Photographic Art, St. Louis, MO. Picture taken at the National Ranching Heritage Center in Lubbock, TX).

FIGURE 1.9 The American Water-Pumping Windmill and storage tank, key to settling the Great Plains of the United States in the nineteenth century. Texas Tech University, National Ranching Heritage Center, Lubbock, TX (Charles Norland, Norland Photographic Art, St. Louis, MO. Picture taken at the National Ranching Heritage Center in Lubbock, TX).

FIGURE 1.10 Windmill repair, nineteenth century; Photo reproduction from wall mural at the American Wind Power Center, Lubbock, TX (Charles Norland, Norland Photographic Art, St. Louis, MO. Picture taken at the American Wind Power Museum in Lubbock, TX).

flood, killed more than 50,000 people. In the thirteenth century, windmills were introduced on a large scale to pump water from low-lying areas, called polders—the name for an area of reclaimed land, usually enclosed by dikes, to form an artificial lowland suitable for agriculture and other uses. Without continual pumping and drainage, however, this land can flood, destroying any development on the reclaimed land (Fig. 1.12).

To maintain these lowland areas and to keep them productive, the Dutch continued to innovate and improve windmill drainage and pumping technology and by the year 1850, there were more than 50,000 windmills operating in the Netherlands. Many of these windmills were for the purpose of drainage and flood control, as described earlier, but they were also used for milling grains and for sawing wood. Even today, although the drainage pumps are mainly powered by electric motors and engine-driven pumps, windmills are often used for backup

FIGURE 1.11 Nineteenth century western town painting, Photo reproduction from wall mural at the American Wind Power Center, Lubbock, TX (Charles Norland, Norland Photographic Art, St. Louis, MO. Picture taken at the American Wind Power Museum in Lubbock, TX).

FIGURE 1.12 Wind Mills in the Netherlands (Photo Credit: Michiel1972, http://en.wikipedia.org/wiki/File:Zandwijkse_Molen.jpg).

power to maintain the required draining of the polders. Engineers have estimated that a typical Dutch windmill operating in a reasonable breeze produced 50 kW of power; thus, in the mid-nineteenth century, with 50,000 windmills operating, the Dutch had installed approximately 2,500 MW of mechanical windmill capacity to power their country—all at a time before the steam engine, the internal combustion engine, and the electric motor.

The Dutch Golden Age is typically considered to be during the seventeenth century, when Amsterdam was the wealthiest trading city and hosted a full-time stock exchange. The country had more than 16,000 merchant sailing ships to conduct trade and commerce worldwide bringing enormous wealth to the country. The Dutch West Indies Company had developed worldwide trading partners and during this time, the population of the country increased from 1.5 to 2 million people. Dutch wealth created an atmosphere in the country that supported famous artists, such as Johannes Vermeer, whose most famous painting called *The Girl with A Pearl Earring* is shown in Figure 1.13 in the "Mural of Dutch History." The painting inspired a modern book and movie of the same name. Without wind power to drive its commercial fleet of sailing ships, provide domestic power for mills, and for flood protection for the country, Holland would have not achieved the Dutch Golden Age.

The development of wind power and hydropower innovated over the centuries by the Dutch not only provided the energy base to drain the large polders but also provided the power and energy for ship building, grinding grain, and other uses to support its robust economy. The social impact of wind power on Dutch civilization and development during this period is remarkable. Even today, leadership in wind technology continues in the Netherlands. Delft University is a European leader in wind energy development. The group "DU Wind" at Delft is a wind energy research and technology organization focused on research, development, and education. The country has also hosted the installation of modern wind turbines with a rated capacity of several thousand megawatts.

FIGURE 1.13 Mural of Dutch History—Trade, Commerce, Tulips, Windmills, and Johannes Vermeer's *The Girl with a Pearl Earring* (Photo Credit: Anne97432, http://en.wikipedia.org/wiki/File:Hollande04.jpg)

1.2.5 The English Experience

The large-scale advent of windmills in England predated the expansion in the Netherlands by several centuries. The food economy of twelfth century England, like many societies throughout Europe and the world, depended heavily on grains. These were typically cereal grains such as wheat, barley, and oats. About half the food expenses of the typical household was for grain products to make bread, oat porridge, and so forth. For locations where waterpower was available to drive gristmills, flour from the grains could be produced mechanically. However, for other areas where water was scarce or not available, hand milling had to be done. Hand milling using a quern, or handheld grinding stone, took approximately two hours per day for a person to produce the flour needed for a household for a day. In the year 1080, a survey of water mills in England was conducted and indicated there were 5624 water mills operating, grinding grain, and producing flour at 3000 locations. That survey, called the Domes Day Survey, made no mention of windmills.

Historical note on terminology: Water mills, which had been developed and operated during Roman times, were sometimes called corn mills, which can be confusing since corn as we know it, also called maize, was originally cultivated in the Americas and was not known to the Europeans until many centuries later. Corn, however, is also an English term for a small granule of grain (like a peppercorn), so a corn mill (also called a gristmill) could be used to grind peppercorns to a fine powder to season food.

Terminology notwithstanding, the gristmill or corn mill was an important device, whether operated manually in small, hand-driven devices or in a mechanized fashion with a water mill, windmill, or horse-driven mill. The milling process of the time was accomplished by two stones, one the bed stone with grooves in it and the other a runner stone that rotated on top, crushing the grain kernels and producing flour. The flour was driven to the outside of the bed stone by centrifugal force and captured in bags. Although the exact date is unknown, sometime in the twelfth century, the English post windmill was developed to grind grain. The post mill typically consisted of four wind vanes with canvas sail cloths, following technology similar to that of sails on a sailing vessel, which would drive a shaft and through gearing drive the millstone to produce flour. The mill was constructed on a "T" that could swivel, and a housing structure around it contained the mechanical equipment. A pole sticking from the back was used to align the windmill with the prevailing wind direction. On later models, a fan tail was introduced to mechanically do the same thing.

Records from the time show that, by the year 1200, the number of post windmills had grown rapidly in England. For example, on the Estates of Bishopric of Ely and located in a region of high grain productivity but low rainfall, a survey of the estates in the year 1222 showed four windmills. A subsequent survey done in 1251 showed 32 post windmills, while the number of watermills on the estates remained constant at 20. Although an inventory of the number of windmills throughout the country is not available, the growth was widespread and rapid (Fig. 1.14).

(a)

(b)

FIGURE 1.14 (a) Restored Post Mill, (b) Millstones, both from the American Wind Power Center, Lubbock, TX (Charles Norland, Norland Photographic Art, St. Louis, MO. Picture taken at the American Wind Power Museum in Lubbock, TX).

From an operational point of view, the water mill was actually preferred to the windmill because they were more easily controlled and could deliver power to the mill wheel on demand. Water mills had a sluice gate that could be opened and closed, controlling the flow of water to the water wheel and thus controlling the power delivered to the mill. The windmill relied on the speed of the wind, which was variable, and therefore capable of providing only intermittent power. However, many areas did not have water resources available and the only alternative was a horse-driven mill or the hand mill discussed earlier. In these locations, the adoption of the windmill was a welcomed alternative, and once introduced, grew rapidly.

The development of the post mill impacted communities in England in a number of interesting and probably unexpected ways. First were legal issues. During this time, if land owners had a stream or river on their property, they controlled the rights to that flowing source of power and any water mill erected was under the total ownership and control of the owners. However, in a well-documented case using church records, in 1190, a local church magistrate Dean Herbert erected a post windmill on his property to mill his own grain. Abbot Samson, another church official in the region but of higher authority (both the word "Dean" and "Abbot" are church titles from medieval times), had control of all milling operations and fees associated with those milling operations, and the fees were to be given to his monastery. According to church records, Abbot Samson became furious with the construction of Dean Herbert's post windmill on the idea that he, the Dean, would grind his own grain without due compensation to the Abbot and the monastery. He demanded that the windmill be torn down and that people be sent from the abbey to carry out the task. Dean Herbert, according to documents at the time, made the first-known argument about the "wind rights" stating that, in fact, wind was free, owned by God and not man, and therefore he had the right on his property to build this device to capture the free power of the wind to grind his own grain without paying any fees. Although eloquent in his speech, Dean Herbert gave in and did in fact take down the post mill—but only after he had made these very strong arguments—with long-lasting effects. The references given in this section give excellent accounts of this exchange of Dean Herbert and Abbot Samson documenting one of the first legal arguments for the legal aspects of wind law (Source: Gimpel [2]).

With the number of post windmills growing, the impacts on the economy were substantial. Millwrights and land owners could build a post windmill and charge for grinding grain, with fees typically being 1/16 of the units of flour produced. Over time, the millwrights and windmill operators became people of prominence in their towns. Some of the windmills were so lucrative that land owners made enough money from the windmills to provide donations to the poor and to hospitals. In his book *Harvesting the Air, Windmill Pioneers in 12th Century England*, Kealey [3] points out that tithes on the profits of windmill operations were considered very suitable gifts for charitable and religious institutions of the period. For example, in about 1170, Reginald Arsic made a grant of a windmill to the monks of Saint Mary of Hatfield Regis. The original document exists with its original wax seal in the British Library. It states the following:

> Reginald Arsic, to all his men in the present and in the future, greetings. Know that I have given and granted and by this my charter have confirmed to the monks of Saint Mary of Hatfield Regis the whole **tithe of my windmill** which is in Silverly....

Windmills also became objects of art. Twelfth century windmill art was quite basic, using simple drawings and usually not very accurate. In later centuries, windmills became the subject for many romantic and pastoral scenes, especially for the Dutch landscape painters who found inspiration in both the large Dutch windmills and the English post mill.

Europeans were among the first to focus on using mechanical power for social development. They were concerned with efficiency, convenience, and entrepreneurship associated with the development of mechanical devices. In earlier times, the Romans who, although they had access to mechanical contrivances, mostly relied on human power to accomplish tasks with little regard for the idea of using mechanical devices to simplify and make easier the life of a common person. The European/English view is the one that is prevalent today with most people understanding that research in science and technology leads to advances in civilization and the concepts of efficiency, convenience, and entrepreneurship are well established. In addition to this overall social view of science and technology, there are a number of parallels in the development of wind power in twelfth century England to that of today.

As mentioned before, water mills for milling grain were directly controllable devices as long as there was water available. Windmills did not have that method of control and were subject to the variances in the wind. They could be shut down but never started up or increased in power beyond that available by the wind. The wind would change direction, requiring the millwright to change the orientation of the rotor in the wind. For these reasons, the water with its controllable power was preferred. Additionally, the fact that the mill and owner fully owned the water resources on the property meant that the landowner had complete monopoly on the water-milling operation.

By analogy, electricity-generating sources from the burning of fossil fuels can be controlled by changing the rate of fuel delivery to the power plant. It can be started, shut down, and controlled as desired. Wind turbines, on the other hand, can be shut down at any time, but the power level and the ability to deliver power are dependent on the wind speed, and yaw motors must be installed to align the wind turbine with the incoming wind direction to extract maximum power. Additionally, the fuel sources are owned and controlled by private enterprise whereas the wind turbine does not rely on any purchased fuel.

Also analogous is the use of twelfth century wind power as compared with the use of wind power today. As mentioned previously, flour used for bread and other food staples was a critical household commodity of twelfth century England. Today, electricity is the critical commodity—used throughout the industrialized world and depended upon to run households—from air conditioning and food refrigeration—commerce and industry. It is interesting that wind power, with all of its variability and use of an essentially free resource, was a key in producing a commodity staple, grain, in the twelfth century and a commodity product, electricity, today. The legal framework that wind operates under today in many ways had its roots back in the twelfth century with the examples of Dean Herbert and Abbot Samson discussed earlier in this section showing the origin of legal arguments related to wind power.

As development of the New World began in the seventeenth century, windmill technology from England and Continental Europe was imported for grain milling and water pumping. Because of its location and geography, Cape Cod is relatively flat with little flowing water. It is, however, quite windy, conducive to the installation

(a)

(b)

FIGURE 1.15 (a) Restored Jonathan Young Mill in Orleans, MA. (b) Historical marker located at the site (Source: R. Walker).

of such windmills. Figure 1.15 shows a restored gristmill on Cape Cod in the City of Orleans, Massachusetts, including a short history of the mill shown on an accompanying plaque.

While taxes on revenue earned from windmills (such as that shown in Fig. 1.15) benefitted the community they were located in, the use of wind revenue for social projects, such as schools and hospitals, continues today with modern wind farm development. The photo in Figure 1.16 shows a new school in the West Texas town of Hermleigh, Texas, partially funded by tax revenue from the wind farm project in the background.

1.2.6 Wind Power for Industry

In addition to using the power of the wind to pump water and mill grains, many other repetitive industrial tasks were adapted to use mechanical power derived from the wind to replace human-, horse-, and waterpower. Several examples include the following:

- Sawing: In 1593, the first patent was issued in Holland for a wind-powered saw mill. Cutting timber was a laborious task and important for a number of commercial enterprises. This mill used a vertical saw blade and horizontal cart to carry the log to be sawed. Wind-powered saw mills were rare in Great Britain, with the first being built by a Dutchman in 1663 near London. The demand for wooden beams used in ship building was very strong at this time, and the mill caused a lot of

FIGURE 1.16 Hermleigh Independent School District, Hermleigh, TX (Charles Norland, Norland Photographic Art, St. Louis, MO. Picture taken at the Hermleigh High School in Hermleigh, TX).

interest. However, the local labor force of hand sawyers were strongly opposed and attacked the sawmill with axes. It stood abandoned for several years [2].

- Dyestuff and paper: Production of these materials required the stamping and hammering of wood and other materials. Early windmills converted rotary motion of the wind shaft to vertical stamping and pulverizing operations through a series of gears and shafts for these tasks.
- Oil mills: Oil could be extracted from the seeds of plants such as flax, rape, and olive by first crushing the seeds under edge runners, heating them, and taking them to a stamping operation for oil production.
- Mining operations: Windmills were used to provide power for ventilation and water-pumping systems in mines before the advent of the steam engine.

1.3 IMPACT OF ELECTRICITY ON SOCIETY

1.3.1 National Academy of Engineering: Great Achievements of the Twentieth Century

At the turn of the millennium in the year 2000, the US National Academy of Engineering (NAE), a group of renowned engineers in the United States, formed a panel to list the 20 greatest engineering achievements of the twentieth century. It was not a technical "gee-whiz" criterion that determined inclusion on the list, rather how

TABLE 1.1 National Academy of Engineering: Top Three Greatest Engineering Achievements of the Twentieth Century[a]

1. Electrification: Vast networks of electricity provide power for the developed world.
2. Automobile: Revolutionary manufacturing practices made cars more reliable and affordable and the automobile became the world's major mode of transportation.
3. Airplane: Flying made the world accessible, spurring globalization on a grand scale.

[a]Source: www.greatachievements.org
The complete list of the NAE's engineering achievements can be found at www.greatachievements.org.

much an achievement improved people's QOL. Former astronaut, Neil Armstrong, the first person to walk on the moon, presented the results and noted that "Almost every part of our lives underwent profound changes during the past 100 years thanks to the efforts of the engineers, changes impossible to imagine a century ago" (Table 1.1).

It is interesting that in this assessment by top engineers, electricity was named the technology that had the highest impact on society—not computers or rocket travel but delivery of electric energy to the homes, businesses, and factories of the world. Modern civilization would literally not be possible without electrification—from the preservation of food with refrigeration to heating and cooling of our homes to delivering light. Thomas Edison's development of the long-burning incandescent light bulb allowed people access to convenient, clean, and affordable lighting. As we will present in this course, electricity is provided from a number of sources through a vast and complex network delivering electrical power on demand. Presently, about one-sixth of the world's people do not have access to electricity. It is envisioned that the developments of renewable power sources and other advanced technologies will help bring electricity to these people and their communities. The modern wind power industry is presently the fastest growing source of bulk electric power and is seen by many to be poised to become a major contributor to delivering electric power worldwide.

1.3.2 History of the Early Electric Utilities

The first electric utility station was built in New York by The Edison Electric Company, founded by Thomas Edison in 1882. It was a power plant that produced direct current (DC) power and provided lighting to a small neighborhood in the New York City area. Between the years 1882 and 1886, there were major discussions as to whether the transmission of electric power would occur with DC or alternating current (AC). In 1886, Westinghouse erected in Great Barrington, Massachusetts, the first AC power station and transmission system. For a number of technical reasons, to be detailed later, the AC mode of electric utility transmission turned out to be the technology of choice and is in use today.

1.3.3 Rural Electrification Administration

In 1934, only about 11% of US farms had electricity. The United States was basically an agricultural economy at that time and providing electric power to farmers, ranchers, and livestock producers was seen as a key step in modernizing the

United States. In 1935, as part of the New Deal Act under President Franklin D. Roosevelt, the Rural Electrification Administration (REA) was formed to provide electricity to farmers, ranchers, and livestock producers within the United States. This would be accomplished through a series of rural electric cooperative companies. The REA was not well received by the private electric companies at the time since it was seen as a government intrusion in the area of private enterprise. However, due to the depression and the need for electric power, the REA moved forward and by 1942 50% of US farms had access to electricity mostly through rural electric cooperative companies and by 1952 almost all farms had access to electric power.

1.3.4 Expansion of the Electric Grid

The years following World War II began a time of rapid expansion of installed electric capacity in the United States. Figure 1.17 shows the capacity listed in millions of kilowatts or gigawatts (for reference, one gigawatt of capacity provides the electrical energy for about half million average households) and the growth from 1949 to 2009.

As shown in Figure 1.17, the United States crossed one terawatt, or a thousand gigawatts, shortly after the year 2000. Most of these power station installations are fueled by fossil fuels—coal and natural gas. The bulk of the renewable installations, especially during the 1950s and 1960s, were hydropower and the recent increase is mostly the addition of wind energy since most viable hydropower sites had been exploited by the mid-1970s. Nuclear power began to expand in the 1970s and now represents about one-fifth of the electricity supply in the United States.

Figure 1.18 shows the gross domestic product (GDP) and its increase during this same time period. It is not coincidence that the growth in electric capacity coincides with the increasing economic output of the nation, as represented by GDP. We will discuss in future sections the more complex relationship between these, especially as it has to do with energy efficiency, but in general one can link the increasing supply of electricity almost directly with the output of the national economy as represented by GDP.

1.3.5 Electricity—World View

- **The Relationship between Electricity Consumption, QOL, and GDP**

  We mentioned previously that access to electricity is an indicator of modern civilized society and the direct relationship between electricity supply and economic growth in the United States. In general, it can be shown that QOL indicators are highly correlated with energy use and in particular electric energy use. In recent years, social scientists and political leaders have determined a number of ways to measure QOL, from a "happiness index" in the country of Bhutan to other more quantitative measures that include such things as life expectancy, political stability, and wealth usually determined by GDP per person.

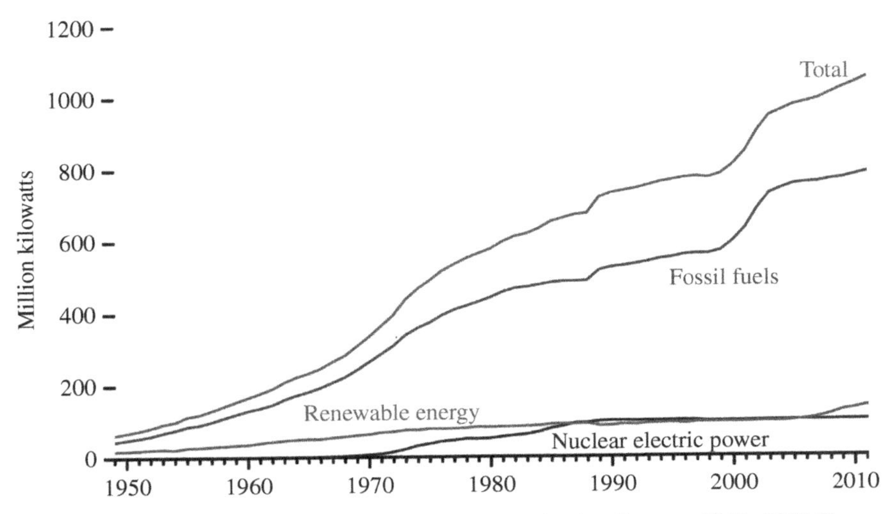

FIGURE 1.17 Installed Electric Generating Capacity by Source, 1949–2011(Source: US-EIA [4]).

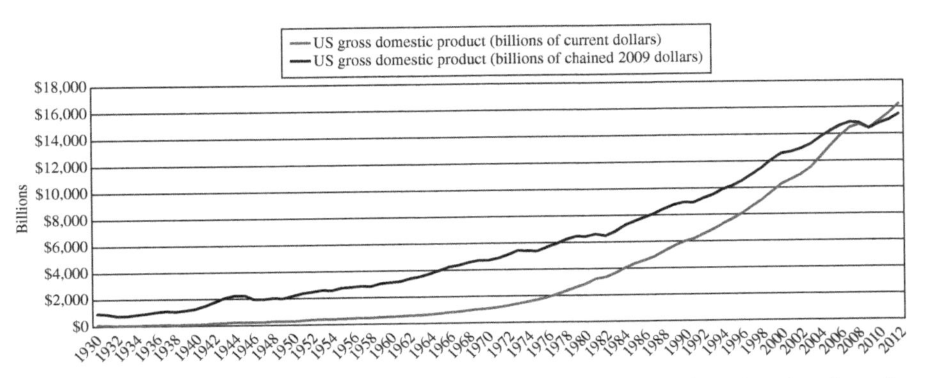

FIGURE 1.18 US GDP, 1930–2012, in billions (Source: R. Walker, based on data from the U.S. Bureau of Economic Analysis [5]).

The QOL index is based on factors including material well-being (based on GDP per person and purchasing power), health (based on life expectancy), political stability and security, family life (based on divorce rate), climate, geography, job security, political freedom, and gender equality. A comparison of this QOL index data with electric energy use per person for a number of countries is shown in Table 1.2, and the related correlation chart is shown in Figure 1.19.

The trend line shown in Figure 1.19 illustrates that, in general, as countries use more electricity per person, the QOL index increases. The data indicate that the increase is rapid at first and then is more gradual as more electrical energy

TABLE 1.2 Quality of Life Rankings Compared With Electricity Consumption[a]

Country	QOL ranking	QOL index	Population (millions)	GDP (at PPP) billion usd ($)	GDP (at PPP) per capita ($)	Electricity consumption (Gwh/year)	Daily kwh per capita	GDP (at PPP) per kwh ($)
Australia	6	7.90	21	803	38,238	257,247	33.54	3.10
Italy	8	7.81	58	1,827	31,500	359,161	16.95	5.10
Spain	10	7.73	41	1,402	34,195	303,179	20.25	4.60
USA	13	7.60	307	14,440	47,036	4,401,698	39.25	3.30
Canada	14	7.60	33	1,303	39,485	620,684	51.50	2.10
Netherlands	16	7.43	17	674	39,647	123,496	19.89	5.50
Japan	17	7.39	127	4,340	34,173	1,083,142	23.35	4.00
Taiwan	21	7.26	23	714	31,043	238,458	28.39	3.00
Germany	26	7.05	82	2,925	35,671	617,132	20.61	4.70
UK	29	6.92	61	2,236	36,656	400,390	17.97	5.60
Korea	30	6.88	49	1,338	27,306	443,888	24.80	3.00
Mexico	32	6.77	111	1,567	14,117	257,812	6.36	6.10
Brazil	39	6.47	199	1,998	10,040	505,083	6.95	4.00
Thailand	42	6.40	66	549	8,318	149,034	6.18	3.70
Philippines	44	6.40	98	318	3,425	60,819	1.70	5.20
Turkey	50	6.29	77	904	11,740	198,085	7.04	4.60
China	60	6.08	1,339	7,992	5,969	3,444,108	7.04	2.30
Vietnam	61	6.08	87	242	2,782	76,269	2.40	3.20
Indonesia	71	5.80	240	917	3,821	149,437	1.70	6.10
Saudi arabia	72	5.77	29	578	19,931	204,200	19.28	2.80
India	73	5.70	1,166	3,304	2,834	860,723	2.02	3.80
Bangladesh	77	5.65	156	226	1,449	35,893	0.63	6.30

Egypt	80	5.60	83	445	5,361	130,144	4.29	3.40
Iran	88	5.35	66	844	12,788	211,972	8.79	4.00
Pakistan	93	5.23	176	431	2,449	91,626	1.43	4.70
Russia	105	4.80	140	2,271	16,221	1,022,726	20.00	2.20
Nigeria	108	4.51	149	336	2,255	21,110	0.39	15.90
World			6,784	70,048	10,325	20,279,640	8.18	3.50

[a]Source: A. Swift based on *The Economist Intelligence Unit* [6], *CIA World Factbook 2009* [7], *Enerdata Statistical Energy Review* [8]; This graph is compiled of data from several sources.

Sources of data

- QOL index from *The Economist Intelligence Unit's Quality-of-Life Index, The World in 2005* [6], http://www.economist.com/media/pdf/QUALITY_OF_LIFE.pdf
- Countries listed are top 20 most populous countries and/or those in top 20 highest GDP as adjusted for purchasing power parity plus Saudi Arabia, based on *CIA World Factbook 2009* [7].
- *Enerdata Statistical Energy Review* [8], electric energy consumption data from 2009 for the 20 most populous countries.
- Productivity as a function of electricity consumption (a concept similar to energy intensity) can be measured by dividing GDP by the electricity consumed. The global average was $3.50 of production for every 1000 kWh of electricity consumed.

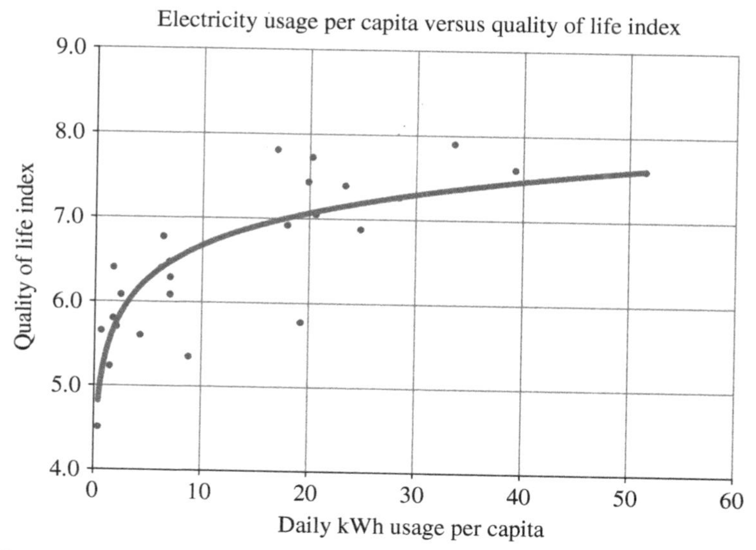

FIGURE 1.19 Chart Showing Quality of Life versus Electricity Usage per person from Table 1.2 (Source: A. Swift).

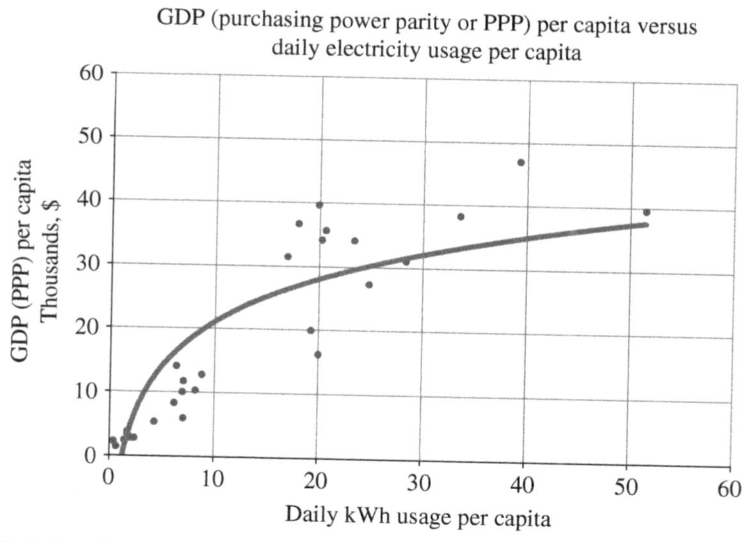

FIGURE 1.20 Chart showing GDP Purchasing Power Parity (PPP) per person and Electric Energy Use per person (Source: A. Swift).

per person is used. The rule is not absolute because such things as political stability can have significant effects on the QOL index, but in general the trend is clear. It can also be shown that GDP on a per-person basis is also directly correlated with electric energy use in a similar fashion as seen in Figure 1.20.

- **Role and Importance of Energy Efficiency**

If one were to forecast the growth in electrical energy requirements with world population growth to provide a basic level of electricity use per person, the need for additional electric-generating capacity would grow very quickly and is probably unsustainable using current electric generation technology. In other words, given our current mix of electric-generation sources, there are just not enough resources to provide all of the electric needs for the growing population of the world. This in fact is one of the motivations driving many people to look at changing the mix of energy supply to include more renewable energy sources, such as wind and solar energy—which are abundant throughout the world. Another way to provide electricity to a growing population is by producing, delivering, and consuming electrical energy resources more efficiently. Traditional energy sources that rely on thermal inputs, either burning of fossil fuels or the fission of nuclear material, are extremely wasteful processes, determined by the laws of thermodynamics. Typically, two-thirds of the heat energy of the source fuel is wasted in conversion to electricity, while one-third is converted to useful electrical energy. Modest increases in thermal power plant efficiencies can have a significant impact. These concepts will be covered more thoroughly in Chapter 15.

Additionally, if one considers the consumption of electricity (in other words, the end use—such as lighting, refrigeration, heating and cooling, and industrial processes) and improves the efficiency of these processes, mainly by delivering the same service for a reduction in energy consumption, this is also beneficial. The first is called supply-side efficiency, and the second is demand-side efficiency. Both of these strategies will allow a growing population access to electricity resources at reduced rates of growth for electric energy supply.

There are two metrics one can use to evaluate efficiency of electric energy production. One is the per capita use of electric energy, as we already looked at in the QOL chart. The other is electric energy use per unit of GDP—in other words, how much electricity is used to drive the economy. Efficiency of electric energy use can be a major determination in calculating the quantities of electric-generating capacity and consumption of resources for electricity generation to meet the given demand. For example, if one uses the maximum electricity consumption per capita number from Table 1.3 and applies population growth from the current time to the year 2050, one can project the estimate of the maximum amount of electrical energy required at that time. Similarly, using the minimal energy per person value at the same QOL, one can project a second, lower-level of required energy capacity. If these limits are combined with increases in electrical-generation efficiency and a transition to renewables, one can begin to understand the range of projected future electricity needs not only for the two billion people who do not have access to electricity but also for providing a reasonable QOL for the growing population of the world.

- **Jevons' Paradox**

  In 1865, William Jevons, an English economist, produced a book called *The Coal Question,* where he examined the use of coal in England. During that period, coal was the primary energy source and Jevons observed that the consumption of coal greatly increased after James Watt introduced the coal-fired steam engine, which was a great improvement in efficiency as compared with Thomas Newcomen's earlier design. At the time, many in England worried that coal reserves were decreasing rapidly and many argued that increasing the efficiency of use would reduce coal consumption. Jevons argued that was not correct. In fact, the increase in efficiency, as with the use of Watt's steam engine, would increase the use of coal since the improved steam engine delivered less expensive power and more people could therefore afford to use it, requiring more steam engines to be built. Modern energy economists have pointed out that Jevons' Paradox applies only to technological improvements that increase fuel efficiency but that the imposition by governments of conservation standards that simultaneously increase costs do not necessarily cause this paradoxical increase in fuel use. Recent examples of such government initiatives include a green tax, a carbon tax, cap and trade, license fees, and so forth, which, it is argued, would counter Jevons' Paradox. However, such regulations have been met with resistance by those who feel energy markets should use a free market approach to set the price of energy.

1.4 HISTORY OF WIND ENERGY FOR ELECTRICITY PRODUCTION

We saw in the previous section the importance of access to electrical power for improving QOL and prior to that the importance of programs such as the REA in providing electricity to rural America in the 1930s. In this section, we will review the history of wind power development for electricity production with a focus mainly on the United States. It should be pointed out that the use of wind power generation proceeded rapidly during the twentieth century in countries such as the Netherlands, Denmark, Russia, and Germany during this same time period, and thus, there are really two histories of wind power development for electricity. The first US wind-driven electric generator was installed in the late 1800s by Charles Brush in Ohio. He built a large wooden wind turbine with a DC generator and a series of batteries to provide electricity for his home. In the 1920s and 1930s, a number of small companies were developed to produce wind generators to charge batteries and power lights and radios for locations without access to grid power. Typically, these locations were in the Great Plains, prior to the REA's program, where the persistent winds provided a reasonable means to provide basic electric power (Fig. 1.21).

The most successful company of this period was the Jacobs Wind Company, which produced thousands of wind generators for home and ranch use. The Jacobs turbine was very rugged and, in fact, was taken to the South Pole to provide electricity for an expedition. It was installed and provided power for the expedition and then left

(a)

(b)

FIGURE 1.21 Advertisement (a) and painting (b) of Zenith Wind Charger at ranch home before REA. Photo—reproduction from wall mural at the American Wind Power Center, Lubbock, TX (Charles Norland, Norland Photographic Art, St. Louis, MO. Picture taken at the American Wind Power Museum in Lubbock, TX).

for several years until the next expedition came back and found it still in working condition. There were plans to put a series of these small turbines on or near electrical lines and provide larger-scale power; however, these ideas never came to fruition as the REA put these companies out of business.

In 1942, a group led by Palmer Putnam designed and installed a large grid-connected wind turbine—the first utility-scale wind turbine connected to the utility grid in the United States. His book, *Power from the Wind,* is an excellent overview of how that project was conceived and developed. The turbine operated for several years, and during a windstorm one evening, one of the blades experienced metal fatigue, broke, and flew off and down the mountain. Since the turbine operator rode in the small cabin at the top of the tower, he experienced a very exciting ride as one blade came off, and due to the imbalance, swung around the top of the tower until he could shut down the turbine. Since World War II was under way at the time, there were insufficient funds and interest in repairing the turbine and thus the project was stopped. After the war, the price of oil and other fossil fuels declined considerably and access to the fuels was readily available. Thus, the idea of using wind power to generate utility electric power was abandoned (Fig. 1.22).

In 1970, there was an oil embargo from the Organization of the Petroleum Exporting Countries that led to an energy crisis in the United States. That event caused a major reexamination as to the dependence of the nation on fossil fuels, especially imported oil, which was a major source of electricity production at the time. During that period, the Department of Energy (DOE) was formed to make the United States independent from foreign energy supplies. One of the technologies that the DOE took very seriously was utility-scale wind generation. In the 1970s and 1980s,

FIGURE 1.22 The 1.25 MW Smith–Putnam Wind Turbine of 1942, the first utility-scale, grid-connected wind turbine in the United States (Photo from the archive of Carl Wilcox in the possession of Paul Gipe).

there were a number of designs produced by researchers both within and outside the DOE. Shown in Figure 1.23 are DOE turbine designs. The first wind farm was located in California.

Wind farms are large groups of wind turbines connected to the grid to generate electrical power. California rapidly grew to become the leading state in installations of wind power in the 1980s. A number of excessively lucrative tax credits in California at the time led to the installation of equipment that was not very reliable and often inefficient, leading to high costs of electrical generation and unreliable production. The California tax credits were tied to equipment installation rather than to how much electricity the equipment produced. In the early 1990s, a federal Production Tax Credit was put in place. At about that same time, a number of Deliberative Polls® taken by utility companies in states including Texas and Louisiana showed that, in general, people were very supportive of wind power and renewable energy. This led to the formation of statewide renewable portfolio standards in several states, including Texas, mandating that a certain fraction of electrical energy be produced from wind power or other renewables. This, along with the entry

FIGURE 1.23 US DOE Wind Turbine Designs. (a) DOE-NASA MOD-0 (1974-75) One-bladed rotor configuration; Sandusky, OH, (b) 200 kW DOE-NASA MOD-0A (1977-81) 38 m rotor diameter; Clayton, NM, (c) 2 MW DOE-NASA MOD-1 (1979-81) 61 m rotor diameter; Boone, NC, (d) 2.5 MW DOE-NASA MOD-2 (1982) 78 m rotor diameter; Goodnoe Hills, WA, and (e) 3.2 MW DOE-NASA; MOD-5B (1991); 97.5 m rotor diameter, variable speed (13–17.3 rpm) Built by Boeing; Kahuku, HI (Source: US DOE).

into the industry of large US companies such as General Electric, combined with rapid advances in Denmark, the Netherlands, and Germany formed the foundation of the modern utility wind industry.

1.5 RENEWABLES AND ELECTRIFICATION IN THIRD-WORLD COUNTRIES

As discussed previously, many people in the world do not have access to electrical power. As solar photovoltaic technology and wind technology have improved, especially on a smaller scale, access to electric power in Third-World countries has been growing. Although solar and wind technologies in remote areas tend to be relatively expensive, the costs are substantially cheaper than the installation of large electric-generating stations and a network of transmission lines to deliver power to remote areas. These stand-alone remote systems have been instrumental in delivering power to many parts of the Third World.

1.6 THE NEXUS OF WIND, WATER, AND ELECTRICITY

Thermal power generation uses large quantities of water for cooling. In fact, the annual withdrawal of water to provide cooling for thermal power plants is the largest use of water in the United States. In arid areas where water is limited or underground wells are used to provide water, these withdrawals (and related consumption of water) can limit the installation of power plants to provide electricity. Additionally, since underground water in arid areas is typically consumed at rates higher than underground aquifers can be recharged, these resources are not sustainable over long periods. Wind energy generates electricity without the use of water, which is a significant advantage when compared with thermal generation sources. These issues will be discussed in greater detail later in the text.

REFERENCES

[1] United Nations. Report of the World Commission on Environment and Development. General Assembly Resolution 42/187. New York, NY: United Nations Department of Economic and Social Affairs; 1987.

[2] Hills RL. *Power from the Wind, a History of Windmill Technology*. Cambridge: Cambridge University Press; 1984.

[3] Kealey EJ. *Harvesting the Air, Windmill Pioneers in Twelfth Century England*. Berkley/Los Angeles: University of California Press; 1987.

[4] U.S. Energy Information Administration. *Annual Energy Review 2009*. Washington, DC: U.S. Government Printing Office; 2011.

[5] U.S. Bureau of Economic Analysis website. Available at http://www.bea.gov/national/index.htm#gdp. Accessed December 29, 2013.

[6] The Economist. 2006. The Economist Intelligence Unit's Quality-of-Life index, The World in 2005. Available at http://www.economist.com/media/pdf/QUALITY_OF_LIFE.pdf. Accessed November 1, 2014.

[7] The Central Intelligence Agency. CIA World Factbook 2009. Washington, DC: Central Intelligence Agency Office of Public Affairs; 2010.

[8] Enerdata Statistical Energy Review. 2010. Electric Energy Consumption data from 2009 for the 20 most populous countries (as listed in Wikipedia).

2

THE BASICS OF ELECTRICITY

2.1 UNITS OF ELECTRICAL MEASUREMENT

A general understanding of electricity and terminology used in the electric power and wind energy industries will be useful as one reads this book. Following are some of the terms and definitions that will be frequently used:

- **Electrical Energy:** Electrical energy is the generation or use of electric power over a period of time expressed in kilowatt-hours (kWh) or megawatt-hours (MWh).
- **Power:** The rate at which work is done, or for electricity, the rate at which electrical energy is generated or consumed (energy per unit of time), usually measured in watts, kilowatts, or megawatts.

The key difference between energy and power that one needs to understand is that power is an instantaneous measurement at a single point in time, whereas energy is measured over some increment of time, such as an hour, a day, or a year. For example, the power requirements of a 60-W light bulb is 60 W, but if one leaves the light bulb on for 24 h, it consumes 1440 watt-hours (60 W × 24 h), or 1.44 kWh. It is likely that your local electric service provider charges you based on kWh usage, at a rate of approximately 10–15 ¢/kWh.

Wind Energy Essentials: Societal, Economic, and Environmental Impacts, First Edition.
Richard P. Walker and Andrew Swift.

- **Ampacity:** The current-carrying capacity of conductors or equipment, expressed in amperes.
- **Ohm or Ω:** The unit of electrical resistance to current flow in a circuit. Resistance is 1 Ω when a voltage of 1 V will send a current of 1 A through (i.e., 1 Ω equals 1 V/A).
- **Volt or V:** The unit of electric force or pressure. The pressure that will cause a current of 1 A to flow through a resistance of 1 Ω.
  - Examples: 12-V batteries
  - The usual voltage in your home is 110–120 V.
  - Clothes dryers, water heaters, and electric ovens may require 220 V.
- **Voltage:** The electrical force or pressure that causes current to flow in a circuit, as measured in volts.
  - Analogy: Pressure in a water hose or pressure created by an elevated water tower.
- **Kilovolt or kV:** A unit of electrical pressure equal to 1000 V.
  - Examples: Large electric transmission lines, such as those extending between a power plant and a city, may be 345,000 V, or 345 kV.
- **Resistance:** A measure of how hard it is for electric current to move through a material; the opposition to the flow of electrical charge. The unit of measurement is ohm, represented as Ω.
  - Analogy: Friction in a water hose causes opposition to the flow of water.
- **Current:** The movement or flow of electrons through a conductor, as measured in amperes; represented by the symbol I. An ampere is a specific number of electrons that move past a given point in 1 s. The capacity of transmission lines or distribution lines is limited by the amount of current the lines can carry without creating excessive heat in the conductors since excess heating causes the line to expand and sag nearer to the ground, possibly violating ground clearance requirements of the National Electric Safety Code.
  - Current = Voltage/Resistance
- **Ampere or Amp or A:** The unit of electric current flow that indicates how much electricity flows through a conductor.
  - Amperes = Watts/Volts
- **Watt or W of Electrical Power:** The unit used for measuring electrical power usage or production. It is the amount of energy expended per second by a current of 1 A under the pressure of 1 V, or the power needed for 1 A to flow through 1 Ω.
  - Watts = Volts × Amperes
  - Watts = Amperes^2 × Resistance
- **Kilowatt or kW:** A unit of electrical power, equal to 1000 W. Electric power is often expressed in kilowatts.
  - Example: 50-kW small wind turbine

- **Megawatt or MW:** A unit of electrical power, equal to 1000 kW or 1 million Watts. The maximum production capabilities of large wind turbines are usually expressed in MW.
  - Example: The Vestas V-90 is a 3.0 MW wind turbine
- **Kilowatt-hour or kWh:** A unit of electrical energy equal to 1000 Watt-hours or a power demand of 1000 W for 1 h. Sales of electricity from the local utility company to one's home or business are expressed in kWh.
- **Megawatt-hour or MWh:** A unit of electrical energy equal to 1000 kWh. Sales of electricity from large wind energy projects to utility companies are usually expressed in MWh.
  - Examples: A 100-MW wind farm in a good wind regime could be expected to produce about 360,000 MWh/year. An average-size home may use about 12,000 kWh/year or 12 MWh/year. Therefore, a 100-MW wind farm would produce enough electricity for almost 30,000 homes.
- **Ohm's Law:** Defines the relationship between current, voltage, and resistance. Three ways that Ohm's Law can be expressed are as follows:
  - Current (in amperes) = Voltage (in volts)/Resistance (in ohms) or $I = V/R$
  - Resistance (in ohms) = Voltage (in volts)/Current (in amperes) or $R = V/I$
  - Voltage (in volts) = Current (in amperes) × Resistance (in ohms) or $V = I \times R$

2.2 DESCRIPTIONS OF COMMON ELECTRICAL EQUIPMENT

The types of electrical equipment referred to frequently throughout this book include the following:

- **Generator:** A general name given to a machine transforming mechanical energy into electrical energy.
- **Wind generator or wind turbine:** A machine that transforms mechanical energy from the wind into electrical energy (Note: Do not call them "windmills" or people knowledgeable about wind energy will know that you are not; windmills are for milling grain or pumping water).
- **Transformer:** An electrical device that raises or lowers voltage to facilitate the connection of electric lines of differing voltages or the interconnection of a generator producing power at one voltage to an electric line of a differing voltage. In a wind farm, padmount transformers are used to increase the voltage coming from the generator to the higher voltage used for the collection system (typically 34.5 kV). Transformers may also be located in the project substation to further increase the voltage to that of the transmission grid (Fig. 2.1).
- **Distribution lines:** Electric lines with operating voltages of 34.5 kV or less; the distribution system carries energy from the local substation to individual households, using both overhead and underground lines.

FIGURE 2.1 Padmount transformer used to step up generator voltage of 575 V to collection system voltage of 34.5 kV (Source: R. Walker).

- **Transmission lines:** The high-voltage electric lines transport generator-produced electric energy to loads usually through multiple paths. Operating voltages of 69, 115, or 138 kV are normally used to distribute energy within regions or areas and consist of overhead lines for the most part. Higher voltages such as 230, 345, 500, or 765 kV provide the "backbone" of most transmission grids in the United States and interconnect the various reliability regions in the United States or interconnect generating stations to large substations located close to load centers. The vast majority of transmission-voltage electric lines are located aboveground due to the significantly higher cost of placing the lines underground, with exceptions typically being in the heart of very large cities where no space is available for aboveground lines (Figs. 2.2 and 2.3).
- **Substation:** An electrical facility containing several types of equipment, usually including one or more transformers and two or more electric lines entering the facility. Substations are used for purposes such as the following: (i) the connection of generators, transmission or distribution lines, and loads to each other; (ii) the transformation of power from one voltage level to another; (iii) controlling system voltage and power flow; and (iv) isolation of sections of an electric line that may be experiencing overloads or faults (Fig. 2.4).

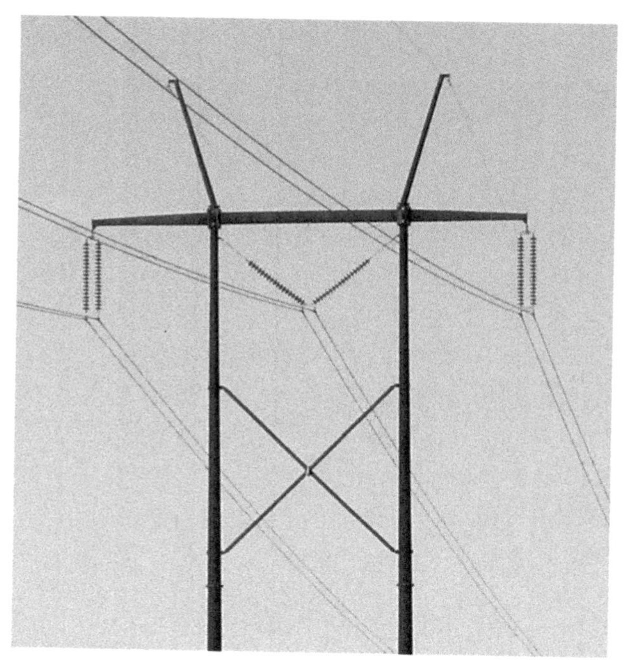

FIGURE 2.2 Single-circuit 345-kV transmission line, which may accommodate 800–1000 MW of wind generation capacity, depending upon conductor size and loading (Source: R. Walker).

FIGURE 2.3 Single-circuit 138-kV transmission line, which may accommodate 150–250 MW of wind generation capacity, depending upon conductor size and loading (Source: R. Walker).

FIGURE 2.4 Electric substation for a wind farm: Underground 34.5 kV lines bring power from turbines to the substation where transformers then convert energy from 34.5 kV to the voltage of the transmission line. The project is connected to a 345 kV line in this case (Source: R. Walker).

2.3 TYPES OF COMPANIES OR BUSINESS UNITS WITHIN THE ELECTRIC INDUSTRY

State regulations applicable to the electric industry vary from state to state. Some states have required traditional utilities to break up into independent business units depending upon the type of service they provide, some of which may be "deregulated," whereas other states have preferred to keep utilities companies as one entity or company. Following are descriptions of various types of companies within the electric sector:

- **Vertically integrated utility:** A utility company that provides generation, transmission, distribution, and retail electric sales for its customers; states typically maintain regulatory authority over the rates and quality of service for vertically integrated utilities.
- **Retail electricity provider (REP):** In some states that have deregulated sectors of the electric utility business, customers may choose between many REPs that have to compete based on price, marketing, and attributes of the energy they sell, such as companies that sell a "green energy" product.
- **Generation company:** Again, in states that have deregulated sectors of the electric utility business, the generation sector of the electric utility industry may have been deregulated. In such cases, power plants have to compete on price and other factors such as the reliability of their power or the environmental attributes of their power.

- **Independent power projects (IPPs):** IPPs are power plants not affiliated with an electric utility within the electric market that they are selling power into. Many wind energy projects are IPPs. IPPs can operate in states where vertically integrated utilities are still commonplace or in deregulated markets.
- **Transmission service provider (TSP or transco):** The company or business unit that owns transmission lines and associated equipment such as transmission voltage substations, and thus would provide transmission services to wind energy projects, other sources of electric generation, and to retail electric providers. TSPs' rates are usually regulated by a state regulatory authority or the Federal Energy Regulatory Commission and are required to provide nondiscriminatory or "open access" transmission service to all generators.
- **Transmission and distribution company:** Similar to TSP, but the company would also own distribution lines and the associated facilities; companies owning the distribution system are normally regulated and are often charged with the responsibility for maintaining highly reliable service to retail customers.

2.4 FREQUENTLY ASKED QUESTIONS

Following are some of the most frequently asked questions about the electric grid and how wind energy interacts with it, along with answers to these questions.

2.4.1 How Do I Tell How Much Electricity I Use and How Much It Costs?

If you are a college student, you may not have ever paid an electric bill before, or perhaps have only recently started paying them, so you may not have ever looked at your electricity bill to figure out how much it costs or how much you are consuming. According to data from the U.S. Energy Information Administration, the average retail price of electricity was 9.74 ¢/kWh during 2008, with West Virginia having the lowest average retail rates at 5.61 ¢/kWh and Hawaii having the highest at 29.20 ¢/kWh. To determine the rate you are paying, divide the total amount of your bill by the number of kWh you used during the billing period (Fig. 2.5).

2.4.2 Can Wind Energy or Electricity Be Stored Until It Is Needed?

Storing electricity in large quantities (such as produced by wind energy projects) is not currently cost-effective due to the cost of the various technologies currently available and due to the energy losses caused by storing and the subsequent use of the energy. However, much research is ongoing in this area and improvements to the technologies are being achieved. Batteries can be used to store small amounts of electricity, and there have been a few megawatt-scale batteries installed in the United States in recent years. However, they tend to wear out within a few years and are expensive relative to the amount of energy they can store. Compressed air energy storage (CAES) is another potential energy storage method under consideration. With CAES systems, energy is stored by using off-peak energy to compress air that is injected into underground

HOMETOWN ELECTRIC COMPANY
5678 MAIN ST.
HOMETOWN, TX 12345

NAME:	JOHN SMITH	ACCOUNT NO.	001-00001-96
ADDRESS:	1234 ELM ST.	CUSTOMER NO.	00001-0
CITY/STATE:	HOMETOWN, TX 12345	SERVICE FROM:	1/01/2010 TO 1/31/2010

BILLING ITEM	METER READING (KWH)		KWH USED	ENERGY RATE	BILLING AMOUNT
	PRESENT	PREVIOUS			
ENERGY CHARGE:	89154	87569	1585	$0.0935	$148.20
FUEL ADJUSTMENT CHARGE:	89154	87569	1585	$0.0125	$19.81
CUSTOMER CHARGE:					$10.00
STATE AND LOCAL TAXES:					$22.25
CURRENT BILLING:					$200.26
PREVIOUS BALANCE:			Your monthly usage (measured in kWh)		$156.23
PAYMENTS:					$156.23 CR
ADJUSTMENTS:					$0.00
PAST DUE AMOUNT:					$0.00
TOTAL AMOUNT DUE:		PLEASE PAY THIS AMOUNT:			$200.26
AVERAGE COST PER KWH:		The average rate you pay in $/kWh			$0.1263 PER KWH
DUE DATE: FEB. 28, 2010					

FIGURE 2.5 Example of an electric bill (Source: R. Walker).

caverns or salt domes. Later, when electricity is needed during on-peak periods, the compressed air can be released through a turbine to produce electricity, although this process may be paired with a natural gas generator for optimum efficiency.

2.4.3 How Much Does It Cost Me to Run the Air Conditioner or Other Appliances in My Home?

This will vary by your own personal habits and the cost of electricity that you pay, but Figures 2.6 and 2.7 show breakdowns of average home energy use and cost by appliance type. Refrigeration typically accounts for 4% of energy use in a home, and space heating and air-conditioning account for about 54% of energy use in an average home. Water heating is another significant contributor to your utility bill, averaging about 18%. Water heating and space heating may be from natural gas, electricity, or some other source.

2.4.4 If We Do Not Store Electricity from Wind Energy, What Happens When the Wind Stops?

Figure 2.8 depicts electric load during the course of a typical weekday in August in a southern state such as Texas, New Mexico, or Arizona and the type of generating resources that an electric utility might use to meet the load. Demand for electricity

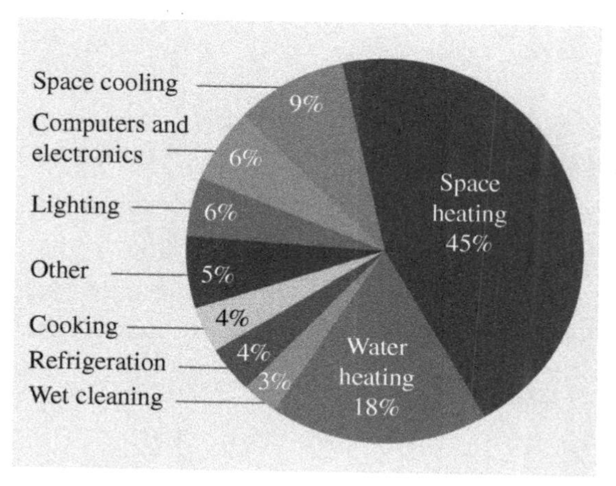

FIGURE 2.6 Residential site energy consumption by end use (Source: US DOE [1]).

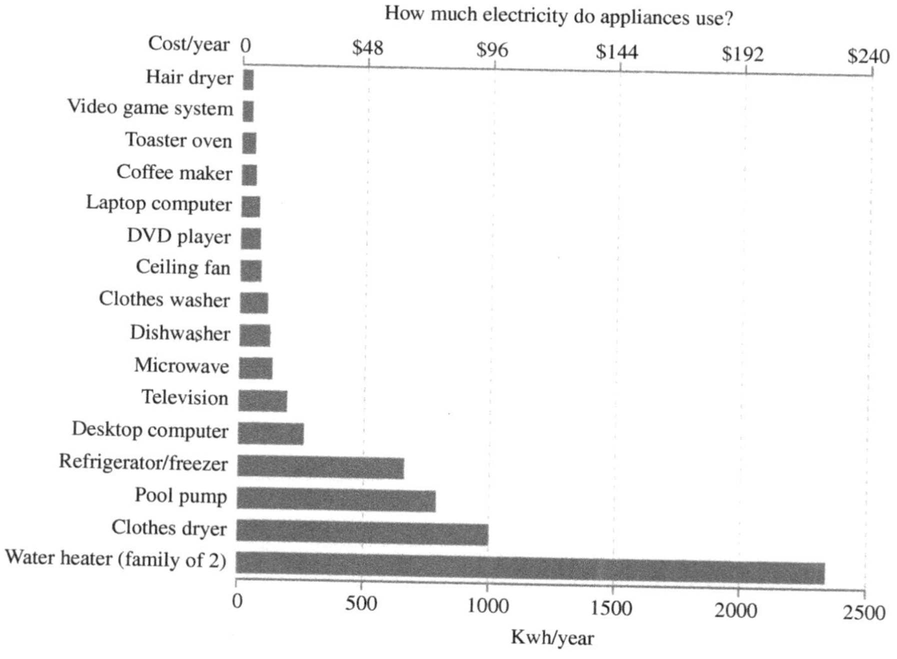

FIGURE 2.7 Average energy cost for appliances (Source: US DOE [1]).

is low during the early morning and late night hours but begins to grow as lights are turned on, people arrive at the office or factory, and the heat of the day requires the use of air conditioners. The electric utility's peak load may occur around 4:00 or 5:00 P.M.

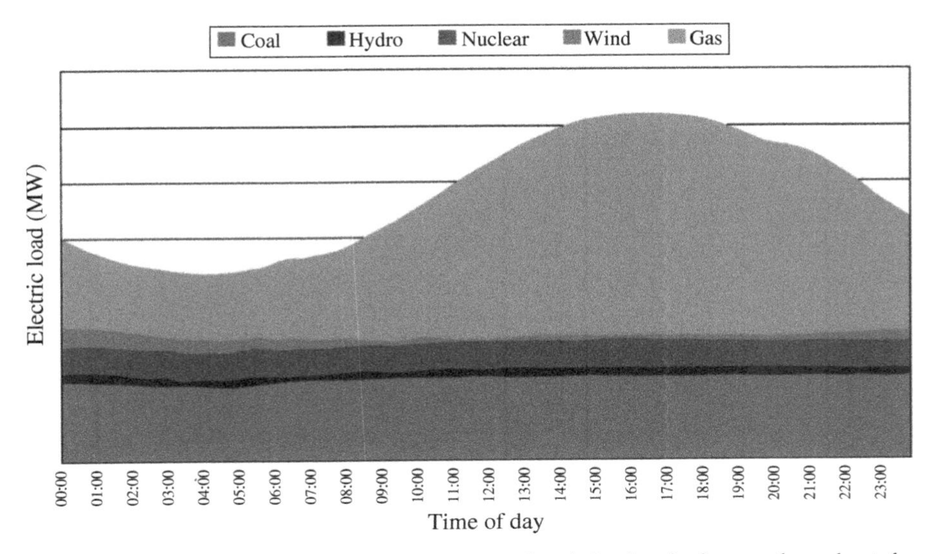

FIGURE 2.8 Generation resources used to meet electric load as it changes throughout the day (Source: R. Walker).

As load grows throughout the course of the day, the electric utility will bring on more and more generating plants. Nuclear and most coal-fired plants are not designed to be frequently cycled; instead, operators prefer to bring them online and keep output levels fairly constant for long periods of time. These are often referred to as "base-load units."

Most modern natural gas generators are designed to allow their output to change either up or down very quickly as demand for electricity changes. While some very efficient natural gas plants may be operated as base-load units (particularly when natural gas prices are low), natural gas plants are generally categorized as "peaking units." Hydroelectric plants, which currently provide about 7% of our nation's electricity requirements, may fall somewhere in the middle between base-load and peaking, dependent on the individual circumstances such as the predictability of rainfall and the intended use of the lakes created by the dam that the hydroelectric facility is attached to, such as flood control or recreational purposes.

In Figure 2.8, production from base-load coal generation and nuclear generation, represented as grey and red, respectively, usually stays fairly consistent throughout the day. The green area represents wind energy, and it will generally be more variable than coal or nuclear generation, and its highest periods of production may not align well with the utility's peak load periods. To meet the remaining demand for electricity, electric utilities will typically use natural gas–fired generators (depicted as the orange area in Fig. 2.8), and these same plants are also called upon to react to fluctuations in wind generation output. This can be done cost-effectively when gas prices are low and if modern, efficient gas-fired units are available but can become very costly in periods of high gas prices or when very high demand for electricity requires the use of older, inefficient units.

2.4.5 How Can Wind Energy Get to My House?

Consumers in many parts of the country now have the ability to choose between multiple REPs, and many of these REPs have a rate plan allowing the consumer to purchase "green electricity" or electricity produced from renewable energy sources such as wind. Green Mountain Energy is an example of such a company. In fact, they are a REP that tries to buy 100% renewable energy, and their website proclaims that the Empire State Building is now powered with 100% renewable energy from Green Mountain.

This leads to the frequently asked question: "Will the electricity delivered to my home really have been produced only from renewable energy sources?" The short answer to this question is "no"—the physics of electricity does not allow us to direct "green electrons" to one home and non-green electrons to another home. Instead, financial instruments called "Renewable Energy Credits" (RECs) are used to track renewable energy purchases of REPs and sales of renewable energy under the REP's green energy tariff, which will be discussed next. Figure 2.9 may help give you a better understanding of how electricity gets to your home, whether it is produced from wind turbines or a nuclear power plant.

The vast majority of electricity produced from wind projects or traditional power plants goes into the electric transmission grid. Typically, electricity is generated at a lower voltage, around 575–800 V, but the electric transmission grid uses higher-voltage lines (69 kV, 138 kV, 345 kV, 500 kV, or greater) to facilitate transfer of large amounts of electricity over long distances while keeping electrical losses at

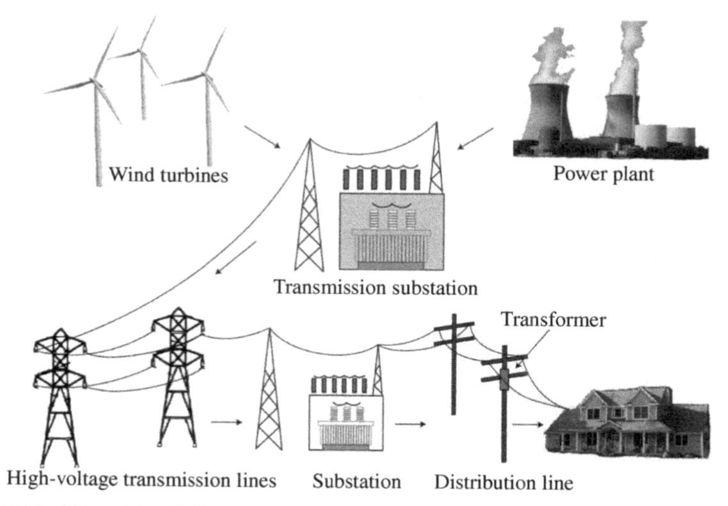

FIGURE 2.9 Electricity delivery from wind turbines and other generation sources (Source: Reproduced by permission of Amy Marcotte).

acceptable levels. Thus, transformers located at a substation are needed to increase the voltage from that of the generator to that of the transmission grid. Once the electricity nears the point of consumption, transformers are again used, but this time to reduce the voltage from that of the transmission grid down to the voltage of distribution system (usually around 12,700 V or 12.7 kV) and then again to further reduce the voltage to that used in one's home. Wind projects using only one or two wind turbines that can be connected directly to the electric distribution grid using smaller transformers than would be required to connect to the transmission grid and some state-of-the-art turbines are being designed so that electricity exiting the turbine is at the same voltage of the distribution system.

2.4.6 If I Am Buying Renewable Energy from My Electricity Provider, How Do I Know That the Electricity I Use Came from a Source of Renewable Energy?

This is a common question from people contemplating electricity purchases under green energy tariffs. As stated earlier, financial instruments called RECs are used to track renewable energy purchases of REPs and sales of renewable energy under the electricity provider's green energy tariff. An REC basically represents the environmental attributes of 1000 kWh or 1 MWh of electricity produced by a qualifying source of renewable energy, such as wind turbines. (Note that an average-size home in the United States will use about 1 MWh/month of electricity.) Thus, if your local electricity provider sells 100,000 MWh of electricity under its renewable energy tariff during a given time period, it will need to show that it has also acquired 100,000 MWh of electricity from qualifying sources of renewable energy, and then it must retire 100,000 RECs. In Texas, the Electric Reliability Council of Texas administers such renewable energy tracking systems, while in other states, it may be the Public Utility Commission, Public Service Commission, or possibly even independent third-party organizations that give their "seal of approval" to green electricity programs meeting the right criteria.

2.4.7 My Utility Charges More for "Green Energy." Does Wind Power Cost More than Electricity from Traditional Sources?

Wind energy can often be purchased by utilities at a lower cost than their incremental cost of generating power from one of their own power plants, although sometimes it is higher. The cost that the electric utility pays for natural gas used in gas-fired power plants and the efficiency of those plants are two of the largest factors involved. When gas prices are high, electric utilities often seek to acquire new sources of wind energy, but when gas prices are low, the utilities' demand for wind energy is reduced. With regard to other sources of electricity, the high cost of nuclear power plant construction has usually been substantially depreciated, so their incremental cost of producing energy is typically low. Similarly, the original cost of many coal-fired power plants has been substantially depreciated, so their incremental cost of producing energy can be low. However, the societal cost of emissions from coal- or gas-fired plants, nor the cost of risk insurance or radioactive waste storage for nuclear plants, is not usually reflected in the cost to the consumer. These costs are called "environmental externalities."

2.4.8 Can I Install a Small Wind Turbine at My House or Business and Reduce My Electric Bill?

The small wind turbine market is quite substantial. Many people want to provide all or part of their power from their own wind turbine or solar modules. These "behind-the-meter" installations are illustrated in Figure 2.10.

As the wind increases and the turbine produces power to meet the facility's electric load, the amount of energy supplied by the utility decreases. If the wind generator output exceeds the facility's load, power flow is reversed and power is supplied to the grid. Depending on the size of the generator and the utility's procedures, the electric meter may either turn backward or the electricity flowing back to the utility is separately metered. Some regions allow this power reversal to be credited to the customer in a process known as net metering.

FIGURE 2.10 Diagram showing a representation of a residence with "behind-the-meter" small wind turbine and photovoltaic system installations (Source: Reproduced by permission of Amy Marcotte).

FIGURE 2.11 One of the largest industrial "behind-the-meter" installations, located at Pyco Industries' cotton seed oil plant in Lubbock, Texas. The project consists of 10 1-MW Fuhrländer units made in China (Charles Norland, Norland Photographic Art, St. Louis MO. Photographed at the Pyco Industries Cotton Seed Oil Plant).

Additionally, large commercial or industrial firms may employ several larger turbines in the same manner, displacing electric energy purchased from the utility as shown in Figure 2.11.

REFERENCE

[1] U.S. Department of Energy, Energy Efficiency & Renewable Energy. *Energy Savers: Tips on Saving Money & Energy at Home*. Washington (DC): U.S. Department of Energy, Energy Efficiency & Renewable Energy; 2011.

3

OVERVIEW OF WIND ENERGY AND OTHER SOURCES OF ELECTRICITY

3.1 DEFINING RENEWABLE ENERGY

Renewable energy is probably defined in hundreds of different ways. The Britannica Online Encyclopedia defines "renewable energy" as "usable energy derived from replenishable sources such as the sun (solar energy), wind (wind power), rivers (hydroelectric power), hot springs (geothermal energy), tides (tidal power), and biomass (biofuels)."

During the late 1990s, the Board of Directors of the Texas Renewable Energy Industries Association developed a definition of renewable energy that was subsequently adopted by the Texas state legislature, and the authors believe this definition to be one of the most accurate descriptions of renewable energy. It is as follows:

> Renewable energy is any energy resource that is naturally regenerated over a short time scale and derived directly from the sun (such as thermal, photochemical, and photoelectric), indirectly from the sun (such as wind, hydropower, and photosynthetic energy stored in biomass), or from other natural movements and mechanisms of the environment (such as geothermal and tidal energy). Renewable energy does not include energy resources derived from fossil fuels, waste products from fossil sources, or waste products from inorganic sources.

The real key to this definition is that the energy is regenerated over a short period of time. Oil, gas, coal, and uranium are all regenerated from natural processes, but it takes a very long time. Thus, the important part of this definition is timescale.

Wind Energy Essentials: Societal, Economic, and Environmental Impacts, First Edition.
Richard P. Walker and Andrew Swift.

Those sources of energy considered to be "renewable" by almost any definition include the following:

- Wind energy
- Solar energy
- Hydroelectric-generated electricity (lakes or run-of-river)
- Biomass (crops or agricultural wastes)
- Landfill gas
- Geothermal energy
- Ocean energy (including energy derived from ocean currents, tides, or waves)

Hydroelectric power can be produced by large dams creating lakes or by run-of-river hydroelectric facilities, which are usually smaller in scale but have significantly fewer environmental issues. Run-of-river hydroelectric projects use the natural flow of rivers or streams created by elevation drop to generate electricity, rather than impounding large bodies of water. In Texas, either form counts as a "qualifying source" of renewable energy, although some states do not include energy from large-scale hydroelectric facilities as renewable energy since they may not want to encourage construction of any more dams or reservoirs due to environmental issues (such as impeding fish from spawning or impacting habitats of endangered species).

The Merriam-Webster Dictionary defines "biomass" as "plant materials and animal waste used especially as a source of fuel." This can include fuel used to generate electricity, fuel used to produce heat, or fuel used for transportation vehicles. Some examples of biomass used for electricity generation include wood wastes, sugar cane bagasse, rice hulls, and manure from cattle feedyards.

Landfill gas refers to gases generated during anaerobic decomposition of organic waste in landfills. The gases produced during decomposition include methane, carbon dioxide, oxygen, nitrogen, and various non-methane organic compounds. Gas from landfills can be used as fuel for an electric generator, or if sufficiently "cleaned up," it can be injected into natural gas pipelines. This raises the question of why landfill gas qualifies as a renewable resource while natural gas from a well does not qualify. First, the content of landfills is mostly organic, and the decomposition process producing the landfill gas is generated over a relatively short timescale. Secondly, if one does not capture landfill gas, which is mostly methane (CH_4, "natural gas"), it will eventually escape into the atmosphere and methane is one of the more significant or problematic greenhouse gases (GHGs). The US Environmental Protection Agency uses a statistic called global warming potential (GWP) to assess the threat posed by various GHGs (see later Chapter 16). GWP measures how much heat one molecule of a gas will trap relative to a molecule of carbon dioxide; thus, carbon dioxide's GWP is 1.0. Methane, on the other hand, has a GWP of 21–25, which means it is 21–25 times more effective at allowing solar radiation to penetrate the Earth's atmosphere and preventing infrared radiation from escaping the planet. While the volume of methane emissions is small compared with the volume of carbon dioxide emissions, they are a significant cause of concern due to their high GWP value.

Therefore, since methane is significantly more potent than carbon dioxide as a GHG, it is far more preferable to capture and to use landfill gas rather than having it escape into the atmosphere. In small landfills, where it may not be economically feasible to use the gas for electric generation or some other source of fuel, the gas may be "flared" or burned converting the methane into carbon dioxide.

Energy sources that are clearly not considered renewable energy include the following:

- Nuclear energy
- Coal
- Natural gas
- Oil

In addition to electricity from large-scale hydroelectric facilities, sources of energy that may or may not qualify as renewable energy sources depending upon how the state or federal government defines renewable energy include the following:

- Municipal solid waste used as fuel for electric generation
- Cofiring biomass in a coal-fired plant
- Tire-derived fuel

Many states do include electricity produced by burning or gasifying municipal solid waste. Typically, the waste is ground up and burned to produce steam, which then goes through a turbine and produces electricity. Texas and some other states may not include this as a qualifying renewable resource since inorganic materials (such as plastics) can be part of the fuel source, producing undesirable emissions when such wastes are burned.

Cofiring biomass refers to mixing biomass or biomass-derived fuel with traditional fossil fuel as a fuel source for a traditional power plant. Most often, this involves mixing agricultural or forestry waste with coal for use in coal-fired power plants. So instead of burning 100% coal, the power plant burns a mixture of 90% coal and 10% wood waste. This is a renewable resource accepted by some states but not by others. For instance, Texas does not include biomass cofired with coal as a qualifying renewable resource.

The importance of clearly defining what types of energy qualify as renewable resources stems from governmental mandates that utilities obtain some portion of their energy from renewable sources. Such mandates may be referred to as renewable portfolio standards (RPSs) or renewable electricity standards (RESs). The obligations of the utilities to purchase renewable energy are tracked by different mechanisms. In Texas and some of the other states that have an RPS or RES, the tracking is done using "Renewable Energy Credits" (RECs). An REC represents the value of environmental or social attributes of the energy produced by renewable energy resources and is typically quantified in megawatt-hours (MWh). For example, if a wind generator produces 500 MWh during a month, its owner can sell 500 MWh of electricity to one company and 500 RECs to another company. Thus, the definition of renewable energy adopted by the state determines whether electricity produced from a generation resource qualifies for RECs. The utilities in Texas have to retire a certain number of RECs based on how much total electricity they sell, which means that they must produce or purchase RECs from some qualifying source of renewable energy whether it be wind energy, solar energy, or landfill gas.

3.2 SOURCES AND USES OF ENERGY

As a nation, the United States uses a lot of energy and in most years, energy use increases. In recessionary periods (such as 1979, 1980, 2009, and 2010), energy consumption may level off or even decline, but in general, when the economy grows, energy use increases. While the efficiency with which the United States uses this energy has increased over time, this improvement has not been able to offset overall growth in energy use, as shown in Chapter 1 and Figure 3.1.

> A British thermal unit (Btu or BTU) is the amount of heat energy needed to raise the temperature of 1 lb of water 1 F°. A kitchen match when burned produces about 1 Btu of heat energy.

As shown in Figure 3.2, petroleum (oil) is the largest source of energy used in the United States, followed by natural gas and coal. According to the U.S. Energy Information Administration, the United States consumed 97.3 quadrillion Btu (or quads) of energy in 2011, with 35.3 quads, or 36%, coming from petroleum.

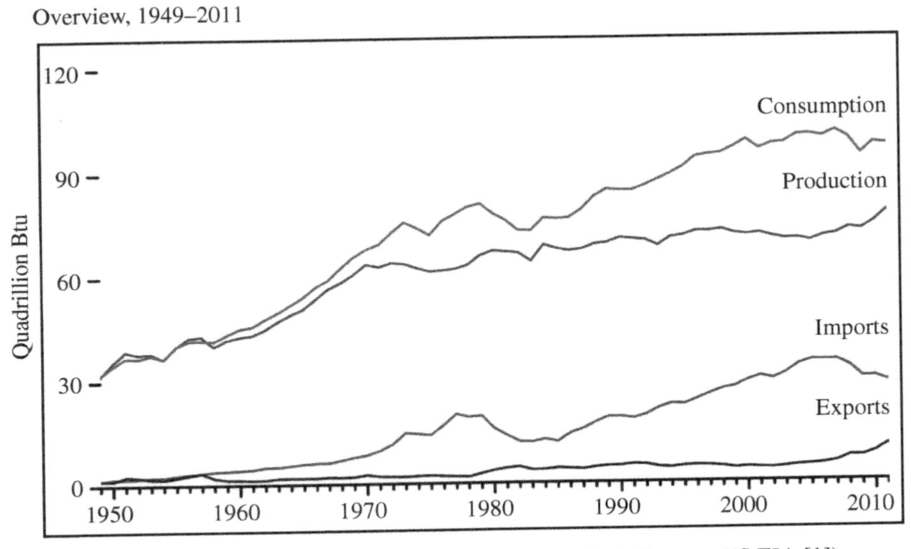

FIGURE 3.1 US energy consumption (Quadrillion Btu) (Source: US EIA [1]).

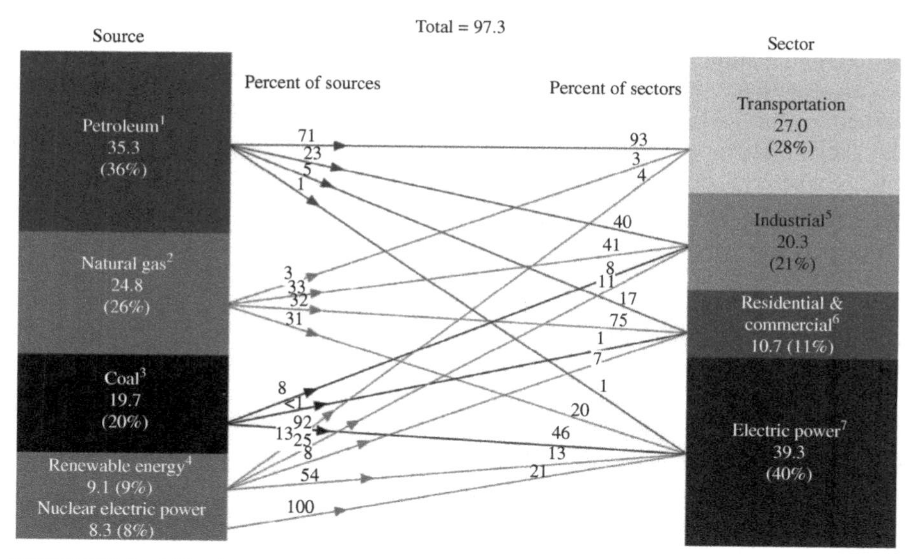

FIGURE 3.2 US primary energy flow by source and sector, 2011 (Quadrillion Btu) (Source: US EIA [1]). Notes: [1]Does not include biofuels that have been blended with petroleum—biofuels are included in "Renewable Energy." [2]Excludes supplemental gaseous fuels. [3]Includes less than 0.1 quadrillion Btu of coal coke net exports. [4]Conventional hydroelectric power, geothermal, solar/PV, wind, and biomass. [5]Includes industrial combined-heat-and-power (CHP) and industrial electricity-only plants. [6]Includes commercial CHP and commercial electricity-only plants. [7]Electricity-only and CHP plants whose primary business is to sell electricity, or electricity and heat, to the public. Includes 0.1 quadrillion Btu of electricity net imports not shown under "Source."

Electrical power production was the largest use of energy, consuming 39.3 quads or 40% of total energy use.

The graph also indicates what the fuel source is used for. In the case of petroleum, one can see that 71% is used for transportation, while only 1% is used to produce electric power. All nuclear power and the majority of coal (92%) consumed in the United States is used to make electric power. While about 31% of natural gas is also used to make electric power, significant volumes of natural gas are consumed directly in the industrial, residential, and commercial sectors. One can also see that renewable energy only accounted for 9.1 quads of energy used in the United States during 2011, or only 9%, with 54% of renewable energy being used for electric power generation, or conversely, only 13% of electric power coming from renewable energy.

Note that sources of renewable energy (such as wind, solar, or hydroelectric power) do not "consume" Btu of energy, and the values represented in the graph shown earlier instead represent the average amount of Btu that would have been consumed at traditional thermal power plants if the renewable energy had not been available.

Turning one's attention from the supply side to the demand side of Figure 3.2, one can also see that petroleum provides 93% of energy used in the transportation sector, with natural gas accounting for 3% and sources of renewable energy (such as ethanol) accounting for 4%. Not surprisingly, nuclear energy does not provide any energy for transportation purposes, although as electric and hybrid-electric cars become more prevalent, one can make the case that even nuclear energy will be used for transportation purposes.

Since this book is focused on social impacts of wind energy, we generally are going to focus on resources used for electricity generation, with much less emphasis on other sectors (such as transportation or industrial use of energy). Of the 39.3 quads of fuel used to produce electric power, coal represented 46% of this amount, followed by nuclear energy's 21% and natural gas' 20%. The United States has abundant reserves of coal and lignite, but the use of coal for electricity generation also presents the largest environmental issues.

3.3 GROWTH OF RENEWABLE ENERGY IN THE UNITED STATES

As previously stated, renewable energy only accounted for 9.1 quads of energy used in the United States during 2011 or only slightly over 9% of total energy consumed based on Btu value. This is in fact only a slightly higher percentage of the nation's total energy use than renewable energy contributed 60 years ago. As shown in Tables 3.1 and 3.2, renewable energy accounted for 8.0% of energy in 1952, compared with 9.3% in 2012.

Hydroelectric generation accounts for a significant proportion of the renewable energy use in the United States but has seen virtually no growth in the past four decades. While use of wind energy and solar energy in the United States has grown rapidly in recent years, the percentage of energy used in the nation coming from renewable sources has only increased from 6.0% in 1972 to 9.3% in 2012. Thus, if the United States is going to markedly increase the proportion of its energy coming from renewable sources, significant growth in the nonhydro renewable energy sectors will be required. However, despite relatively low overall growth in the use of renewable energy, there is hope in some of the numbers. For instance, wind energy has grown at an annualized rate of over 21% since 1992 and 29% since 2002 while solar energy has exceeded a 14% growth rate in the past 10 years.

3.4 USE OF RENEWABLE ENERGY FOR ELECTRICITY PRODUCTION IN THE UNITED STATES

Electricity generation accounts for about 40% of the nation's energy use. As shown in Figure 3.3, renewable resources were used to meet only about 12.2% of US electric needs in 2012, with hydroelectric generation accounting for most of that. Sixty years ago, the United States obtained about one-third of its electricity from hydroelectric generation but virtually none from other sources of renewable energy.

Since growth in hydroelectric generation has been essentially nonexistent in the past 40 years within the United States, the overall percentage of electricity coming from renewable energy declined from around one-third in the 1940s to a low of 8.5% in 2007;

TABLE 3.1 US Energy Consumption (1952–2012)[a]

	U.S. Primary energy consumption by source									
	Quadrillion Btu					Percentage of total consumption				
Year	Coal	Natural gas	Petroleum	Nuclear electric	Renewable energy	Coal (%)	Natural gas (%)	Petroleum (%)	Nuclear electric (%)	Renewable energy (%)
1952	11.31	7.55	14.96	0.00	2.94	30.8	20.5	40.7	0.0	8.0
1962	9.91	13.73	21.05	0.03	3.12	20.7	28.7	44.0	0.1	6.5
1972	12.08	22.70	32.95	0.58	4.38	16.6	31.2	45.3	0.8	6.0
1982	15.32	18.36	30.23	3.13	5.98	21.0	25.1	41.4	4.3	8.2
1992	19.12	20.71	33.52	6.48	5.82	22.3	24.1	39.1	7.6	6.8
2002	21.90	23.51	38.22	8.15	5.73	22.4	24.1	39.1	8.3	5.9
2012	17.36	26.00	34.58	8.05	8.82	18.3	27.4	36.4	8.5	9.3

[a]Source: US EIA [2].

TABLE 3.2 US Consumption of Renewable Energy (1952–2012) in Billions of Btu[a]

Year	Billions of Btu's						
	Hydroelectric	Geothermal	Solar/PV	Wind	Biomass	Total renewables	Total energy
1952	1,465,812	NA	NA	NA	1,474,369	2,940,181	36,747,825
1962	1,816,141	1,061	NA	NA	1,300,242	3,117,444	47,826,437
1972	2,863,865	15,079	NA	NA	1,503,065	4,382,009	72,687,867
1982	3,265,558	50,627	NA	NA	2,663,452	5,979,637	73,099,185
1992	2,617,436	178,699	63,676	29,863	2,931,678	5,821,352	85,782,977
2002	2,689,017	171,164	63,006	105,334	2,700,621	5,729,142	97,645,141
2012	2,686,815	226,643	235,066	1,361,104	4,308,797	8,818,425	94,976,067
10-year growth rate (%)	−0.01	2.85	14.07	29.16	4.78	4.41	−0.28
20-year growth rate (%)	0.13	1.20	6.75	21.04	1.94	2.10	0.51

[a]Source: US EIA [2].

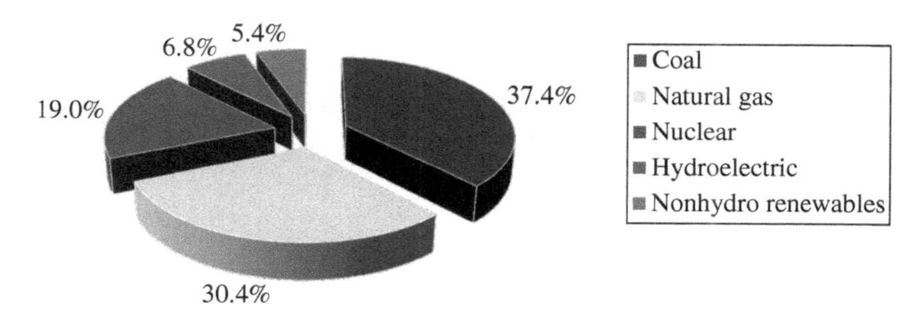

FIGURE 3.3 US electric generation by fuel source in 2012 (Source: R. Walker, based on data from United States Energy Information Administration, *Annual Energy Review* 2011 [2]).

however, the percentage of our electricity coming from renewable energy sources is now on an upswing, accounting for about 12.2% of electricity production in 2012, with the increase primarily due to rapid growth in the wind energy and solar energy industries (Table 3.3).

When Barack Obama and Joe Biden were running for president and vice president, respectively, in 2008, they rolled out the *Obama–Biden Comprehensive New Energy for America Plan* [3], which included the following objectives:

- Provide short-term relief to American families facing pain at the pump
- Help create five million new jobs by strategically investing $150 billion over the next 10 years to catalyze private efforts to build a clean energy future
- Save more oil than is currently imported from the Middle East and Venezuela combined within 10 years
- Put one million plug-in hybrid cars—cars that can get up to 150 miles/gallon—on the road by 2015, cars that are built here in America
- Ensure 10% of electricity consumed in the United States comes from renewable sources by 2012 and 25% by 2025
- Implement an economy-wide cap-and-trade program to reduce GHG emissions 80% by 2050

Since the United States was already obtaining about 9.2% of its electricity from renewable energy sources when the Obama–Biden plan was proposed, that was not really a very aggressive goal. On the other hand, the goal of achieving 25% of electricity from renewable energy by 2025 is a worthy target to shoot for and will no doubt require a concerted effort by the renewable energy industry and elected officials at the local, state, and federal levels.

While the Obama administration has expressed the desire to increase the nation's use of electricity, neither an RPS nor an RES has been passed by the US Congress, although such legislation has been proposed on several occasions. However, as of the beginning of 2013, 29 states plus Puerto Rico and the District of Columbia have

TABLE 3.3 Percentage of US Electric Generation By Fuel Source[a]

	Percentage of electric generation by energy source								
Year	Coal (%)	Natural gas (%)	Other (%)	Nuclear (%)	Hydroelectric conventional (%)	Wind (%)	Solar thermal & PV (%)	Biomass (%)	Geothermal (%)
2003	50.83	16.74	3.48	19.67	7.10	0.29	0.01	1.37	0.37
2004	49.82	17.88	3.44	19.86	6.76	0.36	0.01	1.35	0.37
2005	49.63	18.76	3.35	19.28	6.67	0.44	0.01	1.34	0.36
2006	48.97	20.09	1.93	19.37	7.12	0.65	0.01	1.35	0.36
2007	48.51	21.57	1.91	19.40	5.95	0.83	0.01	1.34	0.35
2008	48.21	21.43	1.41	19.57	6.19	1.34	0.02	1.34	0.36
2009	44.45	23.31	1.25	20.22	6.92	1.87	0.02	1.38	0.38
2010	44.78	23.94	1.17	19.56	6.31	2.29	0.03	1.36	0.37
2011	42.28	24.72	1.02	19.27	7.79	2.93	0.04	1.38	0.37
2012	37.42	30.35	0.84	18.97	6.82	3.46	0.11	1.42	0.41

[a]Source: R. Walker based on data from US-EIA website [2].

implemented RPS or RES policies, and eight other states had goals for increasing use of renewable energy.

3.5 GROWTH OF WIND POWER CAPACITY IN THE UNITED STATES

Installed wind power capacity grew very slowly from 1988 to 2000 but showed very rapid growth beginning in 2001. While newly installed wind capacity additions in the United States slowed during 2010 due to the global recession, transmission congestion in the windiest areas of the country, and low natural gas prices, the US wind energy industry rebounded in 2011 and 2012. In fact, during 2012, for the first time ever, the capacity of wind energy added in the United States exceeded the capacity of natural gas–fired generation added during that year. However, as can be seen, wind power capacity additions in 2010 were about one-half of what they were in 2009 but rebounded in 2011 and 2012, followed by another steep decline in 2013 due to Congress' failure to extend the production tax credit (PTC) applicable to wind energy until the day it was scheduled to expire (December 31, 2012) resulting in almost no US turbine orders during the last one-half of 2012. Despite the ups and downs created by lapses or near lapses in the PTC, wind energy's 27.5% average annual growth rate in the United States between 2002 and 2013 would be a growth rate most industries would covet. In addition, 2007, 2008, and 2009 were all banner years for the wind industry, with total installed capacity increasing by 45%, 50%, and 39%, respectively.

The primary contributors to the improved economics and desirability of wind energy after the year 2000 include the following:

- Larger and more reliable turbines that were developed
- Rising costs of natural gas and coal
- Natural gas price volatility
- Increasing public concern about global climate change caused by carbon dioxide emissions from fossil fuel plants and concern by utility companies that state governments or the Federal government may limit emissions of carbon dioxide or place a tax upon such emission
- Increasing regulation of emissions such as sulfur dioxide, nitrous oxide, and mercury from coal-fired plants or nitrous oxide from gas-fired plants
- Improved understanding of the impacts that large-scale wind generation can have on the electric grid and improvements in wind turbine technology making it more compatible with the electric grid
- Improved understanding of our nation's wind resources, including locations that have the best wind resource and wind forecasting technologies that help utilities plan for fluctuations in wind energy production
- Availability of the PTC during most of these years, particularly in those periods when Congress was able to agree on a multiyear extension of the PTC

3.6 SUBSIDIES OR INCENTIVES FOR WIND ENERGY

Governments wanting to boost some sectors of the economy (such as farming, housing, energy, or manufacturing) may offer financial incentives to attract additional investment into that economic sector. Examples of such incentives can include investment tax credits, income tax credits or deductions, accelerated depreciation for tax purposes, or *ad valorem* tax abatements. Governments may also promote industries by allocating funds for research and development that can make those industries more competitive. The United States has had a long-standing policy of "cheap energy" as a way to help its industries remain cost-competitive globally so that hundreds of financial incentives are available for various energy sectors, wind energy being just one of many.

Since the enactment of the Energy Policy Act of 1992 by the US Congress, the Federal renewable energy PTC has provided an incentive for investors to invest in renewable energy projects. The PTC has been a large reason for the rapid growth of wind energy, but it has also created some issues and problems for the wind energy industry. Referring back to Figure 3.4, note the "down years" of 2000, 2002, and 2004 when wind project construction in the United States was minimal. Each time, this was a direct result of lapses in the PTC. When implemented in 1992, the PTC provided for a tax credit of 1.5 ¢/kWh for wind energy produced during the first ten years of a wind project. The 1.5 ¢/kWh escalates with the rate of inflation and stood at 2.3 ¢/kWh as of December 2013.

Like many Federal tax credits, Congress must periodically reauthorize or extend them, but in 1999, 2001, and 2003, it failed to extend them in a timely manner, causing new construction of wind projects to come to a halt until such time as Congress was able to extend the PTC. While Congress did extend the credit retroactively each time to the date on which the PTC expired, wind project developers were hesitant to

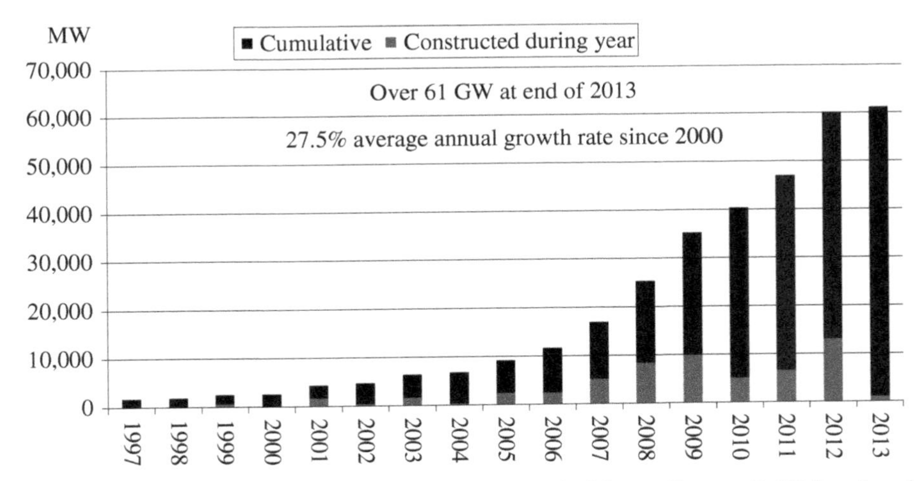

FIGURE 3.4 Installed wind power capacity in the United States (Source: R. Walker, based on data from AWEA Annual Market Reports [4]).

commit to the projects without knowing for certain that Congress would do so, further hampering development.

There are many reasons that governments may want to support the increased use of wind energy and other renewable sources of energy including reducing dependence on energy imports, promoting rural economic development, reducing energy price volatility, improving air quality, and addressing concerns about global climate change. In the United States, virtually every energy technology is supported in one way or another by the federal government. Wind energy is no exception, nor should it be.

Federal support and/or subsidies of the various forms of energy will be further discussed in later chapters of this book, including Chapter 18 covering the Economics of Electricity Generation and Chapter 21 State and National Energy Policies.

In addition to direct subsidies or incentives offered by governments, there are other societal costs of electricity generation that may not be reflected in the cost of electricity paid by the consumer. For example, owners of fossil-fueled power plants rarely are required to pay for the release of emissions into the atmosphere provided they stay below any government-mandated rate of emissions. These emissions, however, can impose a cost on society. Particulate matter from a coal-fired power plant goes out of the smokestack and into the air but eventually comes back down near the Earth's surface where it may be inhaled by an asthmatic person, further complicating his or her medical condition. Thus, there is a societal cost, or "environmental externality," such as in this case (increased medical expenses) associated with releases of emissions into the atmosphere. However, in the vast majority of cases, the company releasing the emissions does not have to pay for the right to emit.

The economic concept of "environmental externalities" refers to unintended and uncompensated environmental effects of production and consumption that do not accrue to the parties involved in the activity. In the case of electricity from a coal-fired power plant, the parties would be the owner of the power plant and the consumer of the electricity. The asthmatic breathing in the particulate matter may have an increase in his or her medical costs, but these costs are not transferred back to either the emitter of the particulate matter or the consumer of electricity. But even electricity-generation sources considered to be very "green" or environmentally friendly (such as wind or solar energy) have some environmental externalities associated with them. This concept will be addressed in greater detail in Chapters 13 and 14, which discuss sources of electricity generation other than wind energy.

Instead of asking why wind energy needs a subsidy, perhaps a better question to ask would be "what are the reasons that governments want to increase the use of renewable energy sources such as wind energy?" Later in this book, both the benefits of wind energy to society and the environmental externalities it may impose will be addressed.

3.7 POTENTIAL FOR INCREASED USE OF WIND ENERGY IN THE UNITED STATES

There are many reasons for the United States to want to increase its use of renewable energy sources, but there are also barriers to doing so. In 2006, then-President Bush, like several previous presidents, identified the need to increase the efficiency with which the country uses energy and the need for greater diversification in the nation's energy portfolio through increased use of renewable energy sources. This led to a joint effort of industry, government, and the US national laboratories, led by the U.S. Department of Energy (DOE) and the National Renewable Energy Laboratory (NREL), to examine scenarios in which wind energy could provide a much greater share of the electricity used in the United States. At that time, wind energy met only about 1% of the nation's electricity needs, yet the DOE/NREL effort resulted in the July 2008 report entitled *20% Wind Energy by 2030: Increasing Wind Energy's Contribution to U.S. Electricity Supply* [5]. This report identified the many issues, costs, challenges, and barriers that will have to be addressed in order to achieve the rather ambitious target of obtaining 20% of the nation's electricity from wind energy by the year 2030. One of the largest barriers is the nation's electric transmission grid. The most populated areas of the United States tend to be along the nation's coastlines, whereas the windiest areas in the United States extend from Texas through North and South Dakota. The transmission lines in that windy "heartland" area of the United States were sized to serve electric load in moderately sized cities, small communities, and rural areas and thus were not constructed to send large amounts of wind energy to more populated areas of the nation.

Based on continued growth in the nation's demand for electricity and accounting for retirements of older wind energy projects over time, the authors of the *20% Wind Energy by 2030* report estimated that about 306,000 MW or (306 GW) of wind energy–generating capacity would need to be installed by the year 2030. As previously shown in Figure 3.4, installed wind capacity reached 60 GW by the end of 2012, which was ahead of the pace envisioned by the authors of *20% Wind Energy by 2030*. Reaching 306 GW will require about 7.5 times as much wind capacity in 2030 as exists now, and since wind generators have an expected useful or economic life of 20–25 years, one could assume that many of the wind turbines currently in operation will be retired by 2030. This leads some to question the likelihood of achieving 20% of the nation's electricity from wind by 2030. The authors of the *20% Wind by 2030* report also estimated that the annual rate of wind capacity installations in the United States would need to grow to about 16 GW by the year 2022. As reflected in Figure 3.5, actual installations of wind capacity in the United States have exceeded the projection in the *20% Wind by 2030* report in each of the years 2006 through 2012, and installations in 2012 were over twice that amount anticipated in the *20% Wind by 2030* report.

There was a significant decline in the number of wind installations in 2010 as compared with the previous year, with only about 5115 MW of new wind capacity installed, but even that exceeded the projection required to reach the goal of 20% wind by 2030. The years 2011 and 2012 saw increases of 6,647 MW and 13,078 MW, respectively, although 2013 is anticipated to be another down year since Congress waited until the last day of 2012 to extend the PTC, resulting in very few turbine orders during the last months of 2012. Despite such down years, the level of support for cleaner sources of energy and increased use of domestic sources of energy remains strong, and given the rapid growth of the wind energy industry, which occurred between 2007 and 2009, and again in 2012, the authors contend that the 20% target is very achievable.

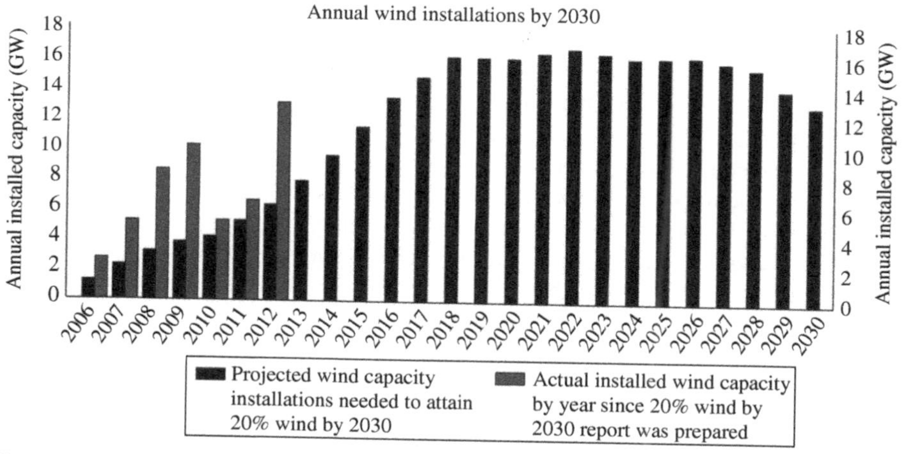

FIGURE 3.5 Annual installations of US wind capacity compared with projected growth needed to meet 20% of electricity from wind energy by 2030 (Source: Reproduced by permission of Amy Marcotte based on data from US-DOE [5] and AWEA Annual Market Reports [6]).

It can be expected that continued rapid growth of the wind energy industry will lead to escalation of existing issues or societal impacts of wind energy and create new issues as more projects are located in proximity to heavily populated areas or offshore from some of the nation's largest cities. This highlights the need to understand the type and magnitude of impacts that wind energy projects can have on society.

3.8 WIND RESOURCES IN THE UNITED STATES

In 2009, the NREL contracted with AWS Truepower to estimate and map the wind energy resources of each US state. As seen in Figure 3.6, states with large amounts of excellent onshore wind resource potential include Texas, Kansas, Oklahoma, Nebraska, North Dakota, South Dakota, Wyoming, Montana, Iowa, Minnesota, Colorado, and New Mexico.

As part of their study of national wind energy resources, AWS Truepower estimated that wind sites in the United States that could expect to have net capacity factors (NCFs) of 35% or greater based on estimated wind speeds at 80 m aboveground could support over 8400 GW of wind energy capacity. Obviously, this far exceeds the 306 GW that the *20% Wind by 2030* report estimated would be needed to in order to obtain 20% of the nation's electricity from wind energy by the year 2030. Since turbine hub heights continue to increase, the study also looked at 100-m wind speeds, which increased the potential wind capacity to over 10,000 GW, again only looking at sites that could be expected to achieve an NCF of 35% or greater. The study also excluded over 20% of land area deemed inappropriate for wind development

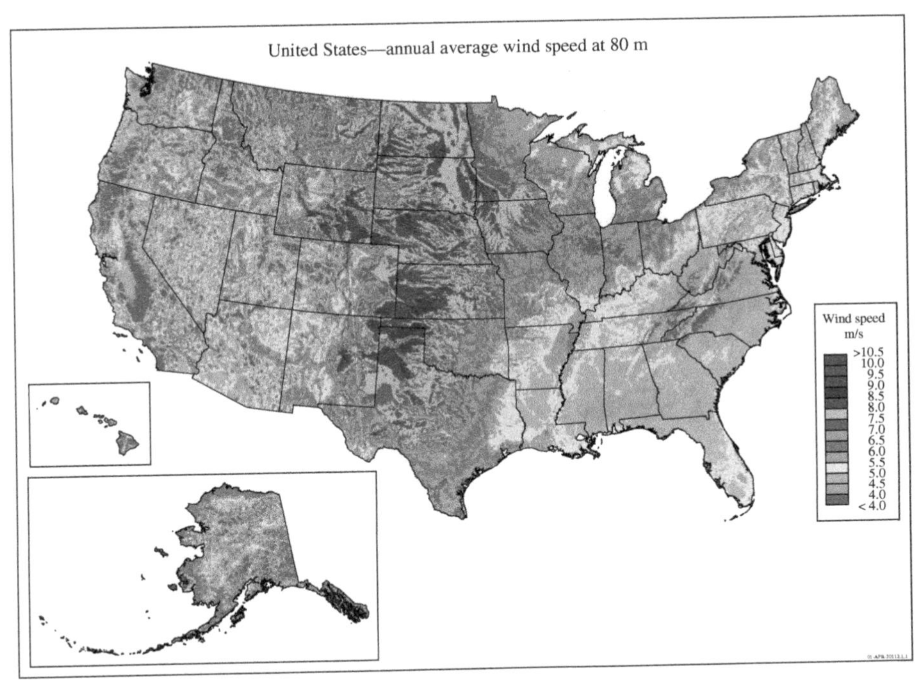

FIGURE 3.6 Annual average wind speed at 80 m aboveground (Source: NREL [7]).

(such as state and national parks, wilderness areas, urban areas, areas near airports, and wetlands).

Net Capacity Factor (NCF): the ratio of actual energy production during a given period of time (including losses such as turbulence, wake effect, planned or forced maintenance, blade degradation, and electrical losses) to potential maximum production if all turbines ran at their rated capacity for the entire period of time.

As shown in Table 3.4, Texas has the greatest potential for wind energy generation, with more than 1300 GW of potential based on winds at 80 m, compared with the 10 GW of wind-generating capacity existing within the state as of the end of 2010. Electricity consumption in the United States is currently around 4 million GWh/year; 1300 GW of wind generation with NCFs averaging 35% would produce about four million GWh, so it is clearly not a lack of windy land area that would prevent wind energy from providing 20% of the nation's electricity by 2030. Much more likely barriers to achieving the 20% goal are constraints in the electric transmission grid, impacts on the electric grid that integration of large amounts of wind capacity can create, and the type of social and environmental impacts of wind energy described in this book.

As previously mentioned, the AWS Truepower study for NREL looked at wind speeds at 80 and 100 m aboveground. The hub height and rotor size of wind turbines

TABLE 3.4 Wind Energy Potential By State[a]

Rank	State	80 m hub height w/≥35% NCF		100 m hub height w/≥35% NCF	
		Potential capacity (MW)	Annual generation (GWh)	Potential capacity (MW)	Annual generation (GWh)
1	Texas	1,360,186	4,989,570	1,757,356	6,696,500
2	Kansas	914,202	3,535,480	951,475	3,933,090
3	Nebraska	888,696	3,455,480	916,159	3,832,600
4	South Dakota	852,372	3,325,230	875,059	3,645,620
5	Montana	718,503	2,581,510	868,012	3,247,260
6	North Dakota	760,034	2,954,260	769,479	3,224,180
7	Iowa	482,415	1,772,460	566,965	2,224,020
8	Oklahoma	400,674	1,457,740	505,259	1,923,820
9	Wyoming	432,083	1,601,240	497,132	1,900,530
10	Minnesota	388,907	1,392,590	483,459	1,832,230
11	Alaska	291,121	1,051,210	410,923	1,489,752
12	New Mexico	325,002	1,170,490	407,760	1,509,690
13	Colorado	266,911	945,484	327,461	1,198,210
14	Missouri	78,917	256,650	247,470	831,473
15	Illinois	120,621	391,737	233,463	799,298
16	Indiana	49,005	160,827	130,218	437,101
17	Wisconsin	20,741	66,171	86,368	284,709
18	Michigan	11,662	37,619	39,312	129,467
19	Ohio	1,163	3,662	24,823	78,712
20	New York	4,805	15,826	20,418	67,619
21	California	14,032	49,073	18,452	64,398
22	Oregon	8,094	27,517	15,588	53,040
23	Washington	6,298	21,289	11,283	38,403
24	Maine	3,450	11,961	8,777	30,106
25	Arkansas	2,213	7,215	8,188	27,217

26	Idaho	3,226	10,938	6,882	23,041
27	Utah	1,517	4,939	3,390	10,988
28	Hawaii	2,483	10,179	2,762	11,434
29	Arizona	972	3,100	2,528	8,075
30	Vermont	1,195	4,243	2,186	7,710
31	Nevada	1,290	4,263	1,917	6,328
32	Pennsylvania	811	2,685	1,863	6,195
33	New Hampshire	962	3,405	1,707	6,046
34	West Virginia	820	2,822	1,270	4,394
35	Virginia	615	2,070	1,111	3,747
36	Maryland	189	607	855	2,753
37	Massachusetts	536	1,945	841	3,074
38	North Carolina	196	695	348	1,220
39	Tennessee	65	220	143	479
40	Georgia	31	101	78	259
41	New Jersey	15	47	53	171
42	Alabama	13	42	48	154
43	Rhode Island	28	99	41	148
44	Kentucky	8	24	26	82
45	South Carolina	3	11	19	60
46	Connecticut	1	3	9	29
47	Delaware	0	0	1	3
48	Florida	0	0	0	0
49	Louisiana	0	0	0	0
50	Mississippi	0	0	0	0
	U.S. Total	8,417,082	31,334,728	10,208,933	39,595,435

[a]Source: NREL and AWS Truepower [7].

have both increased significantly over the years, making wind energy viable at an increasing number of locations. In 1995, hub heights were around 40 m and rotor diameters were also around 40 m; by 2000, turbines might have a 65-m hub height and a 70 m rotor diameter. Currently, 80 m is probably the most common hub height in the United States, and as of the end of 2010, the largest turbines in the United States had a 105-m hub height and a 90-m rotor diameter. In Europe, there are even larger turbines in operation, and most wind turbine vendors continue to develop larger and larger turbines.

As turbines get larger, more and more areas of the country become viable for wind energy generation. Figure 3.7, also the result of the AWS Truepower study for NREL, shows the relationship between the potential for viable wind generation and hub height.

Texas leads the United States in installed wind capacity with 12,214 MW as of September 30, 2013, surpassing California in 2006 as the leading state for installed wind capacity. The next highest state is California with 5587 MW, followed by Iowa with 5133 MW. The American Wind Energy Association tracks wind energy projects in the United States, both completed and under construction. Figure 3.8 reflects operating wind energy projects as of December 31, 2013. Note that several states still do not have any wind generation installations, other than small turbines used to produce energy for one's home, farm, ranch, or small business, with most of these being located in the southeastern United States.

One of the reasons that Texas has been a leader in the installation of wind generation is that it enacted an RPS about 10 years ago. Other reasons for the large amount of wind generation in Texas include a great wind resource, large amounts of open land, and the high percentage of electricity coming from natural gas–fired plants,

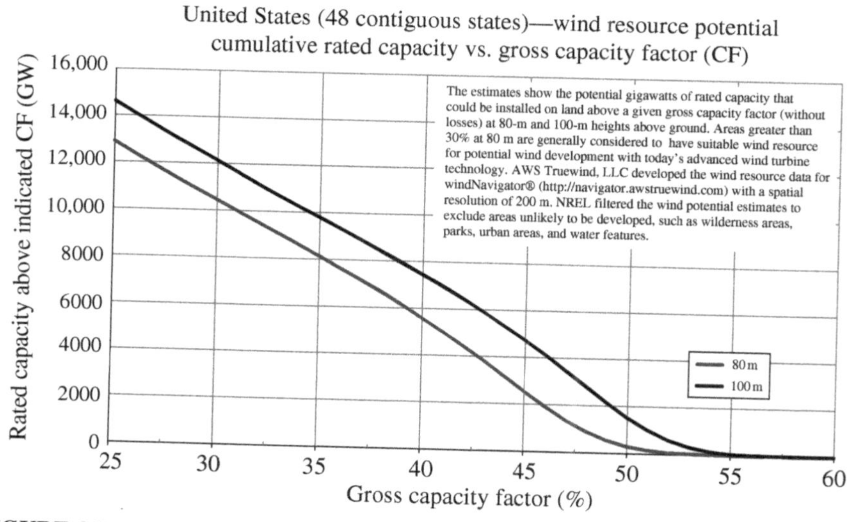

FIGURE 3.7 Wind resource potential in the United States (48 contiguous states) (Source: NREL and AWS Truepower [7]).

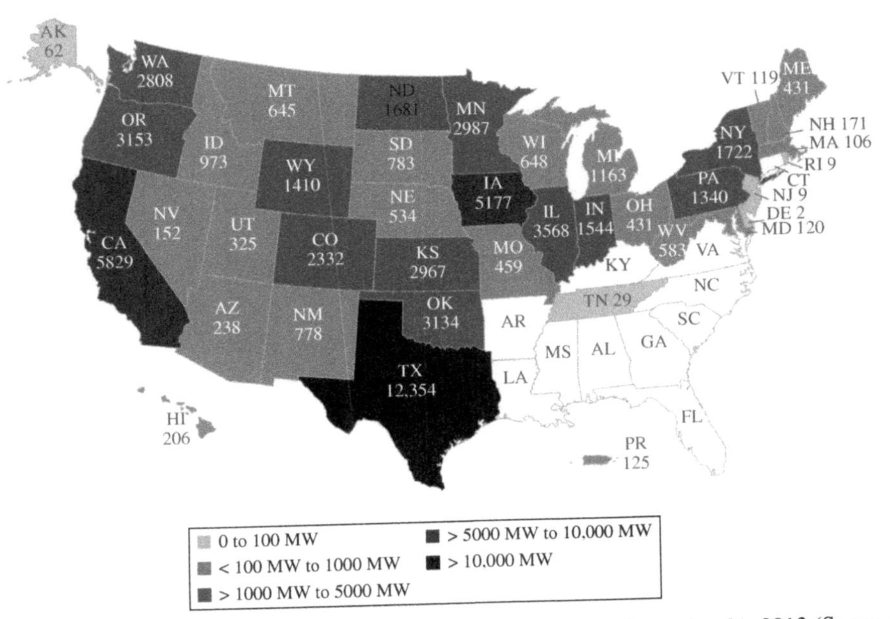

FIGURE 3.8 Installed wind generation capacity by state as of December 31, 2013 (Source: AWEA U.S. Wind Industry Annual Market Report 2013 [6]).

which adjust quickly to fluctuations in both electric load and wind production, but which also can become very expensive to operate during periods in which natural gas prices escalate rapidly.

An RPS, sometimes referred to as an RES, requires that electricity utilities obtain some portion of their electricity production or purchases from qualifying renewable sources of electricity. As shown in Figure 3.9, 29 states plus Puerto Rico and the District of Columbia had enacted an RPS or RES by the end of 2013. An additional ten states have set goals to attain some level of renewable energy resources in the state. Such goals generally differ from an RPS or RES in that there is not a firm legislative mandate requiring a set percentage or amount of purchases by utilities but instead look to incentives (such as tax abatements or tax credits) to encourage development of renewable energy projects in the state.

3.9 OVERVIEW OF OTHER SOURCES OF ELECTRIC GENERATION IN THE UNITED STATES

The United States has been blessed with an abundance of natural resources, and wind is only one of these. The nation also has vast quantities of coal, natural gas, oil, hydroelectric resources, solar energy potential, and biomass materials. Each of these resources will be discussed in detail in later chapters of this book, but the following paragraphs try to address some of the more frequently asked questions about some of these resources.

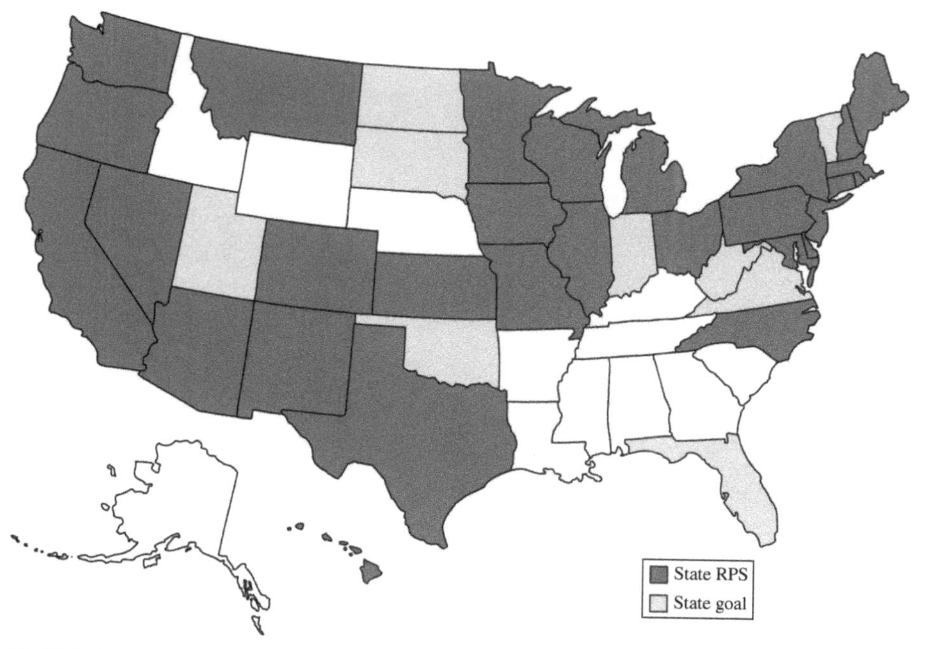

FIGURE 3.9 States with RPSs, RESs, or renewable energy goals as of December 31, 2013 (Source: Reproduced by permission of Amy Marcotte, based on data from Database of State Incentives for Renewables & Efficiency [8] and US EIA [9]).

3.9.1 Coal

Does not the United States have vast amounts of coal reserves? Why not just use it to produce electricity?

The United States does have large amounts of coal reserves, but the biggest issues with the use of coal for electricity generation are the air emissions and the solid waste coal produces when burned. Many state regulatory agencies have denied permits for new coal plants in recent years. Even in politically conservative states (such as Texas, Oklahoma, and Kansas), it is becoming increasingly difficult to obtain all necessary permits needed to construct a new coal plant, primarily due to atmospheric emissions and the large volume of water required for cooling purposes.

Emissions and resulting effects created by burning coal can include the following:

- Carbon dioxide (CO_2) emissions: global climate change
- Nitrous oxide (NO_x) emissions: smog
- Sulfur dioxide (SO_2) emissions: acid rain
- Mercury emissions: neurotoxicity, birth defects
- Particulate matter: asthma, respiratory problems

- Radioactive trace elements such as uranium and thorium: According to a report produced by the Oak Ridge National Laboratory in February 2008 entitled *Coal Combustion: Nuclear Resource or Danger* [10], coal plants can emit much more radioactive materials than comparable nuclear energy plant due to the presence of radioactive materials in the coal being burned.

3.9.2 Nuclear Energy

What about using nuclear energy since it does not produce air emissions?

While nuclear energy does not produce air emissions, the United States still has not concluded how it will permanently store the radioactive waste produced by nuclear power plants. While billions of dollars have been spent evaluating and preparing the Yucca Mountain site in southern Nevada as the national nuclear waste repository, the Obama Administration and Secretary of Energy Steven Chu ruled out its use for this purpose in March 2009, soon after the inauguration of President Obama. Currently, an estimated 60,000 tons of radioactive waste is stored on-site at the nation's 104 nuclear power plants. In addition to the issue of storing radioactive waste, many people are concerned about the safety of nuclear energy, remembering accidents at Chernobyl (Ukraine), Three Mile Island (Pennsylvania), and the recent incident at Japan's Fukushima Daiichi plant as a result of the earthquake and related tsunami in March 2011. Additionally, as with all thermal power plants, nuclear plants use large amounts of water for cooling.

3.9.3 Natural Gas

Why not just use natural gas to produce our electricity—isn't it a clean-burning fuel?

Natural gas power plants react quickly to changes in customer electric load and wind generation, so they actually facilitate the use of wind generation. Natural gas is a much cleaner-burning fuel than coal when used for electricity production but still produces NO_x, a contributor to smog in urban areas, and CO_2, a GHG. Natural gas plants may also require large volumes of water for cooling purposes, which can be a significant issue in the dry western regions of the United States. In addition, the US DOE projects that demand for natural gas will exceed supply around the year 2025. But perhaps the most significant issue with heavy reliance on natural gas for electricity generation is its price variability.

Figure 3.10 shows how volatile the price of natural gas can be, with prices doubling or tripling over the course of a few months. Electric utilities and their customers all like to have stable prices, so this type of volatility in natural gas prices can be troublesome. While natural gas generation can benefit wind generation through its ability to react quickly, wind energy can benefit companies relying heavily on natural gas generation by mitigating some of the impact of natural gas price fluctuation. When wind energy is being generated, it results in fossil fuels not being burned, and in most cases, that fossil fuel is natural gas. This is good for the environment, helps preserve fossil fuels for other uses or future uses, and reducing the demand for natural gas helps keep the price of natural gas lower. Thus, wind generation and gas generation are complementary.

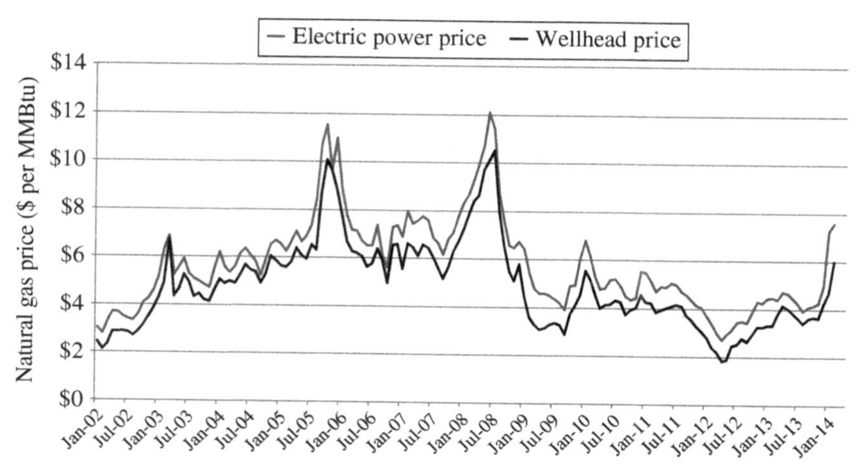

FIGURE 3.10 Monthly average natural gas price: wellhead price and delivered price for electric power generation (Source: R. Walker, based on data from U.S. Energy Information Administration, Monthly Average Prices of Natural Gas for Electric Power [11]).

REFERENCES

[1] U.S. Energy Information Administration. *Annual Energy Review 2011*. Washington, DC: US Govt Printing Office; 2012.

[2] U.S. Energy Information Administration website. Annual energy review; primary energy consumption estimates by source, 1949–2012. Available at http://www.eia.gov/totalenergy/data/annual/pdf/aer.pdf. Accessed December 14, 2013.

[3] Change.gov website. 2008. Obama-Biden comprehensive new energy for America plan. Available at http://change.gov/agenda/energy_and_environment_agenda/. Accessed August 12, 2012.

[4] American Wind Energy Association. *Annual Wind Industry Reports*. Washington, DC: American Wind Energy Association; 2000–2012.

[5] U.S. Department of Energy, Energy Efficiency and Renewable Energy. 2008. 20% Wind Energy by 2030: Increasing Wind Energy's Contribution to U.S. Electricity Supply, DOE/GO-102008-2567. Washington, DC: U.S. Department of Energy, Energy Efficiency and Renewable Energy.

[6] American Wind Energy Association. AWEA U.S. Wind Industry Annual Market Report 2013, Washington, DC: American Wind Energy Association; 2014.

[7] AWS Truepower. 2010. Study for the National Renewable Energy Laboratory, Wind Powering America website. Available at http://www.windpoweringamerica.gov. Accessed October 28, 2014.

[8] Database of State Renewable Incentives for Renewable & Efficiency (DSIRETM). Available at http://www.dsireusa.org/. Accessed August 10, 2012.

[9] U .S. Energy Information Administration website. Today in energy. Available at http://www.eia.gov. Accessed December 16, 2013.

[10] Gabbard A. *Coal Combustion: Nuclear Resource or Danger*. Oak Ridge: Oak Ridge National Laboratory; 2008.

[11] U.S. Energy Information Administration website. Monthly average prices of natural gas for electric power. Available at http://www.eia.gov/dnav/ng/ng_pri_sum_a_epg0_peu_dmcf_m.htm. Accessed November 24, 2013.

4

CONVERSION OF POWER IN THE WIND TO ELECTRICITY

4.1 WIND POWER PLANTS AND WIND TURBINES

4.1.1 How a Wind Turbine Works

This section discusses the following:

- Review of how a wind power plant is interconnected with an electric grid to deliver power to the load
- How a modern utility-scale wind turbine operates
- What its major components are
- How it converts energy in the wind into electrical energy delivered to the utility grid

Chapter 2 explained and demonstrated how electricity is generated at a power plant and then delivered through the transmission grid and a substation to the final point of use. The electricity is routed through a transformer and meter that measures the consumption of electrical power at the load for billing purposes. Chapter 2 also explained that a wind power plant can be incorporated into the system by connecting into the transmission system in parallel with existing power plants. Sometimes, these wind power plants are also known as wind farms, the designation that was first used in wind development in California in the 1980s. As electrical energy is produced by the wind farm, it is delivered to the grid in parallel with electricity produced by the conventional power plant.

Wind Energy Essentials: Societal, Economic, and Environmental Impacts, First Edition.
Richard P. Walker and Andrew Swift.

FIGURE 4.1 Modern utility-scale wind turbine (Source: R. Walker).

Figure 4.1 shows a picture of a modern utility-scale wind turbine. Notice the size of the turbine relative to the pickup truck at the base of the tower. This modern wind turbine consists of three rotor blades operating in an upwind fashion (i.e., the rotor operates upwind of the tower) with rotor blade pitch variation to control the power output of the turbine as a function of wind speed, which will be further described later.

When a number of these large-scale wind turbines are connected together, they are typically called a wind power plant or wind farm. Figure 4.2 shows a modern 3.0-MW wind turbine with some of the key elements shown along with the height of the tower (in this case 105 m).

4.1.2 Wind Turbine Anatomy

Figure 4.3 shows the architecture of the modern wind turbine. Wind comes into the rotor of the wind turbine (shown by the green arrow) where it interacts with the three blades of the turbine. The blades are typically constructed of fiberglass and can be rotated to change their pitch, which helps to control the power output of the turbine. The rotor blades on modern utility-scale wind turbines are specially designed and constructed for the wind industry so that they are able to withstand high wind loads, while capturing maximum energy reliably over the life of the turbine at reasonable cost (see Figs. 4.4, 4.5, and 4.6).

FIGURE 4.2 A modern 3.0 MW wind turbine installed in a wind power plant (Source: R. Walker).

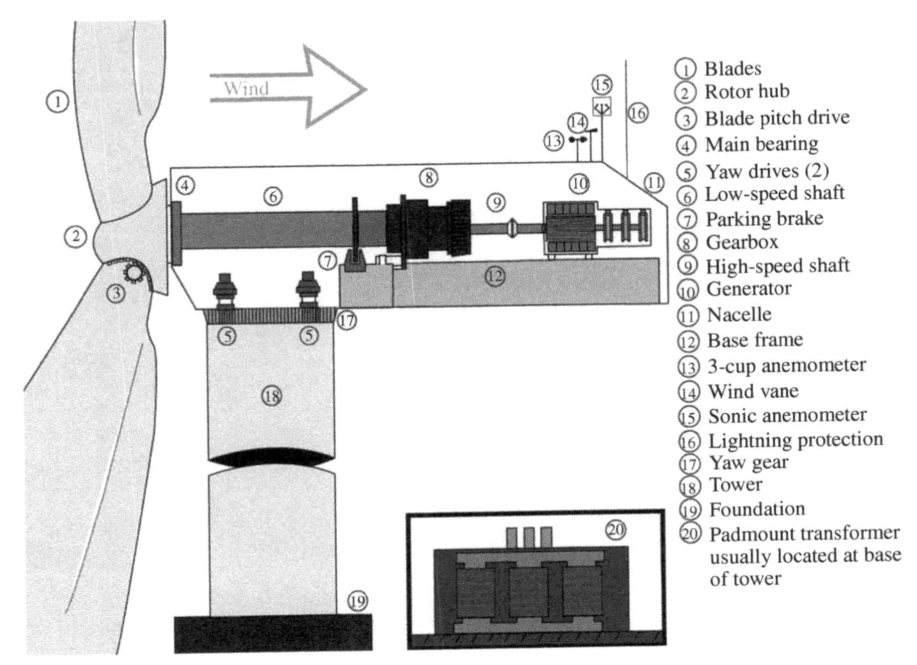

FIGURE 4.3 Wind turbine components (Source: Reproduced by permission of K. Jay).

FIGURE 4.4 Voith generator. notice the high speed shaft which will attach to the high speed output of the gearbox of Figure 4.5 (Source: Reproduced by permission of K. Jay).

FIGURE 4.5 ZF gearbox. Power from the rotor is transmitted to the gearbox by the low speed shaft (pictured on the right) through the gearbox to the high speed output shaft (Source: Reproduced by permission of K. Jay).

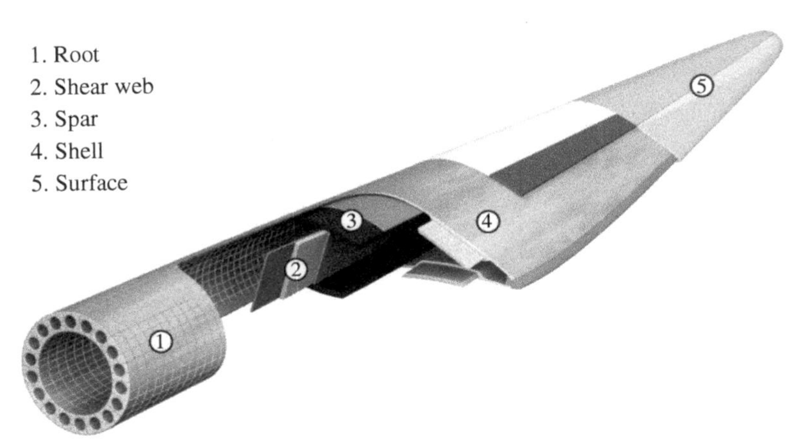

FIGURE 4.6 Wind turbine blade cross section (Source: Reproduced by permission of Gurit).

The rotor is then connected to a low-speed shaft—named that due to the fact that the rotor typically runs at a much lower speed than the generator. The low-speed shaft is connected to a gear box (indicated as item 8 in Fig. 4.3), which is then connected through the high-speed shaft to an electrical generator (item 10 in Fig. 4.3).

The turbine is kept aligned with the wind by yaw motors and a yaw drive system, which is shown directly under the low-speed shaft and on top of the tower. In order to control the wind turbine, an anemometer, which measures the wind speed, and a wind vane, which measures wind direction, are mounted on top of the nacelle (the name for the housing of the wind turbine parts on top of the tower). A controller signals the pitch of the blades and controls the yaw motor as part of the yaw drive system to keep the wind turbine aligned with the wind direction. On many newer turbines, a sonic anemometer (described later in this chapter) is used as the primary source of wind information to the turbine controller, with the cup and wind vane providing backup data.

Figure 4.7 is a another diagram of a wind turbine showing how the wind comes into the rotor, shown by arrow 1, which drives the turbine, as shown by arrow 2. The rotor then drives the low-speed shaft, shown by letter "C"; the gearbox shown by "D"; the high-speed shaft, shown by the black arrow; and finally the generator, creating electricity and delivering it to the utility grid.

Traditional electric generators for wind turbines must operate at high speed, while the rotors of large wind turbines rotate relatively slowly. The rotor speed is in fact inversely proportional to the size of the turbine—the larger the turbine, the slower the rotor turns for optimal performance. As a result, a gearbox is necessary to increase the rotor speed to the generator for it to operate efficiently. Some more recent turbine designs, however, use permanent magnets and direct-drive generators, which eliminate the need for a gearbox, as shown in Figure 4.8.

Most people think that modern utility-scale wind turbines operate in a fashion similar to a child's pinwheel, where wind is captured or collected in cups to drive the

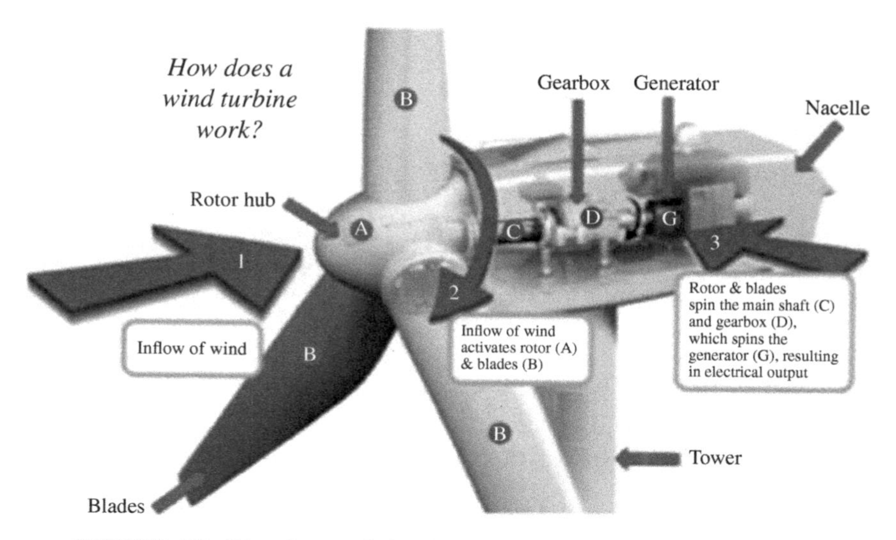

FIGURE 4.7 How does a wind turbine make electricity? (Source: NREL).

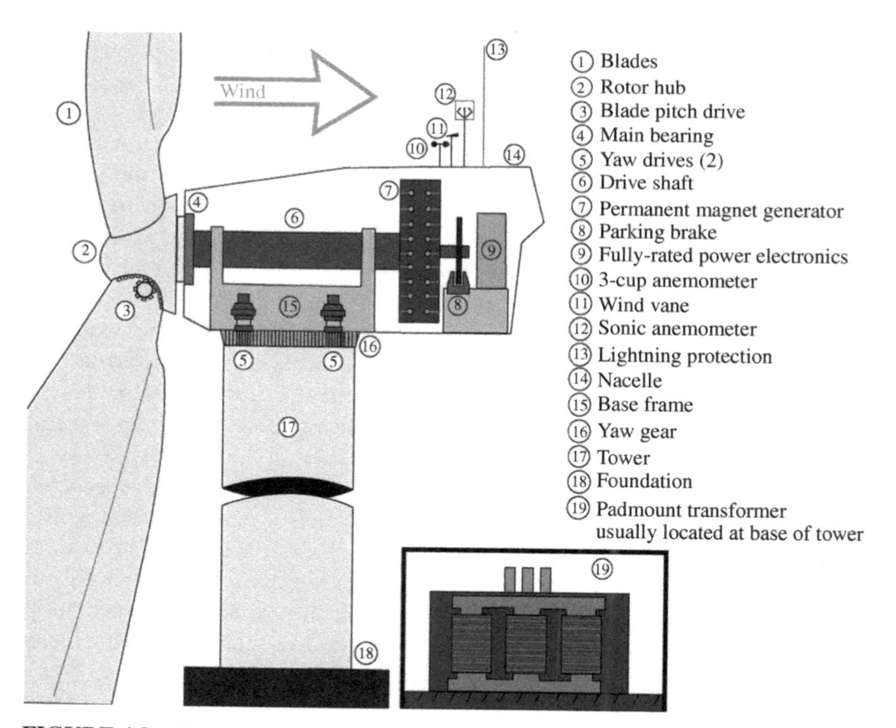

FIGURE 4.8 Direct drive architecture (Source: Reproduced by permission of K. Jay).

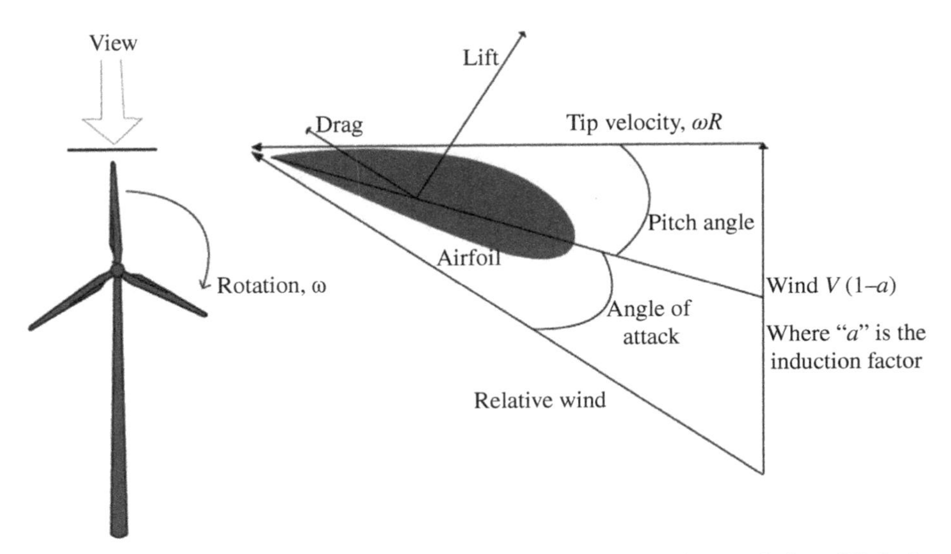

FIGURE 4.9 Lift forces make the rotor turn (Source: Reproduced by permission of K. Jay).

wind turbine. This is actually not correct. Figure 4.9 shows the aerodynamic forces generated by the rotor of a wind turbine that cause it to operate. The rotor blades on a wind turbine are actually airfoils just like the wing of an airplane. These airfoils generate lift-and-drag forces due to pressure differences on the surface when air passes over them. Because the outer portions of the rotor move at a high speed (usually between 50 and 100 m/s (~100–200 mph)), large forces are developed due to this high rotational velocity combined with the natural wind as shown by the Relative wind vector in the diagram. Relative wind is further explained in the example accompanying Figure 4.21 later in this chapter.

In the cut-away view of Figure 4.9, one can see the shadowed indication of the airfoil and how its pitch angle is shown relative to the rotation of the rotor. One can also see that the wind comes into the rotor as indicated by the arrow with some slowdown in the wind velocity $V(1-a)$, where "a" is the induced velocity factor, in order to extract the energy from the wind. It is the combination of the free-stream wind and wind produced by the rotational speed that allows modern wind turbines to be very efficient, by creating large lift forces that drive the tip speed of the rotor at speeds much faster than the actual wind speed. For an ideal rotor, it can be shown that the maximum wind extraction will occur when the induced velocity factor, a, is one-third—meaning that the wind speed at the rotor is slowed by one-third of its free-stream value upwind. This operating point is called the *Betz Limit* and will be further discussed later in the chapter.

Wind turbine sizes have increased dramatically in the last 15 years (Fig. 4.10). Enercon produces some of the largest turbines being offered commercially, such as their Enercon E-126 with a 126 m rotor diameter and a hub height of 135 m. However, as of the writing of this book, several manufacturers are working on the development of even larger turbines. Figure 4.11 shows the increasing size of turbines from 1995, when the state of the art was a turbine of only 40 m diameter on a 60 m tower, to

FIGURE 4.10 Both onshore and offshore wind turbines continue to increase in size (Photo Credit: Phil Hollman, http://commons.wikimedia.org/wiki/File:Off-shore_Wind_Farm_Turbine.jpg).

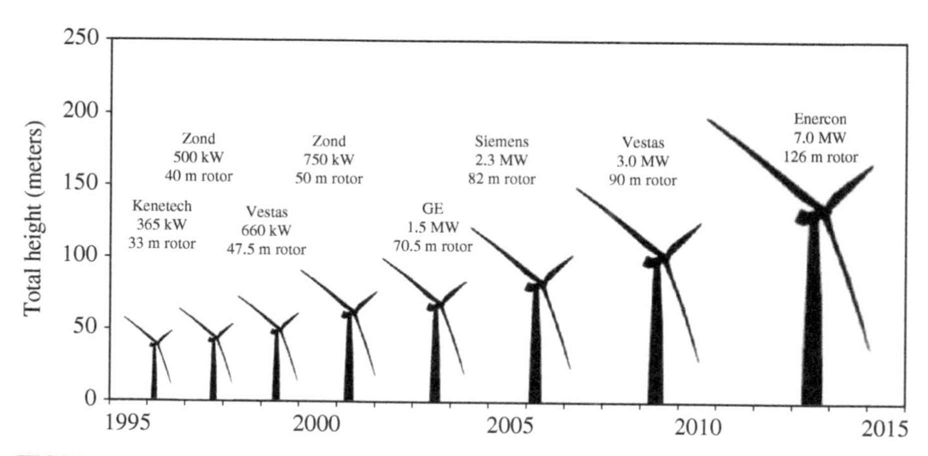

FIGURE 4.11 Rapidly increasing size of commercially available wind turbines (Source: R. Walker).

today's turbines in which hub heights of 100 m or greater are becoming common as are rotor diameters in excess of 100 m. Total height to the tip of the blade can reach almost 200 m. For comparison, the Statue of Liberty, from the bottom of the base to the top of the torch, is only 93 m.

4.1.3 The Power Curve

Wind turbines produce power as a function of the wind speed. The relationship between the power output and wind speed is shown on what is called a "power curve," as indicated in Figure 4.12. Wind speed is plotted along the bottom axis with power output on the vertical axis, here measured in kilowatts. (2300 kW is equivalent to 2.3 MW of power.) To understand the magnitude of these power ratings, a 2.3-MW turbine (as indicated in this power curve) would produce annually sufficient electrical energy to supply approximately 700 average households, so this is a significant amount of power. Figure 4.12 also gives some details about the power curve showing what is called the "cut-in" wind speed, measured in meters per second—typically on the order of 3.0–3.5 m/s. The power output increases with wind speed until rated wind speed and rated output power are reached.

For the wind turbine indicated in Figure 4.12, rated wind speed would be 15 m/s and rated output power is 2300 kW, or 2.3 MW. Beyond rated wind speed, turbine power is controlled by pitching the blades and is held at rated power until a maximum wind speed is reached (typically at around 25 m/s, 56 mph) when the wind turbine is shut down to avoid damage in high winds by pitching the blades to align with the wind, called "feathering," which stops the wind turbine.

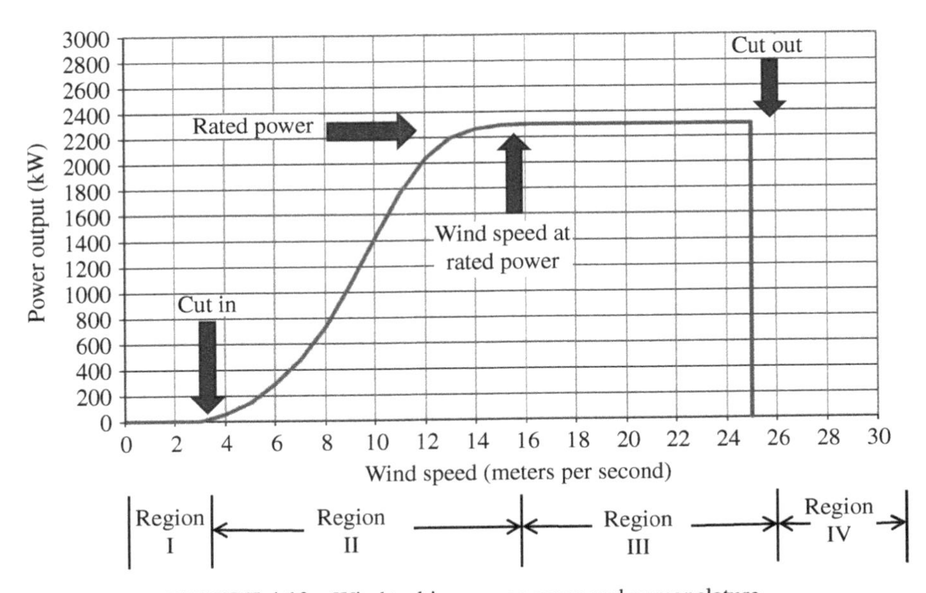

FIGURE 4.12 Wind turbine power curve and nomenclature.

4.1.4 The Wind Power Formula

A question often asked is how efficient are modern wind turbines at converting wind energy into electrical energy. To answer that question, we have to first understand how to calculate the power in the wind and then to estimate how much of that power can be extracted on a theoretical maximum basis and subsequently how much power is actually extracted by modern wind turbines. It was mentioned previously that modern wind turbines amplify the wind speed by using the speed of the wind as well as the speed of the rotation of the rotor to generate high lift forces and high driving forces for the wind turbine; they can in fact be quite efficient. As we will see in this section, modern wind turbines extract almost 50% of the energy available in the wind. The theoretical maximum is 59%.

Figure 4.13 is a schematic that shows the wind power equation. We will go through the equation in a step-by-step fashion in this section along with the physical units that are used. In describing the equation, we will use standard international units, which are usual for the wind industry for several reasons.

First, there is a very strong European element to modern wind turbine production and design where standard international units are used. These include the kilogram, meter, second, watt, and so forth. When one uses standard international units, the power is calculated in watts, rather than in English units of horsepower. One can then make a direct conversion of mechanical power in watts to electrical power in watts. This is convenient since modern wind turbines are generating electricity.

As shown in Figure 4.13, there are several key elements in the wind power equation. The power, as mentioned previously, is calculated in watts and can then be divided by 1000 to get kilowatts (kW) or by 1,000,000 to get megawatts (MW). Air density (the mass of the air per unit volume) is represented by the Greek symbol ρ. The rotor-swept area can be calculated as the area of a circle, or πr^2, where r is the length of one rotor blade, measured in meters and then squared. The wind velocity is represented in meters per second. Later in this section, we will consider the power density of the wind, which uses the wind power equation with $A = 1\,\text{m}^2$, giving the wind power density in watts per square meter. This is often used in wind resource mapping.

We will now consider each of the elements of the equation. The most important parameter is the wind velocity, since it is cubed. For example, if the wind is blowing at 10 m/s. The cube of 10 is 1000. If the wind increases to 11 m/s, the cube of 11 is 1331. Thus, a 1 m/s increase in wind speed from 10 to 11 is a 10% increase in wind speed—which gives an increase in wind power available of 33%. This cubic factor in the equation is very powerful and is why the wind power curve increases very rapidly from cut in to rated power.

The next element of importance is the size of the rotor, since the power is a function of the radius, or blade length, squared. A 40-m radius rotor would result in a 5026 m^2 area for the rotor, using the formula for the area of a circle, πr^2. If that is increased to a 44-m radius, the area increases to 6082 m^2. Thus, a 10% increase in rotor radius gives a 21% increase in power output. Again, this is due to the fact that the area is related to radius as a function of the square.

Next in importance is the density of the air, which is the mass or weight per unit volume. It is easy for many people to think of air as having little or no weight, but in

Power (watts) = $P = (1/2)\ \rho A V^3$

where

- ρ = air density (kg/m^3)
- A = swept area of turbine rotor = πr^2
- V = wind velocity (m/s)
- r = radius of turbine rotor (m)

Note: 1 Watt = 1 kg-m^2/s^3

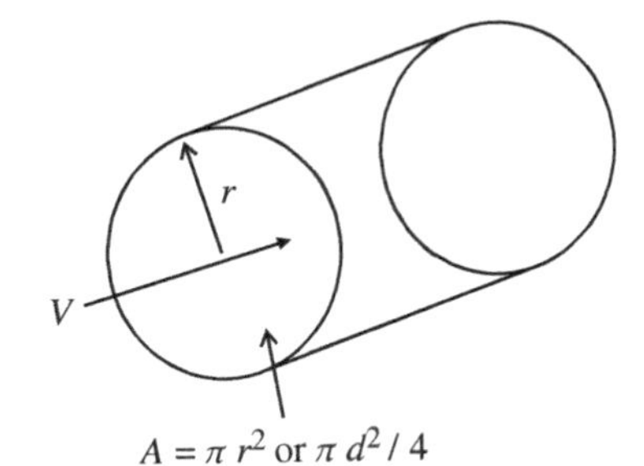

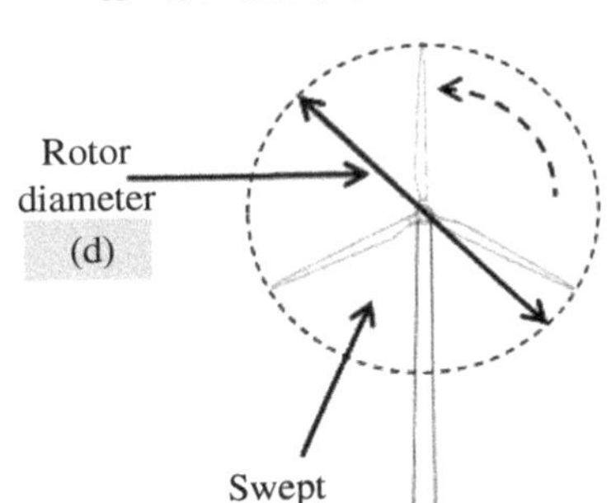

FIGURE 4.13 The power in the wind across the swept area of the turbine rotor (Source: R. Walker).

fact it has significant weight—which results in the wind actually producing a force. If there were no air density, there would be no force due to the wind. The density of the air is directly related to the elevation above sea level as well as to the pressure and temperature due to the gas law. Increasing temperature causes the density to decrease; increasing pressure causes the air density to increase. Due to decreasing atmospheric pressure with increasing elevation, the air density also decreases with elevation. When air pressure and temperature information are unknown, a simplified method to estimate air density is

$$\rho = 1.225 - (0.00011 \times z)$$

where

ρ = air density in kg/m³

z = elevation above sea level in meters

Table 4.1 shows some typical values of air density as a function of elevation and location and the related decrease in power at different locations due to the change in altitude above sea level.

Figure 4.14 shows a graph of the change in air density as a function of altitude above sea level for the Standard Atmosphere (sea level, 15°C, 1 atm pressure). The actual data are very close to the approximation using the simplified formula

TABLE 4.1 Air Density[a]

Standard sea level air density	1.225 kg/m³
Lubbock, Texas Air Density (elevation = 1000 m)	1.112 kg/m³
Denver, Colorado Air Density (elevation = 1610 m)	1.047 kg/m³
Decrease in power output (Lubbock to Denver)	5.7%
Decrease in power output (Galveston to Denver)	14.5%

[a]Source: R. Walker.

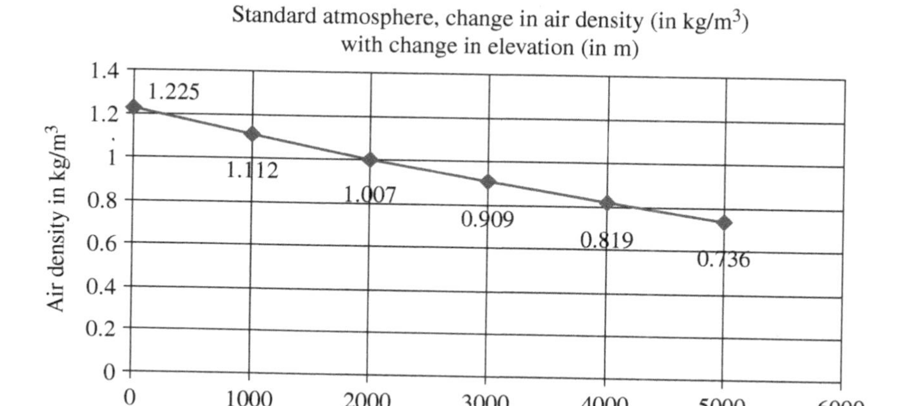

FIGURE 4.14 Air density versus altitude (Source: A. Swift).

presented earlier. One can use this graph to estimate air density and wind power available at a certain location, as compared with operating at sea level. The wind turbine power curve that was introduced earlier is usually presented at sea level. The operator and wind farm project designer must make the appropriate adjustments for altitude of the wind farm to account for these density changes. In addition, turbine output can vary due to decreasing air density on hot days with low pressure, as compared with cold days with high pressure—even at a given location. Thus, when additional data such as air temperature, air pressure, and relative humidity are known, more precise calculations of air density can be made for each increment of time being considered.

Not all wind power available can be converted to shaft power, meaning that all of the energy available in the wind cannot be converted. This operating limit is called the *Betz Limit*, after Albert Betz who first completed these calculations in the 1920s while working on rotor aerodynamics and propeller design. Betz calculated for an ideal rotor that 16/27 or about 59% is the maximum fraction of power that can be extracted from the wind. This number can be derived from a concept called *momentum theory*, used in helicopter, propeller, and wind turbine rotor design.

The fraction of power extracted from the wind by a rotor is called the coefficient of performance, or the efficiency of the rotor. It is dimensionless and usually given the abbreviation C_p. It is defined mathematically as

$$C_p = \frac{\text{Rotor power}}{\text{Power in the wind}}$$

or

$$C_p = \frac{P_{LSS}}{\frac{1}{2}\rho AV^3}$$

where low-speed shaft power (P_{LSS}) is the power delivered to the low-speed shaft from the rotor blades, in watts, and the denominator is the power in the wind, in watts.

The rotor power coefficient, C_p, is determined by the operating state of the rotor—pitch of the blades (for maximum power extraction, modern wind turbines set the blade pitch angle near zero or flat to the wind), speed of the rotor in relation to the wind speed, and the induced velocity factor, a—mentioned earlier. For an ideal rotor with no losses due to friction or aerodynamic drag, the relationship between C_p and induced velocity can be mathematically derived. That relationship is shown graphically in Figure 4.15. Notice that the maximum C_p (rotor efficiency) is produced when the induced velocity is one-third and its value is 59%, *the Betz Limit*, as described earlier.

For actual wind turbine rotors with friction losses and aerodynamic drag, the induction factor, which is not easily measured, can be related to the ratio of rotational tip speed of the rotor to the wind speed, called the *Tip Speed Ratio* (TSR)—which is easily measured. If the TSR is 1, for example, the rotor blade tips are spinning at the same speed as the wind. Using the definitions of C_p and TSR, a rotor efficiency curve for a typical rotor operating at zero degree blade pitch angle is shown in Figure 4.16. Note that the maximum rotor efficiency for this rotor is about 48–50%, which is typical of modern wind turbines, at TSRs near a value of 8. In other words, maximum performance for this rotor occurs when the tips of the rotor are rotating with a speed about eight times the wind speed.

To complete the conversion of power in the wind to electrical power, we next consider the gearbox and generator. In most modern wind turbines, the low-speed shaft is coupled to a gearbox to increase the shaft speed and then the high-speed shaft is coupled to the generator. The generator is constructed internally with coils of wire that rotate relative to a magnetic field and produce electricity. Each of these processes has losses due to friction and power transfer. These losses are often combined into a single efficiency parameter for the gearbox and generator:

$$\varepsilon = \text{combined gearbox and generator efficiency} \approx 90\%$$

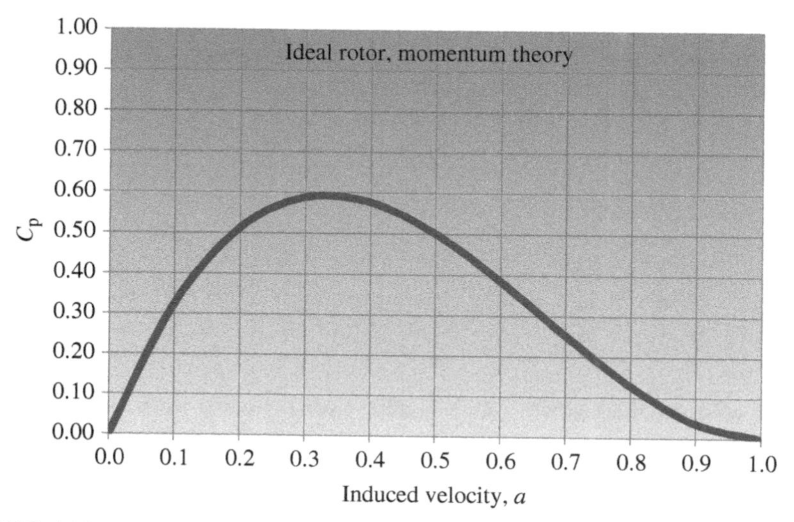

FIGURE 4.15 Rotor coefficient of performance versus induced velocity, ideal rotor and momentum theory (Source A. Swift).

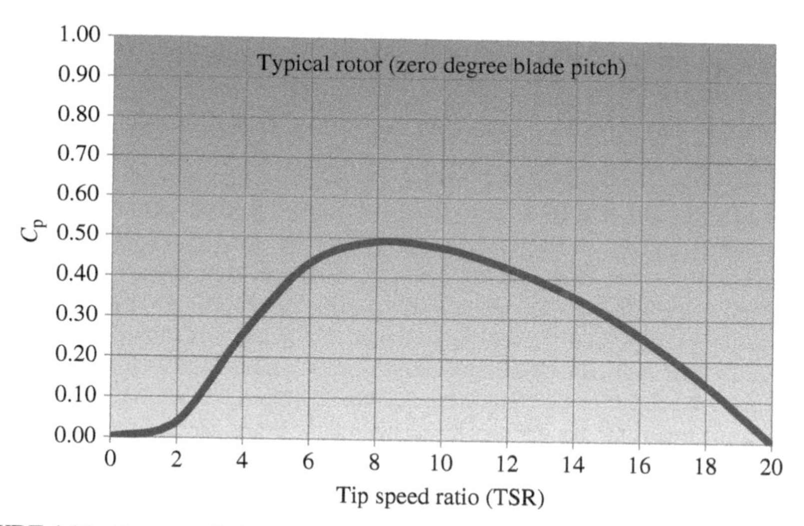

FIGURE 4.16 Rotor coefficient of performance versus TSR, Zero degree pitch, typical rotor (Source A. Swift).

An overall conversion efficiency of power in the wind to electrical power delivered can now be established. The equation for calculating power (in units of watts) is

$$P_{\text{elec.}} = (C_p)\,(\varepsilon)\,(\tfrac{1}{2}\rho \text{AV}^3).$$

Further development of these topics can be found in text books on Wind Turbine Engineering or Wind Turbine Aerodynamics and are beyond the scope of this book.

4.2 ATMOSPHERIC SCIENCE

4.2.1 Wind Characterization

Generally, the wind blows everywhere on the surface of the Earth. It may vary in speed and direction, and it may blow as a steady breeze or in turbulent gusts, but the characterization of wind is important if one has the objective to produce energy from a wind turbine. Wind is the movement of air over the surface of the planet. As we have presented in previous chapters, the wind has inspired mythology; influenced events in history; expanded transport and warfare; and provided a power source for mechanical work, recreation, and electricity. Wind can shape land forms due to erosion and dust storms and drive the spread of wildfire. When combined with cold temperatures, wind has a chilling effect on people and livestock—known as the wind chill factor. Wind has two components, speed and direction, from which the wind is blowing. Short bursts of wind are typically called gusts, while strong winds or wind storms are called squalls and are often associated with thunderstorms. Longer-duration effects of high wind are called gales or hurricanes.

Wind typically occurs when air moves from a high-pressure area to a lower-pressure area, and then its direction is affected by the rotation of the planet in what is called the Coriolis Effect. This effect causes clockwise rotation around a high-pressure area and counterclockwise rotation around a low-pressure area, as shown in Figure 4.17 for the Northern Hemisphere (Fig. 4.18).

Differential heating of the Earth's atmosphere by the Sun between the equator and the poles is a major driving force for wind circulation. These general circulation patterns lead to certain latitudes having what are called the "trade winds," meaning winds of fairly reasonable strength from a given direction, and other areas called the "doldrums," a belt of calm winds north of the equator between the northern and southern trade winds. During the age of sail, understanding of these wind patterns was critical to effective circumnavigation of the globe.

The variations in wind speed, direction, location, and the time of year and day all combine to make the generation of energy from wind very challenging. A study of meteorology is required to understand and to describe the characteristics of the wind, which is important for the effective production of energy from wind resources (Fig. 4.19).

4.2.2 Wind Storms

As mentioned in the previous section, wind can be extremely variable, going from light breezes to gale and hurricane force winds and in extreme cases tornadic winds. Although the energy in a hurricane or tornado is immense, its value as a source of energy is negligible since devices to extract energy from these strong winds are presently unavailable. Even thunderstorms, which typically have winds above 25 m/s or 50 mph,

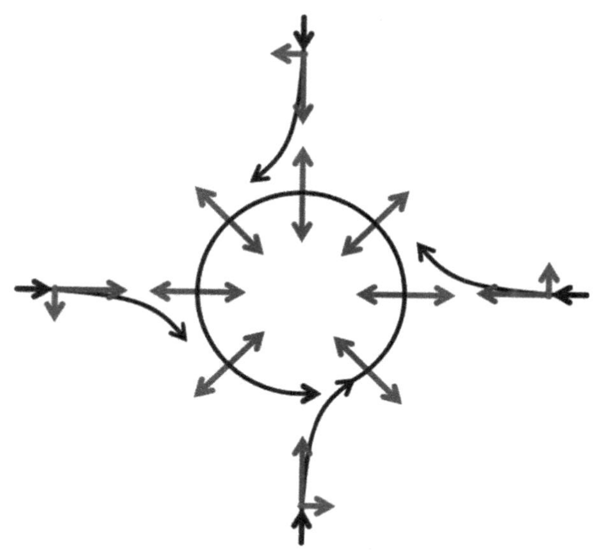

FIGURE 4.17 Illustration of Coriolis Effect; schematic representation of flow around a low-pressure area in the Northern Hemisphere. The pressure gradient force is represented by blue arrows. The Coriolis force, always perpendicular to the velocity, is shown by red arrows. The resulting air flow is indicated by the black lines and arrows showing counterclockwise circulation (Photo Credit: Roland Geider, http://en.wikipedia.org/wiki/File:Coriolis_effect10.svg).

FIGURE 4.18 Low-pressure system rotation due to Coriolis Effect. The image was taken by the Aqua MODIS instrument in September by NASA. Because this low-pressure system occurred in the Northern Hemisphere, over Iceland, the winds spun in toward the center of the low-pressure system in a counterclockwise direction, due to the Coriolis force. In the Southern Hemisphere, the Coriolis force would be manifested in a clockwise direction of movement (Photo Credit: NASA, http://en.wikipedia.org/wiki/File:Low_pressure_system_over_Iceland.jpg).

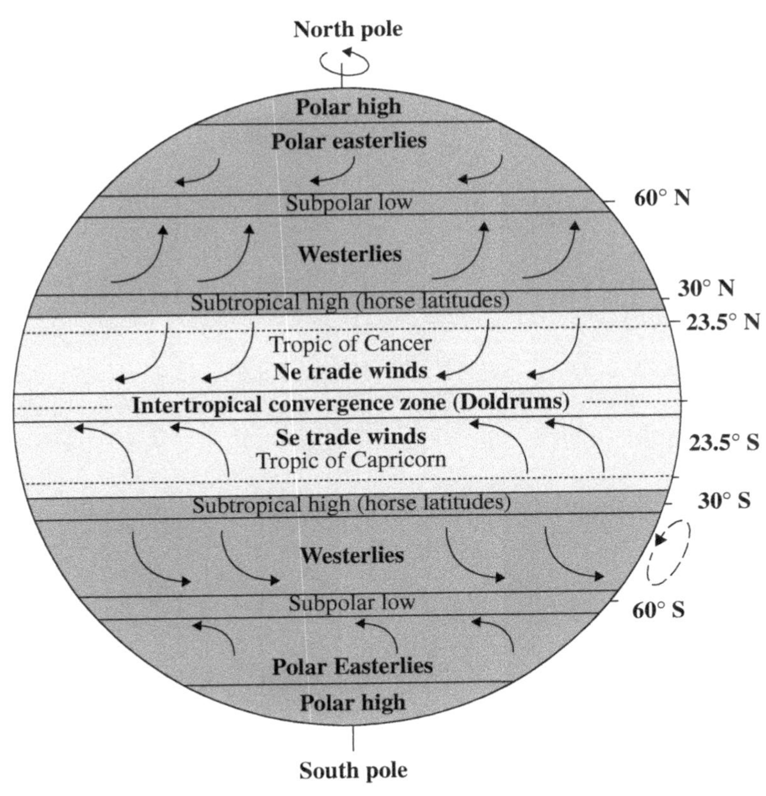

FIGURE 4.19 Trade winds (Credit: U.S. Geological Survey Department of the Interior/USGS).

are not considered useful for energy production due to the short duration of these high winds and their detrimental effect on wind machines. Studies of thunderstorms, hurricanes, and tornadoes and their effects on structures (such as buildings and bridges) are considered under the topic of *wind engineering*. Texas Tech University (TTU) has a well-established wind engineering group, and the reader may have seen demonstrations of the TTU tornado debris air cannon, or news stories of the TTU storm chasing group gathering valuable data from wind storms.

4.2.3 Wind Measurement

Wind direction is always reported by the direction from which it originates. For example, a westerly wind blows from the west toward the east. A southerly wind blows from the south toward the north. A statistical distribution in which the direction the wind blows at a given site and for what fraction of the time is called a wind rose and is used in wind plant siting. Figure 4.20 shows a wind rose for Reese Technology Center in Lubbock, Texas, from Texas Tech University's West Texas Mesonet system. This shows the fraction of time, in percent, the wind blows from a given direction.

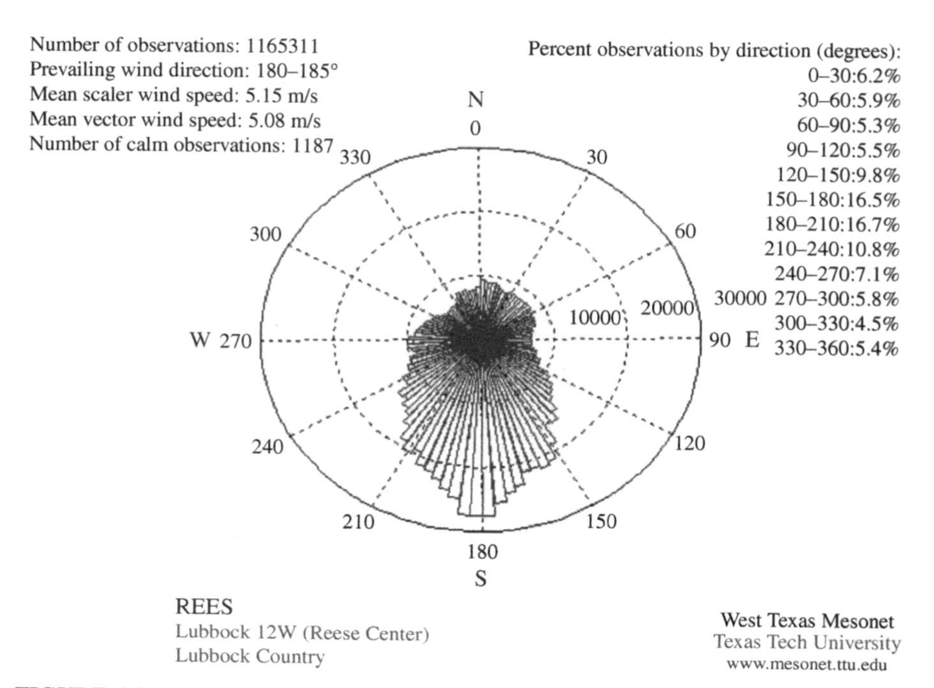

FIGURE 4.20 Wind Rose for Reese Technology Center, Lubbock, Texas (Source: Texas Tech University's National Wind Institute).

Since wind has both a speed and a direction, it can best be represented as a *vector* quantity. Vectors add graphically by combining them in a tip-to-tail fashion, and vector addition is used to calculate total wind effects. A simple example is shown in Figure 4.21. A jogger runs at 3 m/s (6.7 mph) heading East (090°) (see Fig. 4.20 for compass directions). The red vector arrow in Figure 4.21 shows the wind vector opposite to the direction of the jogger, indicating the wind effect on the jogger due to the jogger's running. There is a crosswind from the North (000°), blue vector arrow, at 4 m/s (8.9 mph). The total relative wind, black vector arrow, that the jogger feels is the vector sum of the two wind vectors and is 5 m/s (11.2 mph) from the Northeast—a compass direction of $(090° - 53°) = 037°$. Vector addition can be accomplished graphically, as shown here, or analytically using geometry and trigonometry to solve for the unknown distances and angles.

A windsock, like the one shown in Figure 4.22, is a device that measures both the direction and the approximate speed of the wind. The direction of the sock indicates wind direction, and the angle at which the sock is lifted from the ground indicates approximate wind speed. When horizontal with the ground, most windsocks indicate that the winds are in excess of 15–20 mph. These devices are typically used at airports to assist pilots in judging the wind for takeoff and landing.

Wind speed is more accurately measured by an *anemometer*. Anemometers for wind resource measurement are typically mounted on meteorological towers as close to the expected wind turbine hub height as possible. Anemometers for wind turbine control are typically mounted on top of the turbine nacelle, as described earlier in

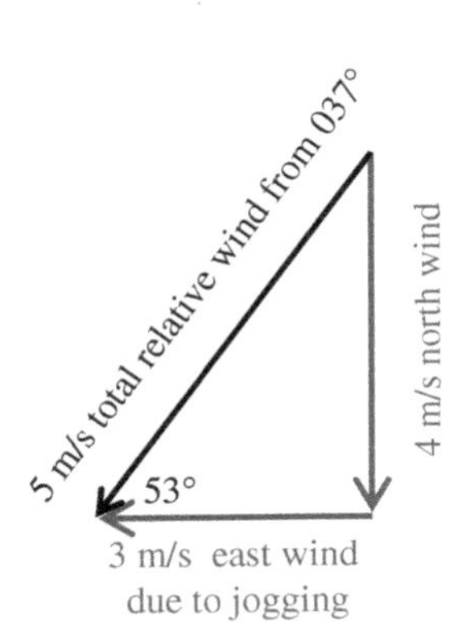

FIGURE 4.21 Vector diagram for total wind effects solution to the jogger problem with a crosswind (Source A. Swift).

FIGURE 4.22 Windsock (Source: Reproduced by permission of K. Jay).

Section 4.1.1. Three types of anemometers are shown in Figure 4.23, as well as a wind vane for measuring wind direction. The sonic anemometer is the most modern instrument and measures both wind speed and direction by measuring the transit time of ultrasound pulses between opposing transducers. By comparing the two sets of transducers, the instrument can measure both the speed and direction of the wind passing through the sensor.

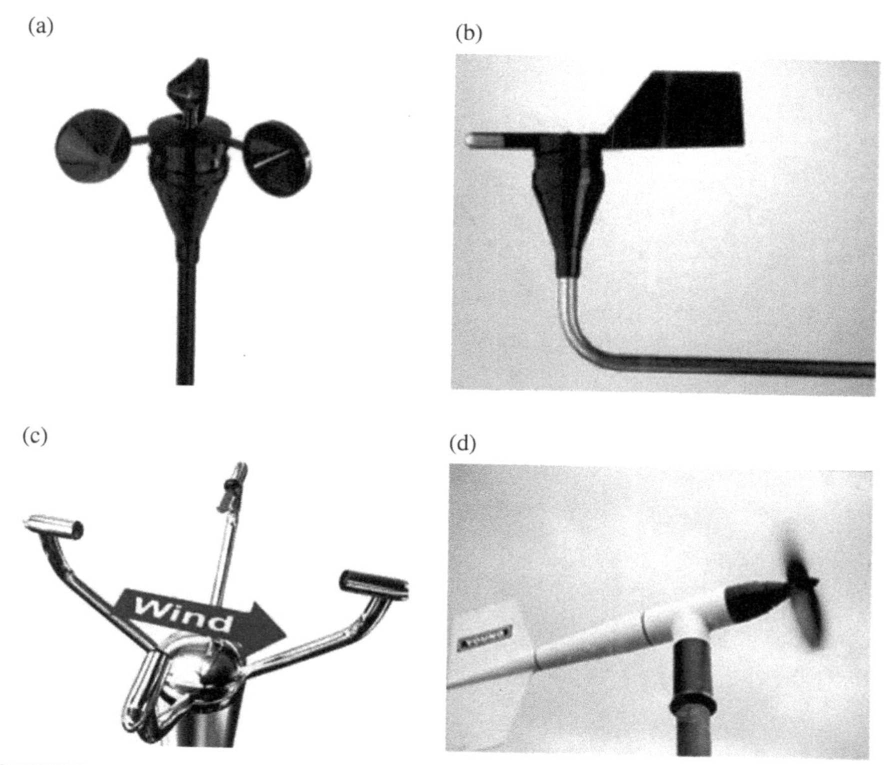

FIGURE 4.23 Three-cup anemometer (a), wind directional vane (b), sonic anemometer (c), and prop-vane anemometer (d) for measuring the wind. The cup anemometer measures only wind speed and the directional vane measures only direction, while the sonic anemometer and the prop-vane anemometer measure both speed and direction (Sources: Two left photos courtesy of NRG Systems; two right photos courtesy of the R.M. Young Company).

Sonic anemometers have become the instrument of choice for turbine control since they have no moving parts. However, the instrument is subject to failures caused by heavy rains, dust, debris, and other environmental factors and thus may require frequent repair and/or calibration.

The cup anemometer measures wind speed by the fact that the cups rotate with the speed of the wind. The wind vane measures the direction of the wind. Currently, these are the industry standard for resource measurements and are often used as backup for the sonic anemometers for turbine control. The wind turbine in Figure 4.24 shows the arrangement of the anemometers on the nacelle of a modern utility-scale wind turbine. Lastly, the prop-vane anemometer, shown in Figure 4.23 (d), measures both speed and direction of the wind and is typically used for wind research and weather data stations.

Weather data are typically collected at 10 m (33 ft) in height from the ground and averaged over 10 min for wind energy analysis. Averaging time for wind speed measurements is important. Wind is often reported with a 5- or 10-min average and a 3-s gust. Additionally, remote sensing technologies are being used to measure wind by collecting

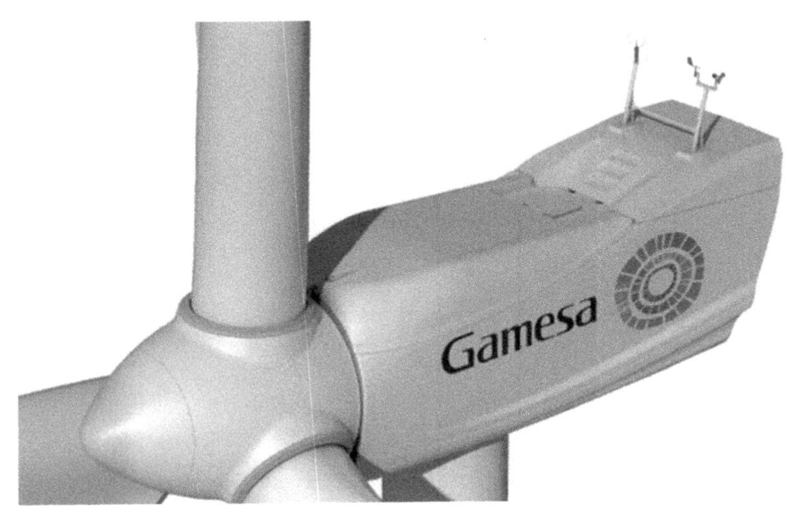

FIGURE 4.24 Arrangement of Anemometers on the Nacelle of a Modern Utility-Scale Wind Turbine (Source: Gamesa G114-2.0 MW 60-Hz Turbine, Courtesy and copyright Gamesa Technology Corporation, Inc., all rights reserved.).

FIGURE 4.25 SODAR wind measuring unit at Reese Technology Center (Source: National Wind Institute at Texas Tech University).

reflected energy from particles travelling in the wind stream, such as dust or water vapor. Ultrasound (SODAR, SOnic Detection and Ranging, Fig. 4.25), lasers (LIDAR, LIght Detection and Ranging), and radio (RADAR, RAdio Detection and Ranging) are now used as remote sensing devices for more sophisticated wind measurements.

FIGURE 4.26 One of two Ka-Band Mobile RADAR trucks for dual-Doppler wind measurement (Source: National Wind Institute at Texas Tech University).

To obtain the full wind field of speed and direction of two radar trucks, Figure 4.26 are positioned and operated simultaneously in a dual-Doppler arrangement. A dual-Doppler sample is shown below for a wind farm in West Texas, Figure 4.27. The hotter colors represent higher wind speeds, as indicated in the scale, measured in m/s. The wind turbine wakes are clearly visible.

Units for wind speed measurement vary, but in the wind power industry, they are typically meters per second. Other common units include miles per hour, kilometers per hour, and nautical miles per hour (knots). Wind direction is measured using the points of a compass, in the direction the wind is coming from and measured in 1° increments starting with 0° for North going clockwise to 90° east, 180° south, and 270° west, as shown in Figure 4.21.

4.2.4 Wind for Energy Production

Using measured data of wind speeds coupled with computer models to fill in data from other locations, analysts are able to produce wind maps that show wind resource availability for the production of wind energy. A typical map is shown in Figure 3.7 of the previous chapter in which the colored scale indicates the class of the wind speed and its characterization for energy production. As pointed out in Section 4.1, wind energy is dependent not only on wind speed but also air density, and thus, wind maps take into account the density of the air and its change with elevation.

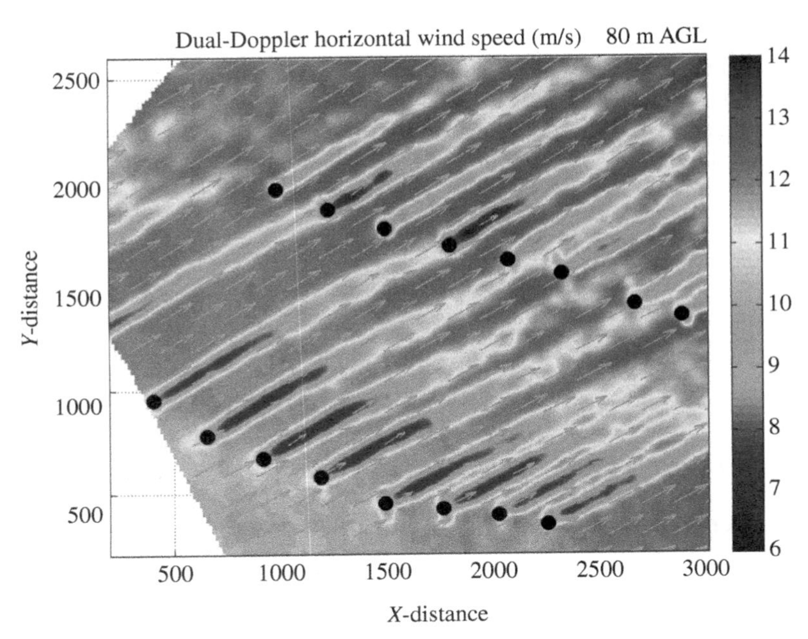

FIGURE 4.27 Dual-Doppler RADAR image from Texas Tech University (Source: National Wind Institute at Texas Tech University).

The central area of the United States is the predominant area for wind resource availability. This wind corridor runs from the South Plains of Texas through the center of the country to North Dakota and on into Canada. This is typically an area of low rainfall and thus has very few trees and limited vegetation as well as sparse population; however, it represents a major area for grain production coupled with the availability of significant wind resources. Wind characteristics in this area can be very unusual in that it is a region of frequent thunderstorms, strong winds and tornadoes, and a phenomenon called the "nocturnal low-level jet."

A nocturnal low-level jet is a "river of wind" that is produced due to a number of large-scale effects but results in higher wind speeds close to the ground during the night, especially in the summer months. These occur frequently and in fact dominate the wind characteristics above 50 m in height. Low level jets contribute to the large wind resource available for wind turbines that have towers above 50 m and result in significant nighttime production of wind energy. Below the 50-m threshold, data collected from 30-m meteorological towers in weather data networks show an increase in wind speed during the day, particularly in the afternoon. These unusual wind characteristics in the Great Plains are a significant factor in the development of wind power plants throughout the region and in the misalignment between the need for peak electrical power production and wind power availability as will be discussed later.

A second US wind resource feature is located in the coastal regions, where daytime heating of the land causes onshore sea breezes. These coastal breezes are often coincident with the need for peak power on hot summer afternoons. This coincidence of wind power production and utility company need for power can significantly add value to wind energy resources in coastal regions.

The third predominant US wind resource area is offshore. A significant fraction of this wind resource is located within 10 miles of the coastline. Access to these wind resources will require the installation of offshore wind turbines to extract the energy and to deliver it by cable to load centers near the coast. Since coastal areas are heavily populated (as compared with inland areas), the availability of wind power to coastal load centers could be significant for the future of wind power development. This will be discussed in more detail in later chapters. A major issue with offshore development, however, especially in the Gulf of Mexico, is the rather frequent occurrence of tropical storms and hurricanes, which can provide a challenge to offshore wind plant designers.

4.3 WIND STATISTICS

4.3.1 Wind Statistical Distributions

If one places a wind anemometer on a meteorological tower at a given location and continuously records the output of the anemometer, one will observe a time record of wind speed, similar to that shown in Figure 4.28.

Time records like this are of interest, but one can imagine that collecting these records for long periods of time at many locations would be extremely cumbersome and result in large quantities of data. An alternative to a time series is the frequency series data collection method. This is called the *method of bins*. Using timed average values, usually 10 min, the average wind speed, within a certain range, is recorded as a histogram. For example, using a 1 m/s bin width, if the wind speed average is between 4½ and 5½ m/s, this will be the 5 m/s wind speed bin. Every time the 10-min average is in that range, a count of one will be added to the total count in the 5 m/s wind speed bin. Over time, one can build up the number of counts in each bin,

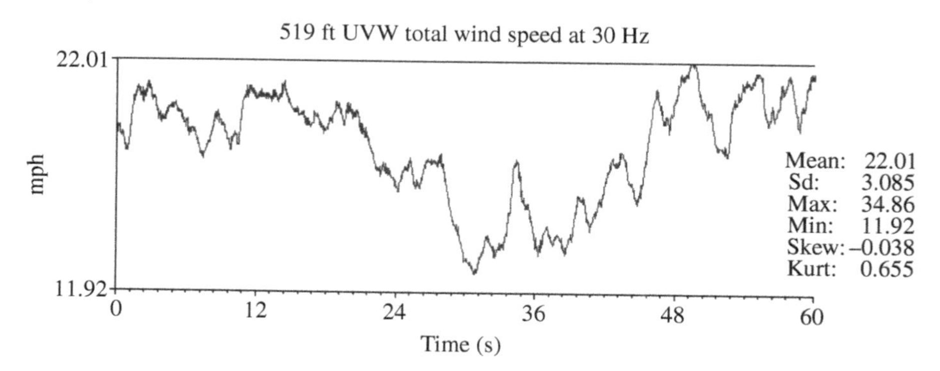

FIGURE 4.28 Time record of total wind speed in mph (Source: National Wind Institute at Texas Tech University).

providing a frequency distribution of the wind at that site, which can then be converted into the number of hours in a month or a year that the wind blows at a given speed for that site.

Notice in Figure 4.29 the difference in the distribution of wind speeds when one measures at 78 m above the ground instead of 10 m. The annual average wind speed at 78 m is 64% greater than that measured at 10 m for calendar year 2005. The increase in wind speed at greater heights is the reason that turbine tower sizes continue to increase.

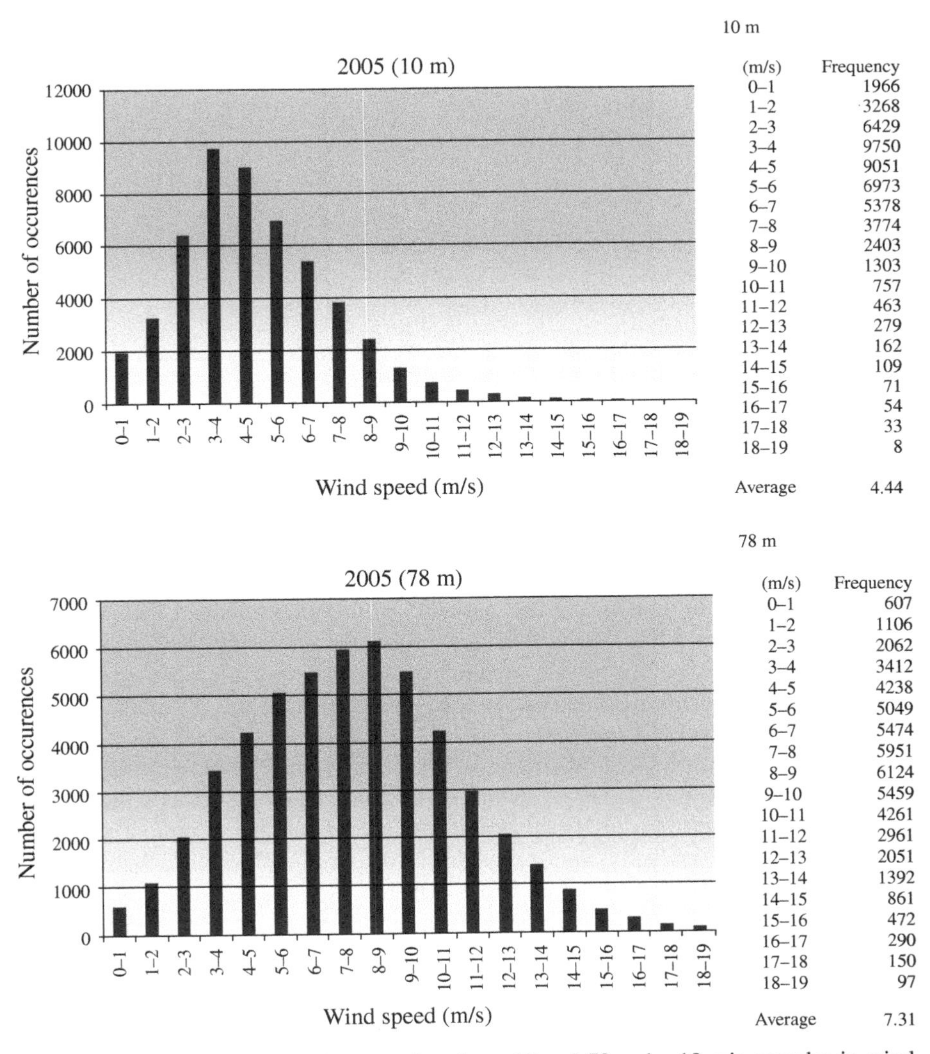

10 m

(m/s)	Frequency
0–1	1966
1–2	3268
2–3	6429
3–4	9750
4–5	9051
5–6	6973
6–7	5378
7–8	3774
8–9	2403
9–10	1303
10–11	757
11–12	463
12–13	279
13–14	162
14–15	109
15–16	71
16–17	54
17–18	33
18–19	8
Average	4.44

78 m

(m/s)	Frequency
0–1	607
1–2	1106
2–3	2062
3–4	3412
4–5	4238
5–6	5049
6–7	5474
7–8	5951
8–9	6124
9–10	5459
10–11	4261
11–12	2961
12–13	2051
13–14	1392
14–15	861
15–16	472
16–17	290
17–18	150
18–19	97
Average	7.31

FIGURE 4.29 Data collected at two elevations, 10 and 78 m, by 10-min samples in wind speed bins (Source: National Wind Institute at Texas Tech University).

Rayleigh distribution function

$$f(u)=\left(\frac{\pi}{2}\right)\left(\frac{\upsilon}{\bar{\upsilon}^2}\right)\exp\left[-\frac{\pi}{4}\left(\frac{\upsilon^2}{\bar{\upsilon}^2}\right)\right]\Delta\upsilon$$

where $f(u)$ = the probability of a given wind speed
υ = the wind speed of interest
$\bar{\upsilon}$ = the mean wind speed
$\Delta\upsilon$ = the bin width in m/s

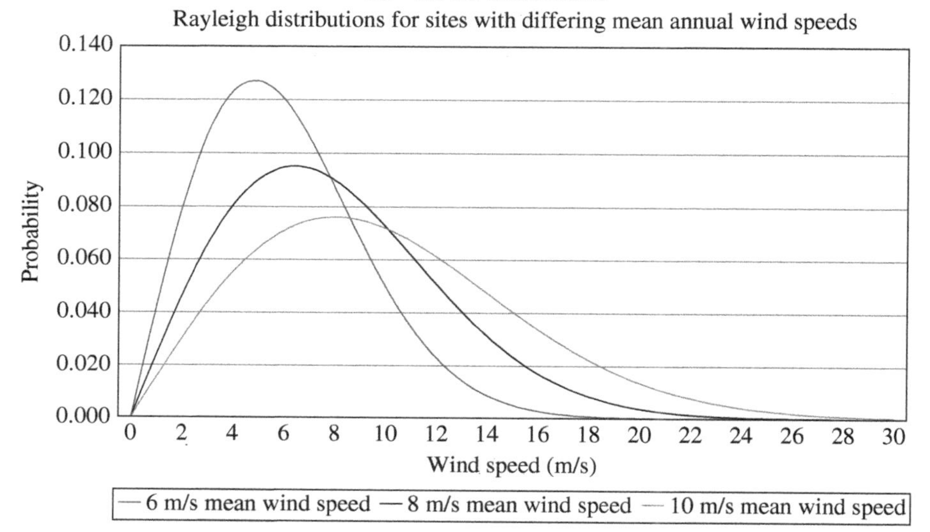

FIGURE 4.30 The Rayleigh Distribution Function or Formula can be used to estimate the distribution of wind speeds when only the annual mean wind speed (in m/s) is known (Source: R. Walker).

Analysis of wind data shows that a statistical correlation called the "Rayleigh distribution" typically fits these data quite well. Familiar statistical distributions (such as the familiar bell curve) are used in the grading of examinations for students. The Rayleigh distribution represents another type of distribution. The mathematical formulation is given in Figure 4.30.

Notable with the Rayleigh distribution is that by simply providing one number, the average wind speed, one can generate an estimated distribution of winds at that site for an entire year. A slightly more complex but related distribution is called the Weibull distribution, which is also used for wind resource analysis. The Weibull wind speed distribution is represented by two parameters, a shape parameter and a mean wind speed parameter, and is therefore more accurate. The Rayleigh is a special case of the Weibull distribution.

4.3.2 Wind Shear

Whenever wind flows over a surface, there is a phenomenon called the "boundary layer," as shown in Figure 4.31. Since friction effects cause the wind speed at the surface to be zero, there is a slow exponential increase in wind speed as one moves

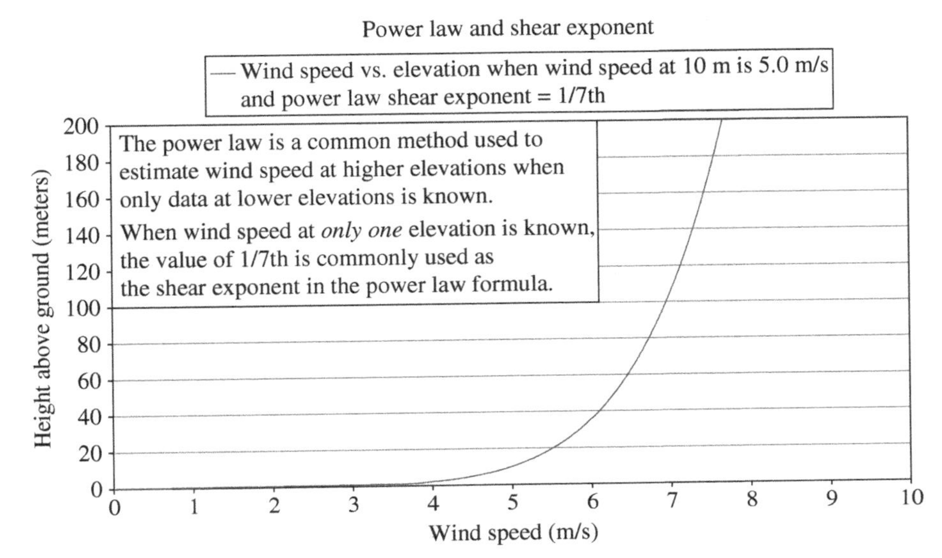

FIGURE 4.31 Illustration of wind shear and use of the Power Law to estimate wind speeds at heights above which wind speeds are known (Source: R. Walker).

to higher and higher elevations. In the box provided is the explanation and the formula that relates the height of interest to a reference height with a wind shear exponent (represented as α), which depends on the atmospheric characteristics. In a neutrally stable atmosphere, for example, on a sunny afternoon when heating causes convection, the value of that exponent is approximately 1/7.

When the atmosphere is stable in the evenings in the Great Plains, for example, these exponents can be quite high, due to strong wind shear, which means much higher wind speeds at higher elevations. The phenomenon of low-level jet that was discussed in Section 4.2.4 can cause very strong wind speed shear and additionally wind direction shear in the lower boundary layer of the atmosphere.

Wind speeds at elevations above where it is being measured can be estimated using the power law formula with an estimated or known vertical wind shear coefficient or exponent (α).

$$\frac{\text{Wind speed}}{\text{Ref. wind speed}} = \left(\frac{\text{Height}}{\text{Ref. height}}\right)^{\alpha}$$

4.3.3 Wind Turbulence

Looking again at the figure of the time history of wind speed (Fig. 4.28) from an anemometer, one can see that the wind speed continually varies. The variability of the wind about the mean is called turbulence, and the larger the variability about the

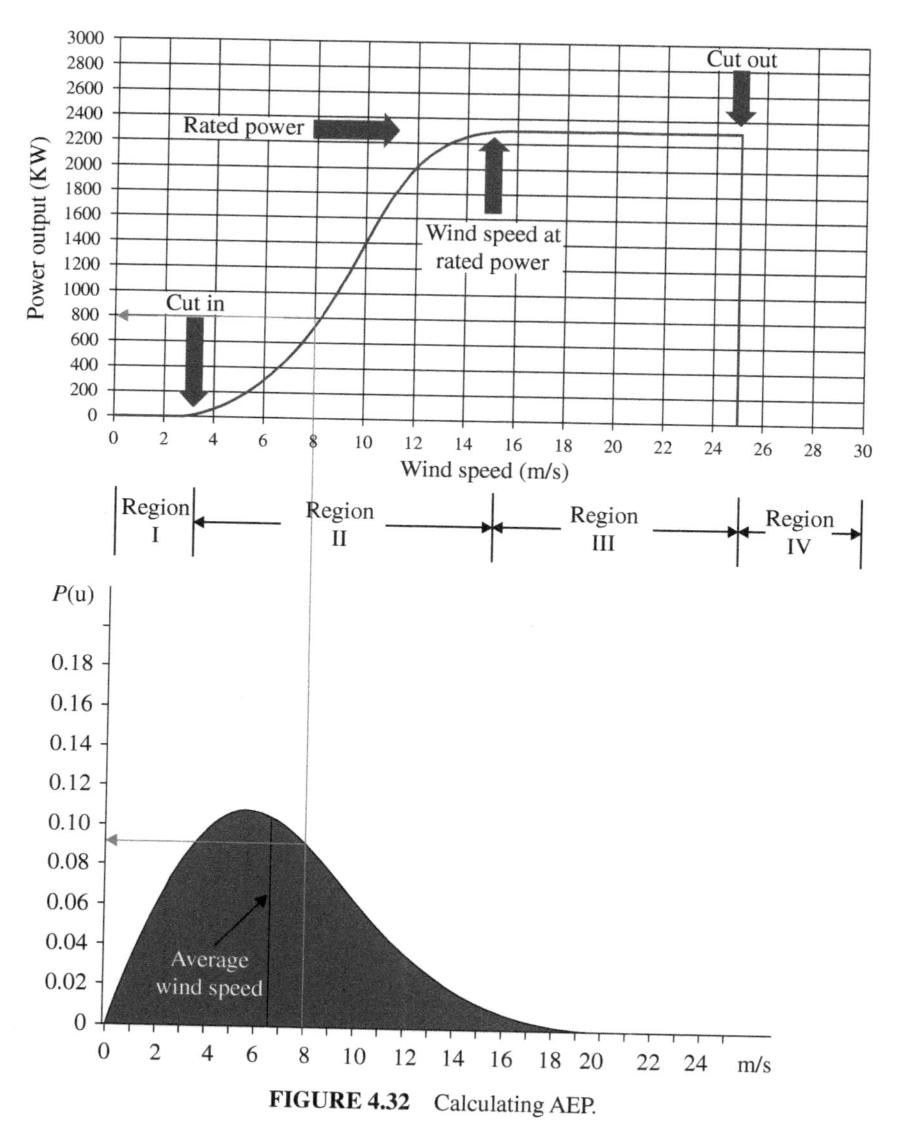

FIGURE 4.32 Calculating AEP.

mean wind speed, the higher the *turbulence intensity*. Sites that have high turbulence intensity can be very destructive to wind turbine equipment. Additionally, even at a given site, the turbulence can change depending on the weather characteristics. For example, low-level jets typically have very low turbulence, whereas the wind associated with thunderstorms can be quite gusty and turbulent.

4.3.4 Calculating Annual Energy Production

In the wind industry, the economics and financial benefits of a wind power plant are strongly dependent on the annual energy production (AEP) of the turbines in the power plant. Because wind is a variable resource, estimating the AEP from a wind power plant can be challenging. However, estimating AEP is necessary to secure financing for a proposed wind project since income will be directly related to the value and the quantity of the energy produced. With the concepts discussed previously, one can estimate the AEP of a single wind turbine or a wind power plant. Since wind speeds can vary from year to year, one also has to take into account the interannual variability in estimating the annual energy output.

The methodology of estimating AEP consists of combining the turbine power curve (as described in Section 4.1) with the statistical distribution of the wind speed at a given location at hub height, with appropriate density corrections. Following are the steps used to calculate AEP, and the accompanying Figure 4.32 helps illustrate this process.

- Align wind speeds between the wind distribution and wind turbine power curve.
- For each wind speed bin, read the related frequency of occurrence.
- Multiply the frequency by the number of hours in a year (8760).
- For the same wind speed bin, read the power output of the turbine, in MWh, and correct for air density changes due to the elevation of the turbine location.
- Multiply the number of hours that occur for that wind speed bin with the power to obtain production, in MWh.
- Complete this for all wind speed bins and total the result to obtain AEP.

The AEP value is usually measured in megawatt hours for a wind power plant. A ratio between the production (AEP) and the maximum possible production (the rated capacity of the turbine times the number of hours in a year) is called the *capacity factor* and is an important determinant in valuing wind power projects. These concepts will be discussed in more detail in later sections.

5

THE WIND ENERGY DEVELOPMENT PROCESS

5.1 REQUIREMENTS FOR AN ECONOMICALLY VIABLE WIND ENERGY PROJECT

Successfully developing and constructing a wind energy project is a highly complex process that may require several years of work and may put several hundred thousand dollars, if not millions of dollars, at risk if the project fails. For example, the proposed Cape Wind Project, whose developers believe it will become the first offshore wind farm in the United States, has been under development for at least 10 years. Following are some of the things needed for a successful wind project.

5.1.1 Property Rights

Property rights obtained by a wind developer need to allow for wind monitoring, soil testing, and environmental studies, plus the right to construct the project and to operate it over a 20- to 50-year period should the site prove to be viable. A wind project typically requires 40–100 acres of land per MW of nameplate capacity, so a 100-MW project may require as much as 10,000 acres of land. It is also wise to have room to expand the project in the future should circumstances result that make a second or third phase of the project viable. Several of the largest wind energy projects in Texas started out in the range of 50–150 MW but ended up as multiphase projects totaling several hundred MW.

Wind Energy Essentials: Societal, Economic, and Environmental Impacts, First Edition.
Richard P. Walker and Andrew Swift.

5.1.2 Good Wind Resource

A successful wind project requires a good wind resource, as represented by the long-term average annual wind speed at potential wind turbine locations across the site. A "good" wind resource will vary significantly by region. For instance, a good wind resource in the Texas Panhandle or in North Dakota might be average annual wind speeds of 8–9 m/s at 80 m aboveground, whereas in Indiana or Pennsylvania, a good wind resource might be 6–7 m/s at 80 m.

5.1.3 Site Wind Data

Potential investors and lenders will require high-quality wind data measured at the site for an extended period of time. The average hub heights of wind turbines being installed in 2010 averaged around 80 m, as opposed to hub heights of about 40 m in the mid-1990s. This trend of increasing hub heights is expected to continue in the future; thus, it is necessary to measure winds at increasingly higher levels, which substantially increases the cost of wind monitoring programs and site assessment. When getting ready to finance a wind project, wind developers should have at least 2 full years of wind data collected at the site and depending upon the size of the site may need data from multiple meteorological towers.

5.1.4 Transmission Access

Wind projects need access to electric lines capable of moving the energy produced at the site to the customer purchasing the electricity and the right to interconnect to these electric lines. Distribution voltage electric lines (under 69 kilovolts or 69 kV) may be capable of handling only a few MW of installed wind capacity, meaning that large wind projects will require access to transmission voltage electric lines (69 kV or above). Many of the largest wind projects in the United States interconnect to 345-kV lines, and the cost to interconnect to such lines can easily exceed $10 million even if the line has ample capacity available. Obtaining the right to interconnect to these lines typically requires many months of studies and engineering design by either the operator of the transmission grid or the company that owns the transmission line. These studies alone can cost $100,000 to $1 million.

5.1.5 A Customer or Market for the Energy

Some of the regional electric grids in the United States have clearly defined rules and protocols allowing interconnection of generation projects that sell into the hourly energy market, as opposed to selling all the output to one defined customer (such as an electric utility). The Electric Reliability Council of Texas (or "ERCOT") region is an example of such a market, and several "merchant" or "market" wind projects were constructed there between 2005 and 2009, allowing Texas to rapidly become the leading state in terms of installed wind generation capacity. In other electricity markets, it is much more difficult to construct a viable merchant wind project, usually necessitating that the energy be purchased by a customer (such as an electric utility) under a long-term agreement called a power purchase agreement (PPA).

5.1.6 Wind Turbines and Related Equipment

When wind development was growing rapidly in 2005–2009, it became increasingly difficult to acquire turbines for proposed wind projects, particularly for smaller wind developers. Lead times for procuring turbines often reached 2 years or longer, and substantial deposits were required by the leading turbine vendors from developers of wind projects. Other equipment (such as power transformers) may also have substantial lead times to procure and require large deposits. As more turbine vendors have entered the US market and more manufacturing facilities have been constructed, lead times have been reduced and deposit requirements may have eased as well.

5.1.7 Permits

Wind generation projects generally require fewer permits than fossil-fueled or nuclear power plants. In many cases, the most critical permit will come from the Federal Aviation Administration (FAA), from which a wind developer will hope to obtain a "Determination of No Hazard to Air Navigation." While the type of permits will vary from region to region, more state or local governments are seeking to add additional regulations or permitting approvals for wind projects. Permitting requirements for wind projects will be explored in more detail in a later chapter.

5.1.8 An Engineering, Procurement, and Construction Contractor

Most developers of wind energy projects do not have an in-house construction department and will therefore look to outside firms specializing in the construction of large projects to handle the engineering, materials procurement, and construction of the project. When there is rapid growth in the wind energy industry (such as during 2006–2009), experienced engineering, procurement, and construction (EPC) contractors having the workforce and construction equipment available to handle such a large project may be in short supply. The huge cranes required to erect the massive wind turbines now being used can be particularly hard and costly to schedule.

5.1.9 Capital

Lots of money, either yours, an investor's, or a bank's, will be needed. Wind projects can cost anywhere from $1.5 million to $2.5 million/MW to construct, with the wind turbines themselves comprising the largest cost component. In order to obtain such large amounts of money, the developer of a wind project will need to provide substantial amounts of accurate information to potential investors or lenders proving that the project is viable and that the vast array of potential environmental issues that may come up have been or can be adequately addressed.

5.2 OTHER FACTORS THAT CONTRIBUTE TO THE SUCCESS OF A WIND ENERGY PROJECT

While the aforementioned items represent the things most critical for viable wind energy projects, there are a number of other things that can increase the potential for the success of a project, and most of these relate to the societal impacts of wind energy as a whole or the societal impacts of a given wind energy project on the region surrounding it. Following are discussions of some of these items.

5.2.1 Federal Renewable Energy Production Tax Credit

The US government frequently uses tax policy to "incentivize" companies to invest in certain business sectors. Renewable energy has numerous societal benefits (such as wind and solar energy's lack of emissions or the preservation of limited fossil fuel resources), and thus, Congress felt it appropriate to enact the renewable energy production tax credit (PTC) as part of the Energy Policy Act of 1992. While Congress has extended the PTC several times in the past prior to its scheduled expiration date, there have been several other times that Congress allowed the PTC to lapse for a period of several months before reauthorizing it. Each time that Congress allowed the PTC to lapse, this caused significant turmoil for the wind energy industry. The PTC clearly increases the competitive position of wind energy versus electricity produced from traditional fossil fuel sources of generation, although fossil fuels themselves also receive billions of dollars of tax incentives each year. Given the increasing difficulty in balancing the national budget, it is possible that the PTC may someday go away completely, and if competing sources of electricity continue to receive subsidies and are allowed to emit pollutants at no cost, this would not bode well for wind energy. In lieu of the PTC, a state renewable electricity standard or renewable portfolio standard mandating a certain amount of renewable energy purchases by utilities would be beneficial to wind energy since it is the most cost-effective form of renewable energy in many states.

5.2.2 County Tax Abatements

Property taxes or *ad valorem* taxes make up one of the largest components of annual operations and maintenance (O&M) expenses for wind project owners. Many counties anxious to attract potential employers and tax base to the county will offer tax abatements to developers of wind energy projects.

5.2.3 School District Tax Abatements or Appraised Value Limitations

School taxes in many parts of the country can be higher than the county taxes. For instance, in most rural areas of Texas, school taxes comprise about two-thirds of all property taxes, with county, water district, hospital districts, and the like making up the other one-third.

5.2.4 Community Support

Community support is very important to a project and may greatly impact the expenses occurred for legal and public relations support during project permitting. Several recent wind energy projects have incurred opposition from sectors of nearby communities. Usually, it is a small percentage of the population opposed to wind projects, but they can be quite vocal and sometimes politically powerful. For example, the Cape Wind Project proposed as an offshore wind project near the coast of Massachusetts was opposed by many wealthy individuals with shoreline homes, including Senator Ted Kennedy. He subsequently threatened to introduce legislation proposing to prohibit or restrict all offshore wind development, and permits were not granted by the U.S. Department of Interior for the project until after Senator Kennedy's death.

5.3 POTENTIAL "FATAL FLAWS"

The *Wind Energy Siting Handbook* [1] (produced for the American Wind Energy Association in February 2008) lists the following 10 items as potential "fatal flaws" or significant considerations for wind energy project development:

- Obtaining (or failure to obtain) required permits; licenses; and regulatory approvals, including federal, state, and/or local regulations
- The presence of threatened or endangered wildlife species or habitat, either federally or state-listed
- The presence of avian and bat species or habitat, particularly involving threatened or endangered species and migratory pathways
- The presence of wetlands and protected water resource areas
- The presence of known archaeological and historical resources
- Proximity to community facilities (such as churches, parks, and recreational areas) and the potential impact on community services (such as police and fire departments)
- Land development constraints, including regulated constraints as well as guidelines suggested by community leaders, which can include noise limits (state and local standards), setback requirements, floodplain issues, height restrictions, or zoning constraints
- Telecommunications interference, including known telecommunications transmissions and microwave paths
- Aviation considerations, including proximity to airports and landing strips, and potential impact on aviation radar and military training routes
- Visual or aesthetic considerations, particularly when the project is near designated scenic vistas, parks, and residences

Other potential fatal flaws or significant environmental issues can include the following:

- Potential impact of wind turbines on weather radar (referred to as "NEXRAD" radar, short for Next Generation Radar)
- Concerns about wind turbine noise or the alleged health effects of prolonged exposure to wind energy projects
- Contribution of wind project construction to soil erosion
- Introduction of noxious weeds into rangeland or any type of weeds into cultivated areas
- Effect on wildlife where game hunting is a significant recreational or economic factor

5.4 COORDINATION OF THE WIND ENERGY PROJECT DEVELOPMENT PROCESS

Depending upon the size and complexity of a planned wind energy project, development of a project (i.e., getting everything ready for construction to begin but not yet commencing construction) can cost anywhere from $100,000 to $10 million. A very complex project (such as the Cape Wind offshore project, which has been years in the planning and permitting stages) would be an example of one on the high end of this spectrum, whereas a small project on one landowner's property adjacent to an existing substation or perhaps even a behind-the-meter project (such as one providing electricity to an industrial or large commercial business) would be an example of a project on the low end of this spectrum.

Most wind developers would prefer to minimize project development expenses until they know for certain that they have both a good quality wind resource and good transmission access. In addition, they should at least do some high-level environmental studies to assure themselves that no major fatal flaws will be encountered. One would not want to have spent several million dollars developing a site and, at the last minute, find out that the site is located in a whooping crane flyway or an area where FAA permit requests will be denied. The wind development process usually entails many activities and moving parts needing to come together at approximately the same time. Ideally, everything should begin to come together within a few days or weeks of each other at some point in time occurring *before* any major financial commitments are made. Such financial commitments are as follows.

5.4.1 Turbine Supply Agreement

The execution of a turbine supply agreement (TSA) may involve a multimillion dollar deposit at the time of execution and needs to occur many months prior to the date that the actual delivery of turbines begins. In periods of high demand for turbines, a developer may need to execute a TSA two or more years in advance. Larger developers can take advantage of "frame agreements" providing for the purchase of many turbines over a long period of time, and which may be directed to those projects ready for turbine delivery.

5.4.2 Transmission Interconnection Agreement

Obtaining access to the transmission grid will require the execution of a transmission interconnection agreement (IA). This can involve putting up funds or security for the design and construction of new transmission lines, substations, and/or interconnection facilities. Interconnection facilities alone can cost as much as $10 million or more, and 345-kV transmission lines may cost on the order of $1 million to $2 million/mile.

5.4.3 Power Purchase Agreements

Utilities purchasing wind energy typically do so through the negotiation of a PPA in which the wind developer commits to provide power to a utility or purchaser of the energy over a long period of time, perhaps as much as 20–25 years, with penalties for nonperformance. In many electricity markets, the project developer will also want to assure himself, investors, and lenders that there is an identified purchaser of the energy before committing millions of dollars to purchase turbines, although several wind energy projects have been constructed as "merchant plants" with no identified power purchaser or PPA.

5.4.4 Financing Agreements

Since large wind projects can cost tens or hundreds of millions of dollars, most wind project developers need to work out agreements with investors and lenders prior to commencing construction or committing to purchase turbines. In doing so, they often are also committing to pay for the cost of legal and financial advisors needed to arrange for equity investors and debt providers.

5.4.5 Signing Day

As depicted in Figure 5.1, the red circle depicts "signing day," or the day on which the developer begins making commitments for spending significant amounts of money, possibly in the millions of dollars.

5.5 TIMELINE FOR DEVELOPING A WIND ENERGY PROJECT

A useful tool for planning a wind energy project is a Gantt Chart, which is typically used to identify all steps or actions that need to occur to develop and to construct a wind project, how long each step or action may take, and which steps or actions are contingent upon completion of a prior step or action. For example, turbines may not be erected until FAA permits have been secured, and FAA permits may not be

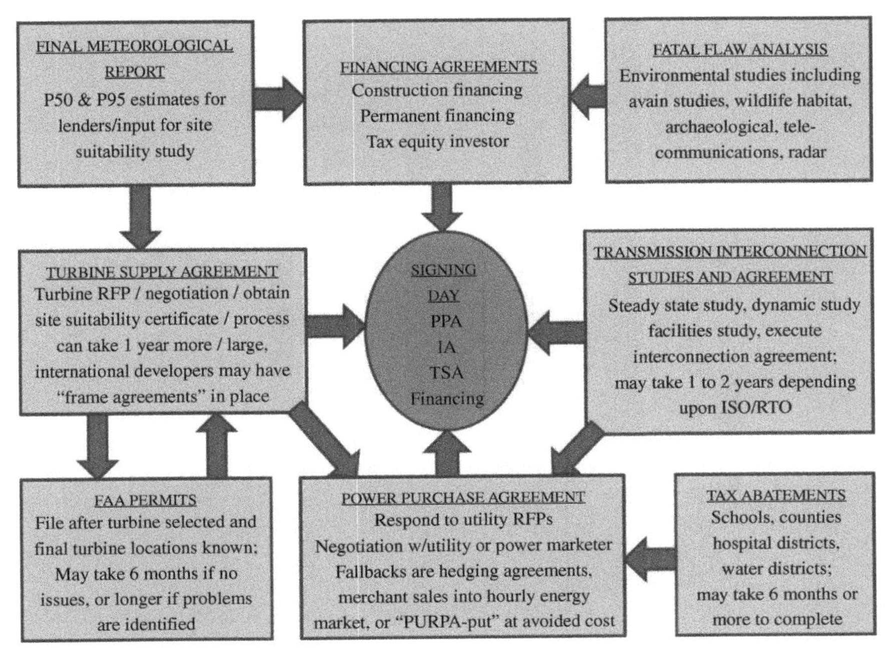

FIGURE 5.1 The wind project development process (Source: R. Walker).

requested until the developer knows what turbine model will be used, how tall the turbine will be, the exact location of turbine, and the groundline elevation of each turbine location. Another example would be turbine foundations, which should not be designed until geotechnical data are obtained from the proposed turbine locations. The Gantt Chart takes its name from Henry Gantt, who designed the chart in the early twentieth century. An example of a Gantt Chart is shown in Figure 5.2.

Note that, due to space limitations, this example only shows a few of the steps in developing and constructing a wind energy project. To be easily legible and to include all the steps of the entire process of development, construction, project commissioning, and achieving commercial operation of a 100^{+} MW wind energy project, the Gantt Chart would probably need to be printed on 24 in. × 36 in. and could cover a time span of 4–5 years.

5.6 MAJOR STEPS IN THE EARLY-STAGE WIND PROJECT DEVELOPMENT PROCESS

Following are some of the major steps in the wind energy project development process:

- Identify the electricity market, transmission independent system operator (ISO) or regional transmission organization (RTO), and relevant market rules. Figure 5.3 shows the various regional electricity markets in the United States.

Abbreviated example of a Gantt Chart

TASK DESCRIPTION	TASK NUMBER	CONTINGENT UPON	START DATE	DURATION	END DATE
SITE IDENTICATION AND CONTROL:	1		07/01/11	335	05/31/12
Determine region and electricity market for potential projects	1.1		07/01/11	30	07/31/11
Identify prevailing wind direction and existing information on regional wind resource	1.2	1.1	07/31/11	31	08/31/11
Identify topographic features that accelerate wind resource	1.3	1.2	08/31/11	31	10/01/11
Perform site reconnaissance and mapping of potential project areas/select preferred location	1.4	1.3	10/01/11	61	12/01/11
Identify landowners and determine if they desire to lease their property for wind energydevelopment	1.5	1.4	12/01/11	61	01/31/12
Execute leases with interested landowners	1.6	1.5	12/01/11	182	05/31/12
RESOURCE ASSESSMENT/TURBINE SELECTION:	2		08/31/11	974	05/01/14
Acquire and analyze data from existing long-term meteorological towers	2.1	1.1	08/31/11	153	01/31/12
Order meteorological towers	2.2	1.1	07/31/11	62	10/01/11
Install on-site meteorological towers	2.3	2.2, 1.5	01/31/12	91	05/01/12
Collect and analyze on-site wind data from meteorological towers	2.4	2.3	05/01/12	730	05/01/14
Estimate energy production from wind turbine(s) being considered for project	2.5	2.3	05/01/13	183	10/31/13
Determine cost and delivery schedule of turbine(s) being considered for project	2.6	2.5	10/31/13	61	12/31/13
Select wind turbine(s) to be used and contract for purchase of turbine and associated equipment	2.7	2.6	12/31/13	61	03/02/14
Obtain site suitability certificate from turbine vendor	2.8	2.7	03/02/14	60	05/01/14
TRANSMISSION STUDIES/INTERCONNECTION:	3		10/01/11	1218	01/31/15
Determine ownership and voltage of nearby transmission lines	3.1	1.3	10/01/11	61	12/01/11
Preliminary assessment of transmission capacity	3.2	3.1	12/01/11	61	01/31/12
File screening request for transmission interconnection with RTO/ISO; await results	3.3	3.2	01/01/13	120	05/01/13
Assuming transmission studies are encouraging, proceed with dynamic study request; await results	3.4	3.3, 2.6	05/01/13	123	09/01/13
Assuming transmission studies are encouraging, proceed with facilities study request; await results	3.5	3.4	12/31/13	182	07/01/14
Execute interconnection agreement and post required security	3.6	3.5	07/01/14	30	07/31/14
Construction of interconnection facilities by transmission service provider	3.7	3.6	07/31/14	184	01/31/15

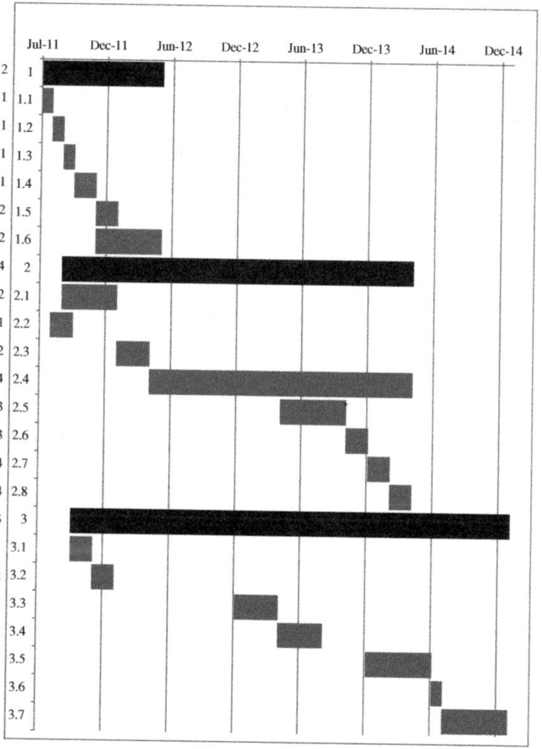

FIGURE 5.2 Abbreviated example of a Gantt Chart (Source: R. Walker).

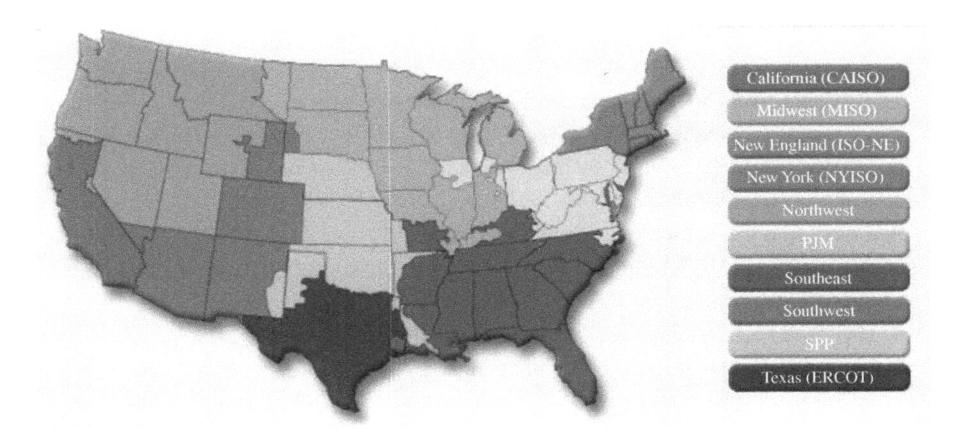

FIGURE 5.3 National electricity market areas (Source: FERC [2]).

- Identify potential energy purchasers and market mechanisms for energy sales. These may include long-term PPAs with utilities, direct sales of energy to large industrial facilities where allowed by state law, or sales into the electricity spot market, hourly market, or sub-hourly market.
- Use available wind maps to identify the windiest areas of the region and topographic or geographic formations that would be likely to accelerate wind speeds in the area. Wind speed maps for all 50 states have been produced as part of the U.S. Department of Energy and National Renewable Energy Laboratory's (NREL) Wind Powering America [3] program. Figure 3.6 showed the wind map for the United States produced by AWS Truepower (previously AWS Truewind) for this initiative. This map and individual maps for each state can be accessed at the Wind Powering America website: http://www.windpoweringamerica.gov/wind_maps.asp.
- Look for other existing wind data or resources to better understand the wind resource of the sites being considered. Potential sources include the following:
  - Data collected by the National Weather Service, usually at airports and usually collected at 10 m above groundline. Such data can be obtained at the following website: http://www.ncdc.noaa.gov/oa/mpp/digitalfiles.html#DIG.
  - Wind data collected by universities. West Texas A&M University's Alternative Energy Institute has installed and maintained many 40- to 100-m tall meteorological towers for several years, and data from many of the towers can be obtained from their website found at: http://www.windenergy.org/.
  - Wind data collected by state-funded programs in efforts to encourage wind energy project development in the state or networks of weather stations such as the West Texas Mesonet system maintained by Texas Tech University or the Oklahoma Mesonet operated by the University of Oklahoma and Oklahoma State University.

  - Simulated data produced using complex computer programs tied to massive databases of weather and topographical data; several meteorological consulting firms specialize in producing such simulations.
  - The Western Wind Dataset or Eastern Wind Dataset produced under contract for the NREL, which can be obtained at the following website: http://www.nrel.gov/wind/integrationdatasets/about.html.
- Determine proximity to transmission lines and the voltage of such lines.
- Perform a desktop review (i.e., information readily available on the Internet) to check for any potential fatal flaw environmental issues.
- Use the FAA/Depart of Defense preliminary screening tool to check for potential issues related to impact on long-range radar, NEXRAD weather radar, or military operations; this tool can be found at FAA's website, https://oeaaa.faa.gov/oeaaa/external/portal.jsp. Examples of the output from the long-range radar tool and the NEXRAD radar tool are shown in a later chapter addressing potential impacts of wind energy on aviation, radar, and telecommunications.
- Perform an on-site inspection of the project area for siting issues and potential fatal flaws (houses, subdivisions, irrigated cropland, wetlands, national or state parks, recreational areas, weather or military radar facilities, airports, etc.)
- Identify owners of the property of interest. A good place to start this process is with the County Appraiser's office or a local realtor specializing in rural property.
- Identify those landowners interested in wind energy project development and leasing their land for this purpose; one strategy to consider is starting with one of the owners of the largest amount of property in the area of interest and try to establish a "foothold" before your interest in the area becomes known, since there are always people out there that may try to preempt your efforts.
- Secure access to the property and the right to develop a wind project on it by developing proposed terms of a property lease agreement; be willing to negotiate terms acceptable to the landowner, particularly if you are unfamiliar with farming, ranching, hunting, or oil and gas development practices in the area.
- Purchase and install one or more meteorological towers; an important part of this is identifying the expected hub height of wind turbines and using meteorological tower heights that are at least 75% of the expected hub height (i.e., if the expected hub height is 80 m, use meteorological towers that are at least 60 m in height). In addition, determining how data are to be collected from the tower is important in determining what type of data acquisition and delivery systems will be used. Choices include manual acquisition of data, cellular phone delivery, or satellite transmission.
- When the project developer feels comfortable that he has secured sufficient property rights to prevent competing developers from coming in and trying to lease property in the same area, it is wise to begin publicizing the project in the area surrounding it and cultivating community relationships. Meetings with

county officials (such as the county judge, county commissions, mayors, and economic development officials) and other community leaders should be arranged, followed by "town hall" meetings in which the general public is invited. At such town hall meetings, the developer may wish to show visual simulations or a representation of what the project will look like in order to alleviate any concerns about the aesthetics of the project.

- Begin collecting and analyzing wind data from the meteorological towers. Since high-quality data will be required for financing the wind project, it may be wise to contract with a well-known, established meteorological consulting firm that is familiar with the data requirements for financing such projects.
- Identify permitting requirements and recommended environmental studies for the region.
- Identify all taxing jurisdictions within the project area (county, school districts, municipalities, hospital districts, water districts, community college districts, etc.) and the current *ad valorem* tax rates for each to determine the expected combined tax rate applicable to areas within the project.
- Determine if there are substantial differences in the tax rates applicable to various areas of the project that may affect turbine siting decisions.
- Perform preliminary transmission studies. Sometimes, it is more cost-effective to use consultants familiar with the region of interest, but at other times, it may be more cost-effective to go ahead and file a screening study request with the transmission system operator covering the region.

At this point in time, a developer may have only spent $50,000 to $100,000 on the project. Beyond this point, things can begin to get substantially more expensive, so it is wise to assess whether your venture can spend much more money on the project or if it should begin to identify potential investors or development partners to move the project forward.

5.7 MAJOR STEPS IN LATTER-STAGE WIND ENERGY PROJECT DEVELOPMENT

Following are subsequent stages in the wind energy project development process that generally cost substantially more than the steps discussed in the previous section:

- Ensure that sufficient capital is available to move forward. Additional costs to get a project ready for construction can total $1 million to $5 million, depending upon the size and the complexity of the project (not even counting the cost of a down payment on wind turbine purchase, which alone can be a multimillion dollar expense on projects as small as 10 MW). It is not uncommon for turbine vendors to require a down payment of 20% of the value of the turbines, and it may be due 1–2 years before turbine delivery, particularly at times when the wind energy industry is booming.

- Have an established environment consultant perform desktop and on-site inspection for potential fatal flaws (endangered species, wetlands, animal habitat, archaeological and historical sites, telecommunications facilities, etc.).
- Initiate environmental studies and permit requests. One of the environmental studies frequently performed for wind energy projects is a "four-season avian study"; thus, as implied by the name, one needs to allow at least four seasons or a full year for the completion of this study.
- Obtain specifications, power curve data, and indicative pricing for all wind turbine models being considered for the site.
- After a full year of on-site wind data are available, have the consulting meteorologist prepare preliminary turbine layouts for each of the turbine models being considered for the site, and prepare an estimate of long-term average annual energy production from the site based on the various turbine models.
- Add additional meteorological towers, some of which may be located at a given place for only a few months (as specified by the project meteorologist to aid in turbine micrositing decisions).
- For each of the turbine models still under consideration, prepare preliminary road and electric collection system designs associated with each turbine layout, and prepare a cost estimate for the balance of plant (i.e., roads, electric collection system, O&M building, permanent meteorological towers, project substation, transmission costs to be borne by the project, etc.).
- Develop or acquire a *pro forma* financial model that can be used to estimate the cost of energy resulting from the project based on each of the turbine models still being considered for the project. Prepare cost estimates including all turbine costs and balance-of-plant costs. Then use the model to estimate cost of energy based on each turbine model and identify the model or models that appear to be most cost-effective for the site.
- Begin negotiations with the preferred turbine vendor to finalize turbine prices, warranty agreements, O&M contracts, and delivery schedules.
- Request pricing proposals from potential EPC contractors for the erection of turbines and the construction of the project balance-of-plant. It may be wise to spend some money to collect some geotechnical information in the project area that can then be used by potential EPC contractors to get a good idea of what types of foundations will be required for the projects, as foundations can represent a multimillion dollar cost component.
- Begin the process of identifying the legal and financial services providers that will be involved in securing debt and equity for the project, and determine the costs that will be associated with these activities.
- Update or refine the cost of energy estimates based on the results of the previous activities and then compare the estimated cost of energy from the site with known indicators of electricity prices in the region or electricity market that the project is located to determine if the project appears to be economically viable. RTOs or ISOs often provide information on their websites regarding average

cost of electricity in the region, locational marginal prices, or the market clearing price of energy.

- Upon identifying the preferred turbine model and optimized turbine layout, submit information regarding the turbine model's and the project's electrical characteristics that are necessary for the transmission service provider and/or regional transmission authority to complete the transmission interconnection studies.
- One of the steps in the transmission interconnection process may be a facilities study (or something similarly named) in which the transmission service provider or regional transmission authority would incur significant amounts of engineering time and expense designing the interconnection facilities and any transmission lines that may be required to interconnect the project. This study can be expensive itself, and the project developer may then be asked to post funds or a letter of credit equal to the cost of materials and construction for the interconnection facilities and transmission lines, which can easily be in the millions of dollars.
- Upon identifying the preferred turbine model and optimized turbine layout, requests for FAA permits can be submitted by filing Notices of Proposed Construction or Alteration (FAA Form 7460-1) for each turbine location. The form can be obtained at the FAA's website: http://oeaaa.faa.gov.
- Begin negotiation of potential tax abatement agreements. As previously discussed, property or *ad valorem* taxes can represent one of the largest expenses for a wind energy project, and abatements can sometimes influence the location of such projects and the economic viability of projects. Successfully negotiating a tax abatement agreement with local government officials will help the project developer determine how low it may go on the proposed price of energy from the wind project when seeking potential purchasers of the energy.
- Identify potential customers for energy from the project and begin negotiations with them or respond to competitive requests for pricing (also sometimes referred to as requests for proposals) from potential energy purchasers.

5.8 FINAL STEPS IN DEVELOPING A WIND ENERGY PROJECT

As previously discussed, the wind development process entails many activities and moving parts needing to come together at approximately the same time, ideally **before** any major financial commitments are made. When it becomes clear that there is a customer for the energy, that there is transmission capacity available for the project, and that the project will be economically viable, the following are some of the steps necessary to bring the project to the point of construction readiness:

- Negotiate and execute the PPA with the intended purchaser of the energy.
- Negotiate and execute the transmission IA with the transmission service provider.

- Negotiate and execute the TSA and other agreements necessary to procure transformers and other long lead time equipment.
- Obtain geotechnical data at each planned turbine location necessary for the EPC contractor and/or turbine vendor to design the foundations for the turbines.
- Contract with the preferred legal and financial services providers, and begin the process of assuring lenders and investors that all property rights are in order, that all permits and regulatory approvals are in place, and that the project is economically viable.
- Contract with the preferred EPC contractor for construction of the project and agree upon the construction schedule. The construction schedule needs to align with the delivery schedule for turbines and other equipment and also needs to include time for delays due to rain, wind-outs (i.e., when the wind speeds are too high to safely erect turbine nacelles, blades, or rotor sets), and unanticipated problems.
- Finalize warranty and O&M agreements with the turbine vendor or other O&M providers. Note that O&M services may be provided by a company other than the turbine vendor as wind energy projects become more widespread.

5.9 CONCLUSIONS

Development of a wind energy project is a complex process that can take several years and many millions of dollars. Thus, it is imperative that wind developers take advantage of tools such as Gantt charts and flow charts to identify all actions that need to be taken without unnecessarily spending or committing funds that may be wasted if unexpected barriers to the project arise. Timing is critical in many aspects of the process, and it is best if all potential fatal flaws to the project are evaluated and understood prior to committing to major purchases (such as turbine purchases or depositing funds for transmission line upgrades).

REFERENCES

[1] Tetra Tech EC, Inc., Nixon Peabody LLP. *Wind Energy Siting Handbook.* Washington, DC: American Wind Energy Association; 2008.

[2] Federal Energy Regulatory Commission Website. 2012. Available at http://www.ferc.gov/market-oversight/mkt-electric/overview.asp. Accessed August 12, 2012.

[3] AWS Truepower Study. 2010. National Renewable Energy Laboratory, Wind Powering America Website. Available at http://www.windpoweringamerica.gov/wind_maps.asp. Accessed August 12, 2012.

6

OVERVIEW OF ISSUES FACED BY WIND ENERGY

6.1 THE NEED TO UNDERSTAND AND RESPOND

Electricity provides many benefits to our society, and it is unlikely that many people in the United States would choose to do without it. However, millions of people on the Earth still do not have access to electricity, and in many developing nations, increased electrification is a top priority of the government. However, all forms of electricity generation have both environmental impacts and economic impacts, and all of them have their proponents and detractors. Wind energy is no exception—there are those in our society who are opposed to the continued growth of wind energy for electricity generation, and their reasons for opposing it range across a wide spectrum. A primary goal of this book is to not only identify benefits, issues, and challenges of wind energy but to contrast them to benefits, issues, and challenges of other means of electricity generation (such as nuclear power or fossil-fueled power plants) and, as nearly as possible, to compare them on an "apples-to-apples" basis.

In 2008, the U.S. Department of Energy, with the support of many experts in fields such as engineering, wind energy, atmospheric science, biological science, and environmental science, produced a report titled *20% Wind Energy by 2030: Increasing Wind Energy's Contribution to U.S. Electricity Supply* [1]. This report and its authors examined the feasibility of producing 20% of the nation's electricity supply by the year 2030, as compared with the approximately 1% contribution of wind energy when the study was initiated. The report examined challenges to achieving the goal

Wind Energy Essentials: Societal, Economic, and Environmental Impacts, First Edition.
Richard P. Walker and Andrew Swift.

and estimated the economic impacts and environmental effects of achieving it. The report stated that "Wind energy is one of the cleanest and most environmentally neutral energy sources in the world today. Compared with conventional fossil fuel energy sources, wind energy generation does not degrade the quality of our air and water and can make important contributions to reducing climate-change effects and meeting national energy security goals. In addition, it avoids environmental effects from the mining, drilling, and hazardous waste storage associated with using fossil fuels. Wind energy offers many ecosystem benefits, especially as compared to other forms of electricity production."

Yet the report also correctly pointed out that "wind energy production can also, however, negatively affect wildlife habitat and individual species, and measures to mitigate prospective impacts may be required."

Since wind energy is held up by those in the wind energy industry as being environmentally friendly, it is important that the industry conduct itself in a manner consistent with this claim and to promote and adhere to high standards for environmental protection. When issues, questions, or challenges are raised, the industry needs to address them, and in general, it has done a good job of doing so. Of course, there are always exceptions to every rule and there are those in any industry who are more motivated by profits than concern for the environment. And there are existing issues in the industry and issues that will arise in the future that are legitimate concerns that must be addressed.

There are also people in competing industries that are quick to jump on any issue related to wind energy no matter how trivial, and they will look for any opportunity

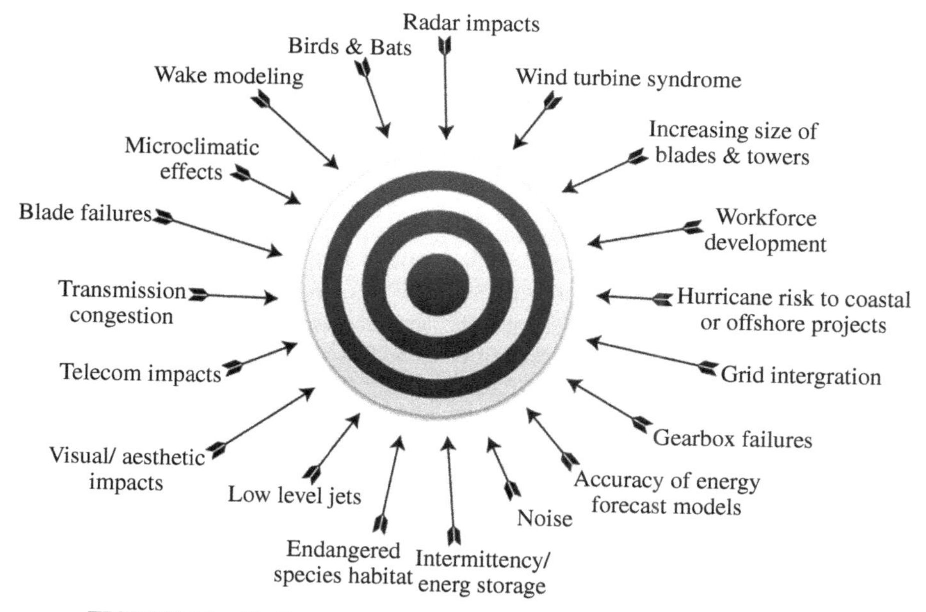

FIGURE 6.1 Challenges for the wind energy industry (Source: R. Walker).

to "hype" it as a reason to stop developing the nation's wind resources. Again, the wind energy industry must be proactive in its response to such challenges and provide clear facts rebutting detractors of wind energy. Another goal of this book is to educate individuals who desire to enter the wind energy industry about the issues or challenges they may be faced with, how to respond to them, and how to avoid repeating mistakes of the past (Fig. 6.1).

6.2 COMMUNITY RELATIONSHIPS AND PUBLIC OUTREACH

Maintaining good relationships with the community surrounding a proposed wind energy project and providing accurate and timely information to the public will mitigate or prevent many potential problems. Admittedly, not all problems, issues, or complaints can be avoided, and almost all wind energy projects will have some type of issue that arises.

Factors that may affect the potential for problems to arise include the following, all of which should be considered when assessing the suitability of a site for development of a wind energy project.

6.2.1 Residential Density

In general, the more homes that are in the area of the proposed project, the more likely it is that resistance will be encountered. This would apply to utility-scale wind energy projects and to small wind turbines alike. With regard to small home-owned wind turbines, areas that have 5- to 10-acre lots may be acceptable locations for a small horizontal axis turbine, but installing such turbines at smaller lots may result in issues with neighbors due to turbine noise. For small lots, vertical axis turbines may be a better solution and most of these would not produce the levels of noise associated with most small horizontal axis turbines.

6.2.2 Socioeconomic Level of Community and Surrounding Residences

Many rural areas that rely on farming or ranching as the backbone of their economy will be very excited about the prospects of a wind energy project; on the other hand, scenic areas near to urban areas may have wealthy individuals who have built "estate homes" or weekend residences in the countryside in order to get away from the hustle and bustle of the city, and many of these would look unfavorably upon wind projects in the area of their homes.

6.2.3 Compatibility of the Project with Existing Land Use

Wind energy projects are compatible with activities such as farming, ranching, oil and gas development, and even hunting. On the other hand, such projects may not be compatible with areas where mining operations or subdivision development is occurring.

6.2.4 Proximity to Parks, Lakes, Recreational Areas, Scenic View Sheds, or Historical Sites

While many people who enjoy the outdoors think of themselves as environmentalists and appreciate the benefits of renewable energy, they may still look upon wind energy projects as industrial development.

6.2.5 Community Desire for Economic Growth in the Area

Many rural areas of the nation are steadily declining in population, and young people growing up in such areas are likely to move to cities where jobs are more readily available. Such communities or areas are typically very appreciative of the jobs, tax base, and economic activity created by wind energy projects. The closer that wind projects are to urbanized areas, the less likely they are to be appreciated for their economic benefits, and the more likely they are to be opposed for their potential negative attributes (such as visual impacts, traffic congestion during construction, or turbine noise).

6.2.6 Effective Community Outreach Efforts

Community outreach is critical to a project's success and should be planned for and initiated fairly early in the development process. Local governmental officials, community leaders, and economic development agencies should be contacted very early in the process. It may be wise to add one or more community leaders to the project development team so that they can provide input on the types of issues or challenges that the proposed project may face since it is likely that the project developer will not be familiar with the community and may be completely unaware of economic or social issues in the community. Project developers who grew up in cities may have little understanding of issues related to farming, ranching, or hunting.

Town hall meetings are a good way to reach members of the community. When such events are held, the project developer should be ready to address the following subjects:

- Local economic benefits including jobs during the construction phase, permanent jobs after the project goes into operation, tax base and tax revenues to counties and schools, landowner royalties, services and equipment that will be obtained locally, and tourism opportunities.
- National or regional economic benefits (such as the use of local resources and promotion of national energy security).
- Environmental benefits (such as reducing emissions from electricity generation and reducing water used for electricity generation).
- Visual impact of the project—an excellent way to address this issue is to have visual simulations prepared so that the public can see how turbines will look rather than speculating on this.

- Noise created by the project both during construction and when in operation.
- Proposed setback practices or planned distances between turbines and homes, other buildings, roads, pipelines, etc.

The number one thing to do at town hall meetings is to listen carefully and to seek to understand the issues or concerns of the public. If answers to some of the questions raised by the public are not known, be sure to obtain contact information of the person asking the question and let him or her know that you will look into the issue and get back to them with an answer or response.

6.3 GENERAL CATEGORIZATION OF ISSUES

The benefits, challenges, and issues for wind energy can be classified into several broad categories, as has been done later, but several of the specific issues may fit into multiple categories. For example, air quality is shown under "Human Health Related" but air quality will also affect wildlife. There will undoubtedly be other issues related to wind energy that will arise following the publication of this book, and others that have not been addressed, but the ones listed as follows are those that will be discussed in subsequent chapters.

6.3.1 Biological or Wildlife-Related Issues

- Threatened or endangered species
- Avian issues (birds and/or bats)
- Wildlife habitat
- Wetlands/waterway crossings
- Vegetation
- Soil erosion

6.3.2 Human Health–Related Issues

- Air quality
- Climate impacts
- Proximity to homes, businesses, roads, etc.
- Audible noise or sound
- Low-frequency sound
- Wind turbine syndrome
- Shadows or shadow flicker
- Ice shedding
- Workforce safety
- Blade drop or throw

- Fire
- Electric and magnetic fields
- Lightning strikes
- Hazardous wastes

6.3.3 Other Human-Related Issues

- Visual impacts/aesthetics
- Ground transportation and impact on traffic
- Impacts on cultural resources
  - Paleontological
  - Archaeological
  - Historical
  - Recreational
- Surface water and groundwater resources
- Soil erosion/stormwater runoff
- Removal of retired or nonoperating equipment
- Restoration of property to original condition

6.3.4 Technology-Related Issues

- Impact on the reliability of the electric grid
- Interference with telecommunications
- Interference with radar signals
- Obstruction to aviation
- Aviation warning lights
- Electromagnetic interference
- Department of Defense and Homeland Security

6.3.5 Economic-Related Issues

- Cost of wind energy compared with other sources of electricity
- Job creation/job losses in competing industries
- Compatibility of wind projects with existing or future land use
- Amount of land taken out of production
- Total amount of land that a project is spread across
- Impacts on property values, either positive or negative
- Effect on tourism, either positive or negative
- Impact on emergency services (fire, police, etc.)
- Property taxes/tax abatements

- Landowner payments for royalties; land rental; land purchase; or purchases of water, gravel, or caliche
- Purchases of goods and services from local businesses

REFERENCE

[1] U.S. Department of Energy, Energy Efficiency and Renewable Energy. *20% Wind Energy by 2030: Increasing Wind Energy's Contribution to U.S. Electricity Supply*, DOE/GO-102008-2567. Washington, DC: U.S. Department of Energy, Energy Efficiency and Renewable Energy; 2008.

7

WIND AND WILDLIFE: SITING ISSUES AND CHALLENGES

7.1 OVERVIEW OF WILDLIFE-RELATED ISSUES FACED BY THE WIND INDUSTRY

As wind energy development expands rapidly within the United States and many other nations, the wind energy industry must be diligent in responding to any potential environmental issues. One of the more significant issues for the industry is the impact on wildlife, and as offshore wind development becomes a reality, the issue is extended to fish as well. While it is virtually impossible for any electric-generating resource to have no impact at all on wildlife, it is possible to mitigate many potential impacts on wildlife by being knowledgeable about the issues and diligent during the siting, construction, and operation of wind plants. The issues and topics to be discussed in this chapter include the following:

- Avian issues
  - State and federal regulations relevant to wind development
  - Collisions with wind turbines
  - Raptors
  - Migratory birds/wetlands
  - Threatened and endangered species
  - Collisions and electrocution with transmission and distribution lines

Wind Energy Essentials: Societal, Economic, and Environmental Impacts, First Edition.
Richard P. Walker and Andrew Swift.

  - Destruction, degradation, and/or fragmentation of avian habitat
  - Organizations studying wind and avian interaction
  - Best practices for mitigation of avian issues
- Bat issues
  - State and federal regulations relevant to wind development
  - Collisions with wind turbines
  - Barotrauma (rupturing of tissues in the lungs of bats due to sudden air pressure changes they encounter when entering air vortices near wind turbines)
  - Organizations studying wind and avian interaction
  - Best practices for mitigation of bat issues
- Other wildlife issues
  - State and federal regulations relevant to wind development
  - Destruction, degradation, and/or fragmentation of wildlife habitat
  - Threatened and endangered species
  - Noise and visual impacts

7.2 STATE AND FEDERAL REGULATIONS RELEVANT TO WIND DEVELOPMENT'S IMPACT ON WILDLIFE

One of the most commonly cited issues with wind energy is the potential for birds to collide with the rotating blades of the turbines. While the rotors may appear to be turning slowly, the tips of these massive blades may be traveling at 150 mph or greater, so there is no question that collisions will almost always be fatal for birds hitting the turbine. But wind development can affect birds and other forms of wildlife in other ways too, including the destruction or fragmentation of their habitat or the disruption of breeding during mating season due to construction noise. Thus, it is imperative that the wind energy industry be diligent during the siting and operation of projects so that the potential for bird collisions and wildlife habitat destruction is minimized, particularly when there may be threatened or endangered species in the region.

Regulations affecting the wind energy industry are constantly evolving, particularly in recent years as annual capacity additions of wind generation have exceeded additions of nuclear power and coal-fired generation and rivaled that of natural gas–fired capacity additions. Therefore, before expending large amounts of time, effort, and money on a wind project, it is wise for the developer to undertake an examination of current local, state, and federal laws relevant to wind development.

While permits for siting wind farms may not be required from the U.S. Fish and Wildlife Service (USFWS) or from state regulatory agencies, there are existing Federal laws that apply to the potential impact of wind energy projects on wildlife of which developers of such projects need to be aware. These include the following:

- The Endangered Species Act, which regulates activities affecting threatened and endangered species

- The Migratory Bird Treaty Act, which is a strict liability statute that prohibits harm or taking of migratory bird species
- The Bald and Golden Eagle Protection Act (BGEPA), another strict liability statute that prohibits harm or taking of bald and golden eagles
- The National Environmental Policy Act, which can be triggered when wind projects are sited on Federal lands
- The Clean Water Act, which regulates the discharge of dredged or fill material into waters of the United States and thus may come into play when wind projects are in the areas of wetlands, navigable rivers, or other bodies of water meeting the definition of "waters of the United States." Wetlands can be a very important habitat or migratory landing spot for many avian species.
- Executive Order 12996 was issued by President Clinton in 1996 to (i) conserve fish and wildlife and their habitat, (ii) further define the mission of the National Wildlife Refuge System, and (iii) provide direction to the Secretary of the Interior regarding management of the Refuge System resources.
- Executive Order 13186 was issued by President Clinton in 2001. It directed federal agencies taking actions having a measurable negative effect on migratory bird populations to develop and implement a memorandum of understanding with the USFWS that promotes the conservation of migratory bird populations.

Violation of any of the aforementioned without having taken appropriate precautions prior to construction (such as four-season avian studies or desktop and on-site analysis of potential endangered species) could lead to serious consequences for the project owner.

7.3 USFWS GUIDELINES FOR WIND DEVELOPMENT

In 2003, USFWS proposed a set of voluntary interim guidelines for mitigating the impacts of land-based wind energy projects to wildlife. In 2007, USFWS established the Wind Turbine Guidelines Advisory (WTGA) Committee, comprising representatives of governmental agencies, wildlife conservation organizations, environmental advocacy groups, and the wind energy industry. This committee submitted recommended guidelines for wind development to the U.S. Secretary of the Interior in early 2010. USFWS then convened an internal working group to review the WTGA Committee's recommendations, which then developed a new set of USFWS wind energy guidelines for utility-scale and community-scale land-based wind energy projects. This resulted in the release of the following two documents, one of which has been finalized and a second that is still a draft version as of the writing of this book.

7.3.1 USFWS' Land-Based Wind Energy Guidelines

These guidelines were developed with the intent of helping industry avoid and minimize impacts to federally protected migratory birds and bats and other impacted wildlife resulting from site selection, construction, operation, and maintenance of land-based wind energy facilities [1]. A final version of the guidelines was released in March 2012 and can be downloaded from the USFWS website at http://www.fws.gov/windenergy/.

7.3.2 USFWS' Eagle Conservation Plan Guidance: *Module 1—Land-Based Wind Energy*

These guidelines were developed to provide interpretive guidance to wind developers, USFWS biologists who evaluate potential impacts on eagles from proposed wind energy projects, and others in applying the regulatory permit standards as specified by the BGEPA and other federal laws [2]. The most current version, as of the writing of this book, was released in April 2013, and it too can be found on the USFWS website.

The USFWS provided the following reasons for their development of the *Eagle Conservation Plan Guidance:* "Of all America's wildlife, eagles hold perhaps the most revered place in our national history and culture. The United States has long imposed special protections for its bald and golden eagle populations. Now, as the nation seeks to increase its production of domestic energy, wind energy developers and wildlife agencies have recognized a need for specific guidance to help make wind energy facilities compatible with eagle conservation and the laws and regulations that protect eagles. To meet this need, the U.S. Fish and Wildlife Service (Service) has developed the Eagle Conservation Plan Guidance (ECPG). This document provides specific in-depth guidance for conserving bald and golden eagles in the course of siting, constructing, and operating wind energy facilities. The ECPG guidance supplements the Service's Land-Based Wind Energy Guidelines (WEG). WEG provides a broad overview of wildlife considerations for siting and operating wind energy facilities, but does not address the in-depth guidance needed for the specific legal protections afforded to bald and golden eagles. The ECPG fills this gap."

7.4 BEST MANAGEMENT PRACTICES FOR WIND ENERGY DEVELOPMENT INCLUDED AS CHAPTER 7 OF THE USFWS *LAND-BASED WIND ENERGY GUIDELINES*

Chapter 7 of the USFWS *Land-Based Wind Energy Guidelines* includes a list of Best Management Practices for wind energy projects, which provides an excellent resource for wind developers, construction companies, and owner/operators of wind projects [2]. Following is their list of Best Management Practices pertaining to the project development. Refer to the *Guidelines* for additional Best Management Practices specific to project retrofitting, repowering, and decommissioning.

- Minimize, to the extent practicable, the area disturbed by preconstruction site monitoring and testing activities and installations.
- Avoid locating wind energy facilities in areas identified as having a demonstrated and unmitigatable high risk to birds and bats.

- Use available data from state and federal agencies and other sources (which could include maps or databases) that show the location of sensitive resources and the results of Tier 2 and/or 3 studies to establish the layout of roads, power lines, fences, and other infrastructure.
- Minimize, to the maximum extent practicable, roads, power lines, fences, and other infrastructure associated with a wind development project. When fencing is necessary, construction should use wildlife-compatible design standards.
- Use native species when seeding or planting during restoration. Consult with appropriate state and federal agencies regarding native species to use for restoration.
- To reduce avian collisions, place low- and medium-voltage connecting power lines associated with the wind energy development underground to the extent possible, unless burial of the lines is prohibitively expensive (e.g., where shallow bedrock exists) or where greater adverse impacts to biological resources would result.
  - Overhead lines may be acceptable if sited away from high bird crossing locations, to the extent practicable, such as between roosting and feeding areas or between lakes, rivers, prairie grouse and sage grouse leks, and nesting habitats. To the extent practicable, the lines should be marked in accordance with Avian Power Line Interaction Committee (APLIC) collision guidelines.
  - Overhead lines may be used when the lines parallel tree lines, employ bird-flight diverters, or are otherwise screened so that collision risk is reduced.
  - Aboveground low- and medium-voltage lines, transformers, and conductors should follow the 2006 or most recent APLIC "Suggested Practices for Avian Protection on Power Lines."
- Avoid guyed communication towers and permanent met towers at wind energy project sites. If guy wires are necessary, bird-flight diverters or high-visibility marking devices should be used.
- Where permanent meteorological towers must be maintained on a project site, use the minimum number necessary.
- Use construction and management practices to minimize activities that may attract prey and predators to the wind energy facility.
- Employ only red, or dual red and white, strobe, strobe-like, or flashing lights, not steady burning lights, to meet Federal Aviation Administration (FAA) requirements for visibility lighting of wind turbines, permanent met towers, and communication towers. Only a portion of the turbines within the wind project should be lighted, and all pilot-warning lights should fire synchronously.
- Keep lighting at both operation and maintenance facilities and substations located within half a mile of the turbines to the minimum required:
  - Use lights with motion or heat sensors and switches to keep lights off when not required.
  - Lights should be hooded downward and directed to minimize horizontal and skyward illumination.

  - Minimize use of high-intensity lighting, steady-burning, or bright lights such as sodium vapor, quartz, halogen, or other bright spotlights.
  - All internal turbine nacelle and tower lighting should be extinguished when unoccupied.
- Establish nondisturbance buffer zones to protect sensitive habitats or areas of high risk for species of concern identified in preconstruction studies. Determine the extent of the buffer zone in consultation with the Service and state, local, and tribal wildlife biologists and land management agencies (e.g., U.S. Bureau of Land Management and U.S. Forest Service), or other credible experts as appropriate.
- Locate turbines to avoid separating bird and bat species of concern from their daily roosting, feeding, or nesting sites if documented that the turbines' presence poses a risk to species.
- Avoid impacts to hydrology and stream morphology, especially where federal- or state-listed aquatic or riparian species may be involved. Use appropriate erosion control measures in construction and operation to eliminate or minimize runoff into water bodies.
- When practical, use tubular towers or best available technology to reduce ability of birds to perch and to reduce risk of collision.
- After project construction, close roads not needed for site operations and restore these roadbeds to native vegetation, consistent with landowner agreements.
- Minimize the number and length of access roads; use existing roads when feasible.
- Minimize impacts to wetlands and water resources by following all applicable provisions of the Clean Water Act (33 USC 1251-1387) and the Rivers and Harbors Act (33 USC 301 et seq.), for instance, by developing and implementing a storm water management plan and taking measures to reduce erosion and avoid delivery of road-generated sediment into streams and waters.
- Reduce vehicle collision risk to wildlife by instructing project personnel to drive at appropriate speeds, be alert for wildlife, and use additional caution in low-visibility conditions.
- Instruct employees, contractors, and site visitors to avoid harassing or disturbing wildlife, particularly during reproductive seasons.
- Reduce fire hazard from vehicles and human activities (instruct employees to use spark arrestors on power equipment; ensure that no metal parts are dragging from vehicles; use caution with open flame, cigarettes, etc.). Site development and operation plans should specifically address the risk of wildfire and provide appropriate cautions and measures to be taken in the event of a wildfire.
- Follow federal and state measures for handling toxic substances to minimize danger to water and wildlife resources from spills. Facility operators should maintain hazardous materials spill kits on site and train personnel in the use of these.

- Reduce the introduction and spread of invasive species by following applicable local policies for invasive species prevention, containment, and control, such as cleaning vehicles and equipment arriving from areas with known invasive species issues, using locally sourced topsoil, and monitoring for and rapidly removing invasive species at least annually.
- Use invasive species prevention and control measures as specified by county or state requirements or by applicable federal agency requirements (such as Integrated Pest Management) when federal policies apply.
- Properly manage garbage and waste disposal on project sites to avoid creating attractive nuisances for wildlife by providing them with supplemental food.
- Promptly remove large animal carcasses (e.g., big game, domestic livestock, or feral animal).
- Wildlife habitat enhancements or improvements such as ponds, guzzlers, rock or brush piles for small mammals, bird nest boxes, nesting platforms and wildlife food plots should not be created or added to wind energy facilities. These wildlife habitat enhancements are often desirable but when added to a wind energy facility result in increased wildlife use of the facility, which may result in increased levels of injury or mortality to them.

7.5 BIRD DEATHS DUE TO COLLISIONS WITH WIND TURBINE BLADES

Several research studies evaluating bird and bat mortality due to wind turbines have been performed. The following sections show results from only a few of these. The American Wind Wildlife Institute (AWWI), the National Wind Coordinating Collaborative (NWCC), and the Bats and Wind Energy Cooperative (BWEC) are all good sources of information for locating new studies since much work is still being done to address this issue.

7.5.1 USFWS's *Migratory Bird Mortality: Many Human-Caused Threats Afflict Bird Populations*

This bulletin, published in 2002 by the USFWS, includes the following points [3]:

- Of the 836 species of birds protected under the Migratory Bird Treaty Act, about one-quarter are known to be in trouble.
- There are 78 bird species listed as "Endangered" and another 14 listed as "Threatened."
- The greatest threat to birds, and to all wildlife, continues to be the loss and/or degradation of habitat due to human development and disturbance.
- For migratory birds and other species that require multiple areas for wintering, breeding, and stopover points, the effects of habitat loss can be complex and far-reaching.

7.5.2 A Summary and Comparison of Bird Mortality from Anthropogenic Causes with an Emphasis on Collisions

This paper by Wallace Erickson, Gregory Johnson, and David Young, Jr., was published in the USDA Forest Service's General Technical Report PSW-GTR-191 in 2005 [4]. The study compared the estimated number of bird deaths caused by various anthropogenic sources, or human activities, including vehicles, buildings, windows, power lines, communication towers, wind turbines, electrocutions, oil spills, pesticides, and other contaminants and deaths due to domestic cats. The authors estimated these sources caused from 500 million to possibly over one billion bird deaths annually in the United States and that bird deaths attributable to wind turbines were less than 0.01% of all bird deaths attributable to human activities.

According to the Erickson et al. study, birds colliding with buildings account for more than one-half of bird deaths due to anthropogenic causes, followed by power lines (over 13%), cats (over 10%), automobiles (over 8%), and pesticides (over 7%). The study was based on the amount of wind turbines that existed in 2003, estimated to be 17,500 turbines totaling 6374 MW, suggesting an average turbine size of 360 kW or 0.36 MW. Wind generation capacity in the United States has grown rapidly since 2003, so one can assume that bird deaths attributable to wind generation have also increased. At year-end 2010, about 35,600 wind turbines were in operation in the United States totaling an estimated 40,190 MW, or about 1.1 MW per turbine, clearly showing that the size of turbines has increased significantly in recent years. Statistics from the American Wind Energy Association indicated that the rated capacity of turbines installed during calendar year 2010 averaged 1.77 MW. Thus, the capacity of wind generation in 2010 was 6.3 times as much as existed in 2003, but only about twice as many wind turbines were in use.

The study by Erickson et al. also found that bird deaths per turbine averaged 3.04 per year, or 2.11 birds per year per MW of installed capacity. If one extrapolated the "per MW" rate based on US wind installations at the end of 2012, bird deaths attributable to wind turbines would be approximately 130,000. Assuming bird deaths in the United States from anthropogenic sources were on the low end of the study's estimate (i.e., 500 million), bird deaths from wind turbines would still only account for about 0.026% of the total, or 1 out of every 4000 bird deaths.

7.5.3 Wind Turbine Interactions with Birds, Bats, and Their Habitats: A Summary of Research Results and Priority Questions

A factsheet produced by the NWCC in early 2010 shows that the incidence of bird collisions varies substantially between projects and regions and that a few wind energy projects around the world have had significant issues with bird collisions [5]. The factsheet includes a comparison of bird mortality rates at 45 wind energy facilities based on the numerous studies that have been performed on the subject. The one real outlier is a small wind project in Tennessee that had a mortality rate of 14 birds per MW per year, but all others ranged from less than 1 bird per MW per year to 7 birds per MW per year, with almost 21 of the 45 project studies indicating a mortality rate of 2 birds per MW per year or less. The factsheet can be obtained at the NWCC's website, http://nationalwind.org/.

7.5.4 The Avian Benefits of Wind Energy: A 2009 Update

In this paper by Benjamin K. Sovacool with the National University of Singapore, Sovacool has attempted to quantify the potential number of avian deaths caused by fossil-fueled power plants, nuclear power plants, and wind energy projects, in addition to some non–power-generating items including communications towers, pesticides, buildings, and feral cats [6].

His analysis, based on 2009, indicates the following numbers of avian deaths in the United States due to these items. He states that "when a range of estimates has been given, the figure presents only data for the lower end of the range."

Feral cats	110,000,000 per year
Buildings/windows	97,000,000 per year
Pesticides	72,000,000 per year
Fossil fuels	14,100,000 per year
Communications towers	4,000,000 per year
Nuclear power plants	332,323 per year
Wind power plants	19,875 per year

He points out that avian deaths from fossil-fueled power plants can be attributed to "coal mining activities, collision and electrocution with operating plant equipment, and poisoning and death caused by acid rain, mercury pollution, and climate change." Dr. Sovacool's conclusions include the following:

- While the avian deaths attributed to fossil fuel, wind, and nuclear power plants do vary, they also imply that there is no form of electricity supply completely benign to birds.
- The first-order estimates of avian mortality per GWh offered here suggest that fossil fuels may be more dangerous to avian wildlife (and nuclear power plants slightly more dangerous) than wind farms, and they remind us that what can sometimes be considered the most obvious consequence of a particular energy system may not always be the most salient.

7.5.5 Additional Impacts with Expected Growth of the Wind Energy Industry

The report produced by the US Department of Energy (DOE) and the National Renewable Energy Laboratory (NREL) titled *20% Wind Energy by 2030: Increasing Wind Energy's Contribution to U.S. Electricity Supply* [7] included an estimate that about 306,000 MW of wind energy would be needed by 2030 in order to provide 20% of the nation's electricity supply (or almost 50 times as much as existed in 2003). But since the rated capacity of wind turbines continues to increase, the total number of turbines may only be five to seven times as many as existed in 2003. Assuming 2.11

bird deaths per MW per year, 306,000 MW equates to about 650,000 birds per year, or about 0.13% of bird deaths from anthropogenic sources, again assuming 500 million total bird deaths per year in the United States.

When expressed as a percentage of total deaths from anthropogenic sources, deaths from wind turbines seem very small, but since the total number of birds killed by wind turbines may now exceed 100,000 per year in the United States, this represents a significant issue for the wind energy industry, even more so if threatened or endangered species are involved. Thus, as stated earlier, it is imperative that the wind energy industry be diligent during the siting and operation of projects so that the potential for bird collisions is minimized. Deaths of raptors at some of the older California wind energy projects in particular has been an issue of high concern to bird advocates, to environmentalists, and to the wind energy industry.

7.5.6 Comparing Impacts of Wind Energy with Other Sources of Electric Generation

Throughout this book, the authors have tried to compare and contrast the impact of electricity production from wind energy with the impact of electricity production from other sources, and there are studies in which researchers have estimated the number of bird deaths from fossil- and nuclear-powered electricity generation. Benjamin K. Sovacool contrasts bird mortality attributable to wind generation to that of fossil and nuclear power plants in his article, *Contextualizing Avian Mortality: A Preliminary Appraisal of Bird and Bat Fatalities from Wind, Fossil-Fuel, and Nuclear Electricity* [8]. His conclusions include the following:

- For wind turbines, the risk appears to be greatest to birds striking towers or turbine blades and for bats suffering barotrauma.
- For fossil-fueled power stations, the most significant fatalities come from climate change, which is altering weather patterns and destroying habitats that birds depend on.
- For nuclear power plants, the risk is almost equally spread across hazardous pollution at uranium mine sites and collisions with draft cooling structures.
- Yet, taken together, fossil-fueled facilities are about 17 times more dangerous to birds on a per GWh basis than wind and nuclear power stations.
- In absolute terms, wind turbines may have killed about 7,000 birds in 2006 but fossil-fueled stations killed 14.5 million and nuclear power plants 327,000.

7.6 WIND GENERATION WILDLIFE ISSUES NOT LIMITED TO AVIAN COLLISIONS

In addition to considering the potential for bird collisions with wind turbines when siting wind plants, wind developers also need to be aware of the potential for disturbing wildlife habitat or the migratory patterns of avian species. Of course, development of any kind (buildings, shopping centers, airports, power plants, etc.) will almost always impact some species of wildlife, so an important issue to always consider is the potential impact on the habitat of threatened or endangered species. Sensitive ecological areas and important areas of wildlife habitat include wetlands, playa lakes, tallgrass prairies, and coastal areas. According to the National Park Service's website [9], tallgrass prairies once covered 140 million acres of North America, but due to development, less than 4% of that acreage remains.

Much of the remaining tallgrass prairie is located in the Flint Hills of Kansas, where some proposed wind developments ran into significant levels of opposition

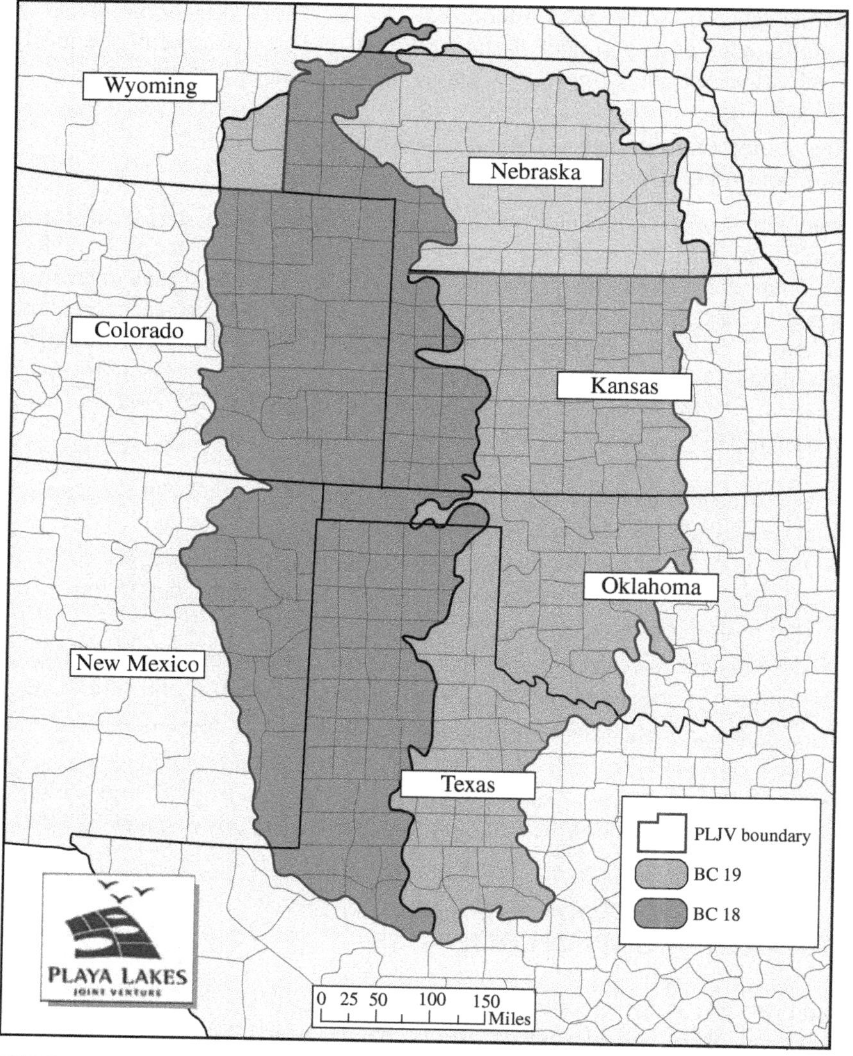

FIGURE 7.1 Playa Lake Joint Venture boundary (in red) (Source: Reproduced by permission of Playa Lakes Joint Venture http://www.pljv.org [10]).

due to the potential for additional destruction of tallgrass prairie. Even the prospect of transmission lines facilitating wind development in regions of Kansas west of the Flint Hills may be opposed if they are routed through the tallgrass prairie. Another ecologically sensitive area of the country called the Playa Lake Joint Venture (PLJV) area also happens to overlap with some of the nation's best wind resources. As shown in Figure 7.1, this area includes parts of Texas, Oklahoma, Kansas, Colorado, New Mexico, and Nebraska. A playa lake is a nearly level area at the bottom of an undrained desert basin, sometimes temporarily covered with water.

The PLJV estimates that there are more than 60,000 playas within the PLJV boundary that serve as habitat for many plants, insects, amphibians, and many species of birds. According to the PLJV website [10], "Playa lakes may be the most important wetland habitat type for waterfowl in the region, hosting about 20 species of waterfowl during wintering and migrating seasons."

One concern about wind development is that construction of turbines and the associated facilities (such as roads and electric lines) can destroy or fragment the habitat of many animal species, including some that may be threatened or endangered. Birds and other animals native to prairies need large contiguous areas of grassland for food and nesting. Some bird species (such as the lesser prairie chicken) may be sensitive to disturbances (such as turbine or vehicle noise) and also have an aversion to tall structures like wind turbines or transmission line towers.

Construction of wind plants in an area containing playa lakes is another concern, either that the playa may be filled in during construction or that increased erosion caused by the project will fill in the playa over time. Since playas may go undetected during dry periods, project designers or construction crews may be unaware of their existence.

7.7 AVIAN SPECIES OF CONCERN

The following avian species are just a few of those that may be impacted by the continued expansion of wind energy. This is not meant to be an exhaustive list but is indicative of the avian-related issues that wind developers must consider when evaluating potential sites for wind energy plants.

7.7.1 Raptors

Raptors are birds of prey such as eagles, hawks, falcons, or owls. Potential deaths of raptors is a significant issue for wind development but is a particularly large problem for wind plants located in the Altamont Pass, lying just east of the San Francisco Bay Area. This is one of the earliest and most widespread areas of wind development in the United States and includes many older, smaller turbine models, many of which use lattice towers.

The NWCC's factsheet *Wind Turbine Interactions with Birds, Bats, and their Habitats: A Summary of Research Results and Priority Questions* [5] also addresses the issue of raptors, including the following insights:

- Studies have indicated that relatively low raptor fatality rates exist at most wind energy developments with the exception of some facilities in parts of California.

- Raptors are known to concentrate along ridge tops, upwind sides of slopes, and canyons to take advantage of wind currents that are favorable for hunting and traveling, as well as for migratory flights [11–15].
- Siting turbines in areas of low prey density may reduce raptor collision rates at wind facilities. A high density of small mammal prey and the conditions favorable to high prey densities [16–18] have often been presumed to be the main factors responsible for the high raptor use, and hence high raptor collision rates at the Altamont Pass wind facility [19–21].
- Using newer monopole tubular support towers rather than lattice support towers associated with older designs may reduce raptor collision rates at wind facilities. Lattice support towers offer many more perching sites for raptors than do monopole towers, and hence may encourage high raptor occupancy in the immediate vicinity, or rotor swept area, of wind turbines [22].
- Newer larger (≥500 kW) turbines may reduce raptor collision rates at wind facilities compared to older smaller (40–330 kW) turbines, but have uncertain effects on songbirds.

7.7.2 Whooping Cranes

Whooping cranes are North America's tallest birds and are also one of the rarest. Hunting and habitat loss caused their populations to decline perilously close to extinction, with only an estimated 15 birds existing in the wild in 1941. Since then, the population has increased somewhat, largely due to conservation efforts. In May 2011, the Whooping Crane Conservation Association [23] estimated that 414 birds were living in the wild. Each year, many of the cranes migrate about 2700 miles from wintering grounds in coastal Texas (Aransas National Wildlife Refuge, Matagorda Island, Islas San Jose, and Lamar Peninsula) to breeding and nesting areas in Alberta, Canada (Wood Buffalo National Park). A few nonmigrating whooping cranes are located in central Florida and Louisiana, while another small group migrates between Wisconsin and Florida.

About 75% of the nation's population of whooping cranes passes through the Salt Plains National Wildlife Refuge in Oklahoma annually. Given the endangered status of these birds and their migratory path across some of the nation's best wind resources, it is critical that wind developers know where the migratory path is in relation to their projects and where the birds are known to stop at during their annual migrations (Fig. 7.2).

7.7.3 Black-capped Vireos

The world's most concentrated area of wind development is centered around Sweetwater, Texas, where more than 6000 MW of wind generation capacity is located within a circle having a 75-mile radius. In this same area, a little-known endangered species of bird can be found. The black-capped vireo, so named due to the black coloring on the top of its head, typically nests in shrublands and open woodlands that have a "patchy" structure.

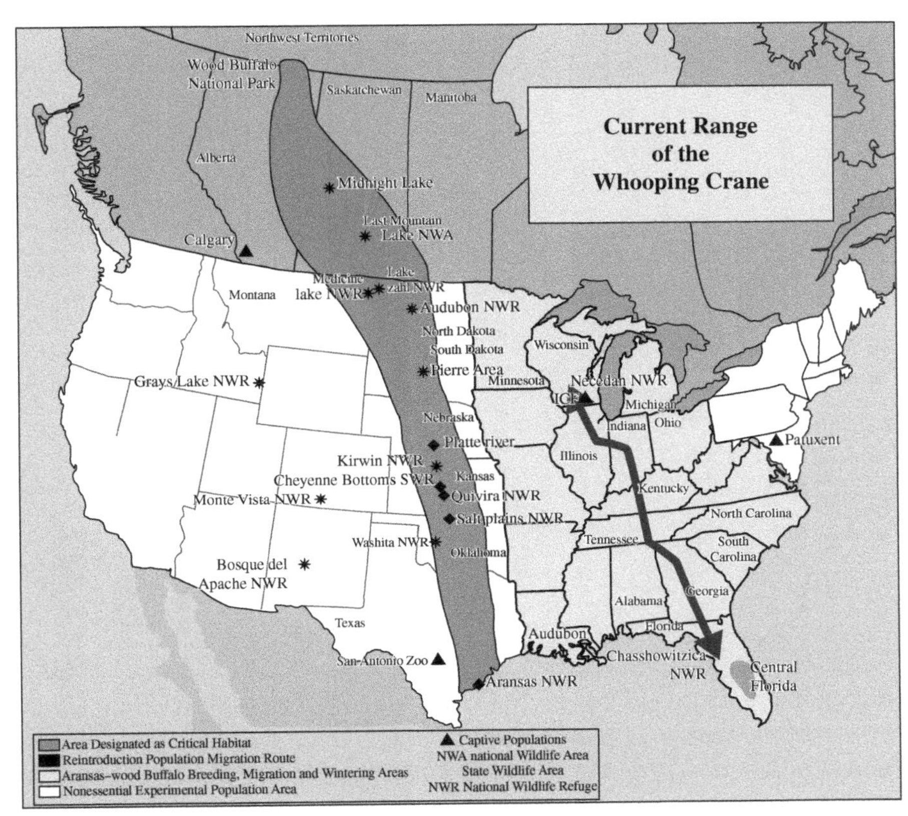

FIGURE 7.2 Whooping crane migratory paths (Photo Credit: Cephas, http://en.wikipedia.org/wiki/File:Grus_americana_map.svg [10]).

According to the Texas Parks and Wildlife Department [24], the decline of the black-capped vireo population can be attributed to habitat loss, natural plant succession, brood parasitism by the brown-headed cowbird, overbrowsing by deer and livestock, and human activities (such as brush clearing, fire suppression, and urbanization).

7.7.4 Lesser Prairie Chicken

The lesser prairie chicken is a medium-sized gray-brown grouse that is being considered by the USFWS for listing as a threatened or endangered species under the Endangered Species Act. A primary reason for its decline is a loss of native prairie as a result of agriculture and overgrazing by livestock. Since this bird flies very near the ground, the potential for collisions with wind turbine blades is not an issue. A study by the George Miksch Sutton Avian Research Center [25] found that more than 40% of lesser prairie chicken mortality in Oklahoma is due to collisions with fences.

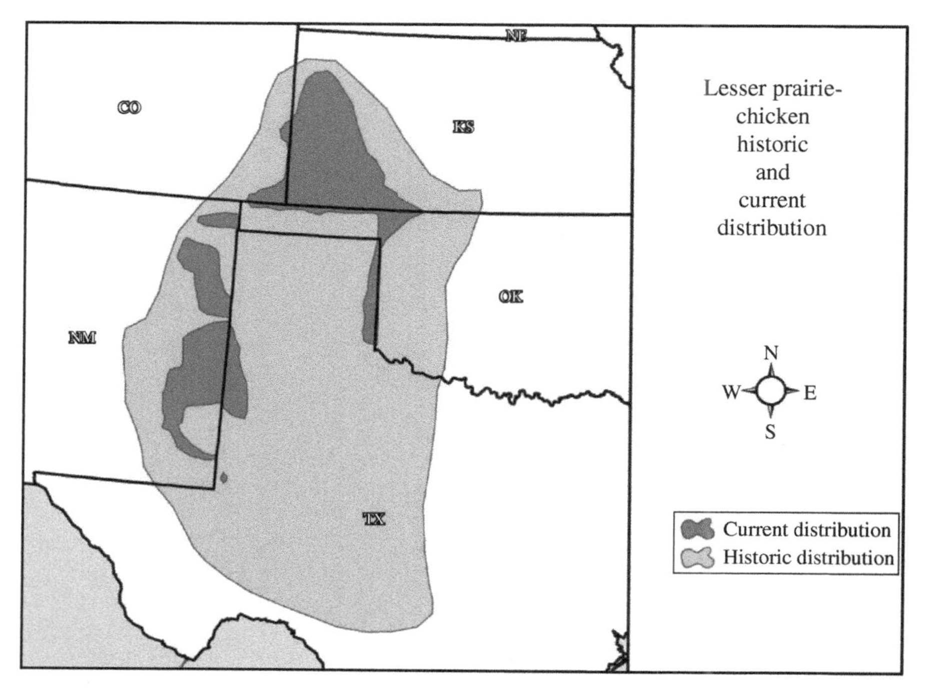

FIGURE 7.3 Lesser prairie chicken distribution (Source: Wild Earth Guardians, *Lesser Prairie-Chicken: A Decade in Purgatory* [26]).

Therefore, with regard to wind energy development, of much greater concern is the potential for habitat destruction or fragmentation during construction of wind plants or transmission lines, as well as the potential disturbance of the bird by noise from wind turbines or traffic or by the height of wind turbines and transmission towers.

In addition, if the lesser prairie chicken is classified as a threatened or endangered species under the Endangered Species Act, siting of wind turbines anywhere in the vicinity of the bird could become significantly more difficult. As shown in Figure 7.3, the habitat of the lesser prairie chicken has declined dramatically, with only small areas of habitat remaining in Texas, Oklahoma, New Mexico, Colorado, and Kansas.

7.8 BAT DEATHS DUE TO WIND ENERGY DEVELOPMENT

Bat deaths at the Mountaineer Wind Energy Center constructed on Backbone Mountain in West Virginia in 2003 first brought this issue to the forefront. A study performed by Jessica Kerns and Paul Kerlinger [27] estimated that more than 2,000 bats were killed during the study period (April 4, 2003 to November 11, 2003), although none of the bat fatalities were identified as being an endangered species.

Since that time, additional concerns have been voiced about the potential impact of wind development on bats including endangered species such as the Indiana bat, which can be found in areas of the central Midwest, Mid-Atlantic and Northeast regions of the United States. The issue led to the formation of the BWEC, discussed later in this chapter. BWEC's website (http://www.batsandwind.org) has links to several excellent reports and scientific publications regarding this subject.

7.8.1 Bat Mortality Rates and Potential Contributors to Bat Deaths from Wind Development

Bat mortality studies undertaken at a number of wind plants indicate that projects located on forested mountaintops or ridgelines in the eastern one-third of the United States tend to have the highest rate of bat deaths, as measured in fatalities per MW per year. NWCC's factsheet titled *Wind Turbine Interactions with Birds, Bats, and their Habitats: A Summary of Research Results and Priority Questions* [5] shows that bat fatalities have ranged from 40 fatalities per MW per year at the same small wind plant in Tennessee that had the highest rate of bird fatalities to less than 1 fatality per MW per year at several wind plants. More than one-half of those projects studied had less than 5 bat fatalities per MW per year.

The NWCC factsheet also includes the following insights about the issue of bats and wind plants:

- The lighting currently recommended by the Federal Aviation Administration (FAA) for installation on commercial wind turbines does not increase collision risk to bats...
- Weather patterns may influence bat fatalities. Some studies demonstrate that bat fatalities occur primarily on nights with low wind speed and typically increase immediately before and after the passage of storm fronts. Weather patterns therefore may be a predictor of bat activity and fatalities, and mitigation efforts that focus on these high-risk periods may reduce bat fatalities substantially [28].
- More adults and more male bats tend to be killed by wind turbines. Although this pattern has been documented at a number of facilities, it may represent an idiosyncrasy of the three species most commonly killed during their fall migration in North America. Furthermore, the pattern of adult fatalities may not necessarily reflect increased susceptibility of adults, but rather a preponderance of adults in the populations. There are notable exceptions, and some studies have reported female and juvenile bias among bat fatalities (e.g., [29–32]).
- It has recently been hypothesized that migratory tree bats (e.g., Hoary and Eastern Red Bats) may exhibit lek mating systems, so that males may be congregating around turbines during autumn in an effort to attract females.
- Bat fatalities in the southwestern United States are poorly understood but the Brazilian Freetailed Bat appears to be vulnerable. The Brazilian Free-tailed Bat comprised a large proportion (41–86%) of the bats killed at developments within this species' range ([28]).

- Bats appear to be attracted to wind turbines [33], and there are several plausible hypotheses that warrant testing as to how and why bats may be attracted to turbines [34], which may prove useful for developing new solutions to prevent collisions. Reasons for apparent attraction may include sounds produced by turbines, a concentration of insects near turbines, and bats attempting to find roost locations. For Hoary and Eastern Red Bats, additional studies need to be performed to better understand lek mating systems in these two species, especially regarding attraction to turbines.

The National Research Council's *Environmental Impacts of Wind-Energy Projects* [21] is another excellent source of information about wind energy development's impact on bats. Some of the more interesting insights included in this book's section addressing bats include the following:

- Migratory tree bats are the commonest reported bat fatalities at wind-energy facilities in the United States.
- The number of bats killed in the eastern United States at wind-energy facilities installed along forested ridge tops has ranged from 15.3 to 41.1 bats/MW/year of installed capacity.
- Bat fatalities reported from other regions of the western and Midwestern United States have been lower, ranging from 0.8 to 8.6 bats/MW/year.
- Bat fatalities at wind-energy facilities appear to be highest along forested ridge tops in the eastern United States and lowest in relatively open landscapes in the Midwestern and the western states.
- One hypothesis is that tall white wind turbine towers may function as a beacon to bats and their insect prey (many insects are attracted to large white objects), especially during nights with sufficient moonlight.
- The turbine and blades produce audible sounds, ultrasound, and infrasonic vibrations, and because some bat species are known to orient to distant sounds [35], it is possible that bats are attracted to sounds produced by turbines or become disoriented and when they are migrating or feeding in the vicinity of wind turbines [34, 36].
- Wind turbines produce obvious blade-tip vortices, and if bats get temporarily trapped in these moving air masses, it may be difficult for them to escape. Rapid pressure changes associated with these conditions may lead to internal injuries, disorientation, and death of bats [34, 36–38].
- Preliminary observations suggest a strong association of bat fatalities with thermal inversions following frontal passage [39]. Thermal inversions create cool, foggy conditions in the valleys with warmer air rising to the ridge tops that remain clear. These conditions would provide strong inducement for both insects and bats, whether migrating or not, to concentrate their activities along ridge tops [34, 36].

7.8.2 Barotrauma

As alluded to in the previous section, researchers undertaking bat mortality studies at wind plants noticed a significant number of dead bats in the vicinity of the wind plants that had no apparent sign of blunt force that would be indicative of contact with a turbine blade. Scientists now believe that bats are dying due to lung damage caused by a rapid reduction in air pressure in the vortices trailing wind turbine blades. This rapid decompression has been compared with "the bends" that deep sea divers can experience if they surface too quickly. When entering an area of low pressure behind the turbine blades, the bat's lungs expand too quickly, resulting in internal hemorrhaging of the lungs.

7.8.3 Minimizing the Impact of Wind Energy Development on Bats

In addition to the Best Management Practices for Wind Energy Development included in the USFWS' *Draft Voluntary, Land-Based Wind Energy Guidelines* [2] and discussed earlier in this chapter, following are some of the ongoing activities of the wind energy industry and the recommended practices for minimizing the impact of wind energy development on bats:

- Perform additional research to better understand and quantify what makes a site more risky for bats and if any design attributes of turbines can affect bat mortality.
- Perform field-testing of potential deterrent devices to warn bats away from wind turbines:
  - White noise generators or "acoustic scarecrows" have been tried as means of discouraging bats but have not worked well.
  - Radar signals emitted from wind turbines appear to reduce bat activity by 30–40% and are being studied as a means of deterrence.
- Use available online resources to identify areas with high concentrations of bats and migratory patterns of bats. Such resources include the following:
  - The United States Geological Service (USGS) Fort Collins Science Center's website (www.fort.usgs.gov) has a tremendous amount of information about the potential impact of wind energy on bats.
  - Endangered and Threatened Animals of Texas from TP&WL: http://www.tpwd.state.tx.us/publications/pwdpubs/media/pwd_bk_w7000_0013.pdf
- Perform field studies of potential wind energy sites under consideration to identify colonies of bats in the area. Techniques that may help to determine the density and species of bats in the area include mist netting, acoustic surveys, night vision or thermal imaging, and portable marine radar systems.
- Try to avoid locating turbines in areas exhibiting higher concentrations of bats or bat activity such as forested ridgelines or near caves containing colonies of bats.

- Minimize the use of outdoor lights around wind plant facilities (such as substations, operations and maintenance (O&M) buildings, and parking areas).
- After operation of the wind plant begins, monitor bat mortality by performing carcass searches and continue to monitor bat activity in the area using technologies such as acoustic surveys, thermal imaging, or radar surveys.
- Consider temporarily shutting turbines off in weather or wind conditions in which bats are more active. A study by Edward Arnett, Manuela Huso, Michael Schirmacher, and John Hayes [40] included the following information: "Relatively small changes to wind-turbine operation resulted in nightly reductions in bat mortality (ranging from 44% to 93%) with marginal annual power loss (<1% of total annual output)."

7.9 ORGANIZATIONS STUDYING WAYS TO MINIMIZE EFFECTS OF WIND DEVELOPMENT ON BIRDS AND OTHER WILDLIFE

Due in part to the concerns raised by biologists, scientists, environmentalists, and bird preservation organizations, the wind energy industry has responded to this issue by funding many studies on the subject of bird collisions and identifying ways to minimize them and by the formation of several organizations or groups that have focused on this subject. Such groups are mentioned as follows.

7.9.1 The American Wind Wildlife Institute

This organization was founded in 2008 by eight wind energy companies and seven environmental and conservation organizations. Its two-part mission is to facilitate timely and responsible development of wind energy while protecting wildlife and wildlife habitat, and its purpose is to help lay the scientific groundwork and best practices for wind farm siting and operations through targeted initiatives including wind–wildlife research, landscape assessment, mitigation, and education. More information about this organization can be found on its website, http://www.awwi.org.

7.9.2 The National Wind Coordinating Collaborative

The NWCC is a consensus-based collaborative formed in 1994 that comprises representatives from the utility, wind industry, environmental, consumer, regulatory, power marketer, agricultural, tribal, economic development, and state and federal government sectors to support the development of an environmentally, economically, and politically sustainable commercial market for wind power. Its website (http://www.nationalwind.org) provides access to several studies on the subject of the interaction of birds and wind turbines, and it has various working groups focusing on a wide variety of issues. One such NWCC working group is the Grassland/Shrub Steppe Species Collaborative.

7.9.3 The USFWS WTGA Committee

In 2007, the U.S. Secretary of the Interior appointed 22 individuals as members of the WTGA Committee. The scope and objective of this committee was to provide advice and recommendations to the Secretary on developing effective measures to avoid or minimize impacts to wildlife and their habitats related to land-based wind energy facilities. Its members represent the varied interests associated with wind energy development and wildlife management.

7.9.4 The Bats and Wind Energy Cooperative

In 2003, American Wind Energy Association (AWEA), Bat Conservation International (BCI), the USFWS, and DOE/NREL formed the BWEC, which conducts research on the issue of bat fatalities at wind plants, including evaluation of technologies, siting assessments, and/or operational techniques that can mitigate impacts on bats.

7.10 CONCLUSIONS

Proper project siting and understanding of wildlife and avian behavior and habitat are critical to the wind project development process. Wind developers must identify regions where it is known that wind projects could pose a substantial risk to wildlife very early in the siting process. Even if habitat has been identified on maps and the project designed so as to minimize the potential impact on wildlife, it is important to clearly mark habitat areas in the field and to thoroughly discuss with all construction personnel their locations and the ramifications of failing to avoid such areas. In many cases, wind development can coexist near habitat of threatened and endangered species by performing detailed studies to identify such habitat, by performing four-season avian studies, by avoiding bisection or fragmentation of habitat areas with roads or electric lines, using tubular steel towers, placing distribution-voltage electric lines underground, and implementing operational protocols such as shutting turbines down during periods of bird or bat migration or activity, which often occur during low-wind periods.

REFERENCES

[1] U.S. Fish and Wildlife Service. *Land-Based Wind Energy Guidelines*. Washington, DC: U.S. Fish and Wildlife Service; 2012.

[2] U.S. Fish and Wildlife Service. *Eagle Conservation Plan Guidance: Module 1—Land-Based Wind Energy*. Washington, DC: U.S. Fish and Wildlife Service; 2011.

[3] U.S. Fish and Wildlife Service. *Migratory Bird Mortality: Many Human-Caused Threats Afflict Bird Populations*. Arlington: U.S. Fish and Wildlife Service; 2002.

[4] Erickson W, Johnson G, Young Jr, D. A summary and comparison of bird mortality from anthropogenic causes with an emphasis on collisions. Bird Conservation Implementation

and Integration in the Americas: Proceedings of the Third International Partners in Flight Conference. 2002 March 20–24; Asilomar, California, Volume 2. Ralph, C. John; Rich, Terrell D., editors. USDA Forest Services General Technical Report PSW-GTR-191. U.S. Dept. of Agriculture, Forest Service, Pacific Southwest Research Station, Albany, CA; 2005.

[5] National Wind Coordinating Collaborative. Wind Turbine interactions with birds, bats, and their habitats: a summary of research results and priority question. NWCC factsheet; 2010.

[6] Sovacool B. The avian benefits of wind energy: a 2009 update. Renew Energy J 2009;49 (2013):19–24.

[7] U.S. Department of Energy, Energy Efficiency and Renewable Energy. 20% Wind Energy by 2030: Increasing Wind Energy's Contribution to U.S. Electricity Supply, DOE/GO-102008-2567. Washington, DC: U.S. Department of Energy, Energy Efficiency and Renewable Energy; 2008.

[8] Sovacool B. Contextualizing avian mortality: a preliminary appraisal of bird and bat fatalities from wind, fossil-fuel, and nuclear electricity. Energy Policy 2009;37: 2241–2248.

[9] National Park Service, U.S. Department of the Interior, Washington, DC. National Park Service website. Available at http://www.nps.gov/tapr/index.htm?showid=74. Accessed July 12, 2014.

[10] Playa Lakes Joint Venture. Available at http://www.pljv.org/cms/our-birds. Accessed August 12, 2012.

[11] Barrios L, Rodríguez A. Behavioural and environmental correlates of soaring-bird mortality at on-shore wind turbines. J Appl Ecol 2004;41:72–81.

[12] Bednarz JC, Klem D, Goodrich LJ, Senner SE. Migration counts of raptors at Hawk Mountain, Pennsylvania, as indicators of population trends, 1934–1986. Auk 1990;107: 96–109.

[13] Curry RC, Kerlinger P. Avian mitigation plan: kinetic model wind turbines, Altamont Pass Wind Resources Area, California. Technical report by Curry and Kerlinger, LLC. McLean, Virginia, USA; 1998.

[14] Hoover SL, Morrison ML. Behavior of red-tailed Hawks in a wind turbine development. J Wildl Manag 2005;69:150–159.

[15] Manville AM II. Towers, turbines, power lines, and buildings – steps being taken by the U.S. Fish and Wildlife Service to avoid or minimize take of migratory birds at these structures. In: Rich TD, Arizmendi C, Demarest D, Thompson C, editors. *Tundra to Tropics: Connecting Habitats and People. Proceedings 4th International Partners in Flight Conference, 13–16 February 2008, McAllen, Texas. Partners in Flight.* 2009. pp. 262–272.

[16] Smallwood KS, Thelander CG. Developing methods to reduce bird mortality in the Altamont pass wind resource area. Final report. P500-04-052. Prepared for California Energy Commission, Public Interest Energy Research Program, Sacramento, CA; 2004.

[17] Smallwood KS, Thelander CG. Bird mortality at the Altamont Pass wind resource area: March 1998–September 2001. Subcontract Report NREL/SR-500-36973. Prepared for National Renewable Energy Laboratory, Golden, CO; 2005.

[18] Smallwood KS, Thelander CG. Bird mortality in the Altamont Pass wind resource area. Calif J Wildl Manag 2008;72:215–223.

[19] Kingsley A, Whittam B. Wind turbines and birds: a background review for environmental assessment. Prepared by Bird Studies Canada Prepared for Environment Canada/ Canadian Wildlife Service; 2007.

[20] Kuvlesky WP, Brennan LA, Morrison ML, Boydston KK, Ballard BM, Bryant FC. Wind energy development and wildlife conservation: challenges and opportunities. J Wildl Manag 2007;71:2487–2498.

[21] NAS (National Academy of Sciences). *Environmental Impacts of Wind-Energy Projects.* Washington, DC: The National Academies Press; 2007.

[22] Orloff S, Flannery A. Wind turbine effects on avian activity, habitat use, and mortality in Altamont Pass and Solano County Wind Resource Areas, 1989–1991. Final Report. P700-92-001. Prepared for Planning Departments of Alameda, Contra Costa and Solano Counties and the California Energy Commission, Sacramento, CA, by BioSystems Analysis, Inc., Tiburon, CA; 1992.

[23] Whooping Crane Conservation Association. Available at http://whoopingcrane.com/ flock-status/flock-status-2011-may/. Accessed August 12, 2012.

[24] Campbell L. *Endangered and Threatened Animals of Texas: Their Life History and Management.* Austin: Texas Parks and Wildlife Department; 2003.

[25] George Miksch Sutton Avian Research Center. Available at http://www.suttoncenter.org/ fence_marking.html. Accessed August 12, 2012.

[26] Wild Earth Guardians. Lesser Prairie-Chicken: a decade in Purgatory. Santa Fe; 2008. Available at: http://www.wildearthguardians.org/support_docs/report_lesser-prairie-chicken_6-9-08.pdf. Accessed June 9, 2008.

[27] Kerns J, Kerlinger P. A study of bird and bat collision fatalities at the Mountaineer Wind Energy Center, Tucker County, West Virginia: Annual Report for 2003. Curry & Kerlinger, LLC for FPL Energy; 2004. Available at http://www.wvhighlands.org/Birds/ MountaineerFinalAvianRpt-%203-15-04PKJK.pdf. February 14, 2004.

[28] Arnett EB, Brown WK, Erickson WP, Fiedler JK, Hamilton BI, Henry TH, Jain A, Johnson GD, Kerns J, Koford RR, Nicholson CP, O'Connell TJ, Piorkowski MD, Tankersley RD Jr. Patterns of bat fatalities at wind energy facilities in North America. J Wildl Manag 2008;72:61–78.

[29] Brown WK, Hamilton BL. Bird and bat monitoring at the McBride Lake Wind Farm, Alberta, 2003–2004. Report for Vision Quest Windelectric, Inc., Calgary, Alberta, Canada; 2004.

[30] Brown WK, Hamilton BL. Bird and bat interactions with wind turbines: Castle River Wind Farm, Alberta, 2001–2002. Report for Vision Quest Windelectric, Inc., Calgary, Alberta, Canada; 2006.

[31] Brown WK, Hamilton BL. Monitoring of bird and bat collisions with wind turbines at the Summer view Wind Power Project, Alberta, 2005–2006. Report for Vision Quest Windelectric, Inc., Calgary, Alberta, Canada; 2006.

[32] Fiedler JK, Henry TH, Nicholson CP, Tankersley RD. *Results of Bat and Bird Mortality Monitoring at the Expanded Buffalo Mountain Windfarm, 2005.* Knoxville: Tennessee Valley Authority; 2007.

[33] Horn JW, Arnett EB, Kunz TH. Behavioral responses of bats to operating wind turbines. J Wildl Manag 2008;72:123–132.

[34] Kunz TH, Arnett EB, Erickson WP, Hoar AR, Johnson GD, Larkin RP, Strickland MD, Thresher RW, Tuttle MD. Ecological impacts of wind energy development on bats: questions, research needs, and hypotheses. Front Ecol Environ 2007;5:315–324.

[35] Buchler ER, Childs SB. Orientation to distant sounds by foraging big brown bats (*Eptesicus fuscus*). Anim Behav 1981;29 (2):428–432.

[36] Kunz TH, Arnett EB, Cooper BM, Erickson WP, Larkin RP, Mabee T, Morrison ML, Strickland MD, Szewczak JM. Assessing impacts of wind-energy development on nocturnally active birds and bats. J Wildl Manag 2007;71:2449–2486.

[37] Durr T, Bach L. Bat deaths and wind turbines: a review of current knowledge and of information available in the database for Germany [in German]. Bremer Beitrage fur Naturkunde und Naturschutz 2004;7:253–264.

[38] Henson MM, Madden VJ, Rask-Andersen H, Henson OW Jr. Smooth muscle in the annulus fibrosus of the tympanic membrane in bats, rodents, insectivores, and humans. Hear Res 2005;200 (1–2):29–37.

[39] Arnett EB. Relationships between bats and wind turbines in Pennsylvania and West Virginia: an assessment of fatality search protocols, patterns of fatality, and behavioral interactions with wind turbines. Final Report. Prepared for the Bats and Wind Energy Cooperative, by Bat Conservation International, Austin, TX. June 2005.

[40] Arnett E, Huso M, Schirmacher M, Hayes J. *Altering Turbine Speed Reduces Bat Mortality at Wind-Energy Facilities*. Washington, DC: The Ecological Society of America's Frontiers in Ecology and the Environment; 2011.

8

ENVIRONMENTAL AND ECOLOGICAL IMPACTS OF WIND ENERGY ON HUMANS: PUBLIC HEALTH ISSUES

8.1 OVERVIEW OF ENVIRONMENTAL AND ECOLOGICAL IMPACTS ON HUMANS

Numerous public opinion polls show that wind energy and other renewable energy sources have very high levels of public support, but like virtually any other method of producing electricity, wind energy does have its detractors. The previous chapter talked about the potential impacts of wind energy on wildlife and the wind energy industry's mitigation efforts. In this chapter and the following two chapters, we begin a discussion of impacts on humans as well as how effective methods are for reducing or mitigating the extent of such impacts.

The issues and/or topics that will be discussed in this chapter include the following:

- Audible noise/ultrasound/low-frequency sound
- Shadow flicker
- Electric and magnetic fields (EMFs)
- Solid and hazardous wastes

Wind Energy Essentials: Societal, Economic, and Environmental Impacts, First Edition.
Richard P. Walker and Andrew Swift.

8.2 WIND TURBINE NOISE: KEY TERMS

Some of the key terms to understand as part of the discussion of wind turbine noise include the following:

- **Sound:** It describes wave-like variations in air pressure that occur at frequencies that can stimulate receptors in the inner ear and, if sufficiently powerful, be appreciated at a conscious level.
- **Noise:** It implies the presence of sound but also implies a response to sound; noise is often defined as unwanted sound.
- **Ambient noise level:** It is the composite of noise from all sources near and far; it is the normal or existing level of environmental noise at a given location.
- **Sound frequency:** The frequency of a sound is measured in vibrations per second, or Hertz (Hz), and determines how the sound may be categorized:
    - **Audible noise** is generally considered to have frequencies between 20 and 20,000 Hz, although some show the range extending down to 16 Hz.
    - **Low-frequency** sound is generally considered to have frequencies in the range of 10–200 Hz.
    - **Infrasound** is sound with a frequency too low to be detected by the human ear and is generally considered to have frequencies below 20 Hz, although there is not always a clear delineation between infrasound and low-frequency sound. In addition, sound frequencies in this range emitted at high sound levels (high decibels) may be audible.
- **Decibel (dB):** It is a unit used to measure the intensity of a sound or the degree of loudness. Decibel measurements use a logarithmic scale, so values do not add up the same as they would for a linear scale. Doubling the sound power increases the sound pressure level by 3 dB. For example, two wind turbines each generating 110 dB of noise would produce a combined noise of 113 dB. Most people subjectively perceive a volume increase of 10 dB as twice as loud.
- **A-weighted and C-weighted sound pressure levels:** The sensitivity of the human ear varies by frequency. For example, a low-frequency sound at a given power level may not seem as loud as a higher-frequency sound at the identical power level. Thus, a weighting scale has been developed that adjusts for the difference in frequency. A-weighted sound pressure is measured in dB(A) on a sound level meter using an A-weighted filter that deemphasizes the very low-frequency and the very high-frequency components of the sound in a manner similar to the frequency response of the human ear and correlates well with subjective reactions to noise. C-weighting is intended for measuring higher or peak levels of noise.
- **L_{max}:** It refers to the maximum sound level measured.
- **L_{eq}:** It is the equivalent continuous sound or an average sound energy over a given time period.
- **L_{10}:** It is the sound level exceeded 10% of the time, which is generally considered to be the sound level that will annoy people.

- L_{90}: It is the sound level exceeded 90% of the time, which is generally considered to be a measure of ambient background noise.
- L_{dn}: Day–night average sound level or the average sound level for a 24-h period.

8.3 ESTIMATES OF NOISE FROM WIND TURBINES

There is no denying that wind turbines do produce sound, as does any other method of producing electricity with the possible exception of solar photovoltaic modules. Sounds from a wind energy project can generally be classified as (i) mechanical noises (such as may be produced by the turbine gearbox, yaw motors, or generator), (ii) electronic noises or "humming" noises (such as may be caused by the generator, power converter, or transformer), and (iii) aerodynamic noise caused by the interaction of the turbine blade with the wind (the "swooshing" noise).

The actual sound level is influenced by many factors including the type of turbine, the wind speed, and the surrounding topography. Much of the sound from wind power plants is often masked by noises in the surrounding environment (such as the sound of wind blowing through trees or bushes, wind blowing around buildings, or highway traffic noise). Some turbine manufacturers include estimated noise levels produced by the turbine in the technical specifications of their product. State-of-the-art wind turbines will generally produce less noise than old turbine designs, as will be discussed in more detail later in this chapter.

Some mechanical noises are an indication of a maintenance issue that needs to be addressed by the project owner. For example, potential problems may include the "clanging" of a nose cone hatch resulting from failure of a hinge or the "whining" noise caused by a tip brake that is not property aligned or sealed. Noise during construction of wind projects can be quite loud, but this is limited to the construction period or sometimes during major maintenance projects. Such noise may include large vehicles delivering turbine components, concrete, or road materials; construction equipment such as cranes, bulldozers, or road graders; or blasting that may be required for foundation construction.

Noise levels drop off rapidly as the distance between the turbine and the location of the measurement increases due to absorption of sound by the environment. Therefore, increasing the distance between turbines and residences and other occupied structures is probably the number one thing that a wind developer can do to reduce a noise problem. Those homeowners wishing to install turbines on their own property need to carefully consider the size of their property, the turbine's location relative to their neighbors' homes, and the type of turbine they plan to use.

The National Research Council produced a book titled *Environmental Impacts of Wind-Energy Projects* [1] in 2007 stating that noise levels from a single turbine are in the range of 90–105 dB(A) at the source but drop to 50–60 dB(A) at a distance of 40 m from the turbine. Small turbines can also produce noise levels in this range, and since some people may install smaller turbines in residential areas, this may create hostilities between neighbors.

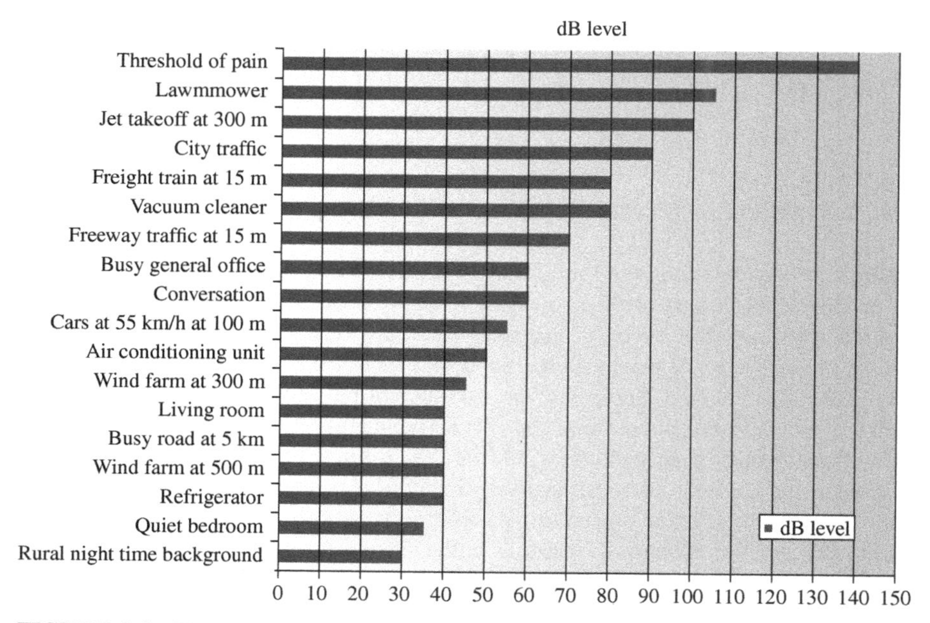

FIGURE 8.1 Typical sound pressure levels measured in the environment and industry (Source: K. Jay based on AWEA/CanWEA Study [3], Scottish Environmental Department [4], and GE Global Research [5]).

The American Wind Energy Association's (AWEA's) factsheet titled *Utility Scale Wind Energy and Sound* [2] indicates that turbine noise drops to 35–45 dB(A) at a distance of 350 m, comparable with noise levels in the living room or bedroom of one's home. Figure 8.1 shows how sound from a wind farm compares to other types of sound. Keeping utility-scale wind turbines at least 350 m (1148 ft) from homes is probably a good idea, particularly if the homeowner is not receiving royalties from the turbine. When a landowner is the one getting royalties from the turbine in question, he or she should be briefed on the issue of noise and, if possible, should be allowed to visit a nearby wind project in order to get an accurate feel for the noise levels.

Since questions about turbine noise are some of the ones most frequently asked of wind developers, AWEA and the Canadian Wind Energy Association (CanWEA) asked that a number of medical doctors, audiologists, and acoustical professionals serve on an advisory panel for the purposes of reviewing available literature about this issue and for preparing a report that could serve as a source of reference for landowners, government officials, and the wind industry. The report titled *Wind Turbine Sound and Health Effects: An Expert Panel Review* [3] was released in December 2009 and can be accessed on the website of the U.S. Department of Energy's Wind Power American program (www.windpoweringamerica.gov). It will be discussed in more detail in the following section of this chapter addressing the potential health effects of wind energy.

8.4 POTENTIAL HEALTH EFFECTS OF AUDIBLE OR SUBAUDIBLE NOISE FROM WIND TURBINES

Dr. Nina Pierpont, a physician specializing in pediatrics, published a paper in 2006 titled *Wind Turbine Syndrome: Noise, Shadow Flicker, and Health* [6], speculating about the potential detrimental health impacts of noise associated with wind generators, labeling these potential health impacts as "wind turbine syndrome." She also testified before the Energy Committee of the New York State Legislature in March 2006. Dr. Pierpont has more recently written an entire book on the subject of wind turbine syndrome in which she has indicated that symptoms of wind turbine syndrome include the following:

- "Sleep problems: noise or physical sensations of pulsation or pressure make it hard to go to sleep and cause frequent awakening
- Headaches which are increased in frequency or severity
- Dizziness, unsteadiness, and nausea
- Exhaustion, anxiety, anger, irritability, and depression
- Problems with concentration and learning
- Tinnitus (ringing in the ears)"

Her paper also states the following:

- "Not everyone near turbines has these symptoms. This does not mean people are making them up; it means there are differences among people in susceptibility. These differences are known as risk factors. Defining risk factors and the proportion of people who get symptoms is the role of epidemiologic studies, which are in progress.
- Chronic sleep disturbance is the most common symptom. Exhaustion, mood problems, and problems with concentration and learning are natural outcomes of poor sleep.
- Sensitivity to low-frequency noise is a potential risk factor. Some people sense low-frequency noise as pressure in the ears rather than heard as sound or experience a feeling or vibration in the chest or throat.
- Neighbors of industrial wind turbines describe the distressing sensation of having to breathe in sync with a rhythmic pulsation from the turbines which is not necessarily audible, especially at night when trying to sleep."

As discussed earlier, AWEA and CanWEA responded to such concerns by asking a panel of medical doctors, audiologists, and acoustical professionals to review available literature about this issue. Their resulting report included the following in Section 8.14:

"Following review, analysis, and discussion, the panel reached agreement on three key points:

- There is nothing unique about the sounds and vibrations emitted by wind turbines.
- The body of accumulated knowledge about sound and health is substantial.
- The body of accumulated knowledge provides no evidence that the audible or subaudible sounds emitted by wind turbines have any direct adverse physiological effects."

"In conclusion:

- Sound from wind turbines does not pose a risk of hearing loss or any other adverse health effect in humans.
- Subaudible, low-frequency sound and infrasound from wind turbines do not present a risk to human health.
- Some people may be annoyed at the presence of sound from wind turbines. Annoyance is not a pathological entity.
- A major cause of concern about wind turbine sound is its fluctuating nature. Some may find this sound annoying, a reaction that depends primarily on personal characteristics as opposed to the intensity of the sound level."

8.5 NOISE ORDINANCES

Many countries, states, or communities have general noise ordinances, but due to the increasing concern about noise specific to wind turbines, some ordinances focused on wind turbines have been enacted and it is very likely that more will be enacted by other governmental entities in the future. Therefore, it is critical that wind developers (or even homeowners wishing to install a small turbine) research this issue for their particular location.

Several European nations and a few others around the world have established ordinances or regulations limiting the level of noise from wind turbines and wind plants. Generally, these seem to be in the range of 35–45 dB(A) during nighttime periods in rural environments, with slightly higher ranges during daytime and in more urban areas. There is significant variation on how the sound measurements are to be taken and how the noise is to be calculated.

For example, some ordinances may require measurements at the edge of the homeowner's property, whereas others may measure at the edge of the home nearest to the turbine. Countries that have implemented such standards include the United Kingdom, France, Germany, the Netherlands, Denmark, Sweden, Norway, Switzerland, Italy, Australia, New Zealand, and Canada.

As of the writing of this book, the authors are unaware of any state-mandated noise standards applicable to utility-scale wind plants in the United States, although several communities or counties have enacted such ordinances. Ohio does require noise studies for proposed wind projects to be submitted to the state Power Siting Board during the permitting process. Some states have established standards for small turbines, although some of these are intended to limit the enactment of ordinances regulating small turbines by communities in the state. State-mandated general noise ordinances, however, may still apply (such as Washington state's limitation of industrial noise, 55 dB(A) during the daytime and 45 dB(A) during the night). Michigan, Illinois, Wisconsin, and Pennsylvania have developed recommended siting guidelines and/or sample ordinance guidelines for local governments to consider adopting and generally recommend limits between 45 and 55 dB(A).

Some states and many counties and communities have chosen to address wind turbine noise by establishing "setback" regulations for wind turbines, meaning that a minimum separation between turbines and residences or other occupied buildings is mandated. The National Renewable Energy Laboratory published its technical report NREL/TP-500-44439 in December 2008 titled *An Overview of Existing Wind Energy Ordinances* [7], which provides a good summary of county and community ordinances regulating wind energy, some of which have specific noise ordinances.

8.6 TECHNOLOGICAL ADVANCES THAT CAN REDUCE TURBINE NOISE

While wind turbines continue to increase in size, technological advances have helped to reduce sounds from wind turbines. Following are examples of how advances in turbine design can affect noise levels.

8.6.1 Blade Pitch Control Versus Stall Regulation

The predominate use of full-span pitch turbines (as opposed to stall-regulated turbines) eliminates the need for tip brakes, ailerons, or seams in blades, all of which are potential sources of aerodynamic noise.

8.6.2 Variable Speed Versus Constant Speed

The use of larger bladed, variable-speed turbines has generally resulted in lower rotational speeds as measured in revolutions per minute. A variable-speed turbine turns slower in low wind speeds than in high wind speeds, thus producing less aerodynamic noise during times when there is less masking by background noise from wind blowing through trees, bushes, or buildings.

8.6.3 Upwind Versus Downwind Turbines

The vast majority of utility-scale turbines are upwind turbines, meaning wind encounters the turbine blade before encountering the tower. Downwind turbines can cause a pulsating noise since each time one of the blades passes behind the tower wind has essentially been blocked by the tower.

8.6.4 Direct Drive Versus Turbines with Gearboxes

A relatively new advent in the wind industry is the use of direct-drive turbines, which eliminate the need for a gearbox. Currently, most turbines still use a gearbox, which can be a source of mechanical noise. Elimination of the gearbox thus may result in quieter turbines.

8.7 SITING PROCEDURES FOR NOISE MITIGATION

The authors of this book are not medical doctors, audiologists, and acoustical professionals and thus have only presented the various sides of this debate rather than opine on their validity. The purpose of this book is to frame the various issues and arguments surrounding the development of wind energy so that its readers can be aware of such issues and have some idea about how to avoid significant opposition to their project development efforts. Following are some of the recommended practices during development of wind projects that can help prevent noise-related issues from becoming a major problem for one's project.

8.7.1 Mapping

Map locations of all buildings prior to turbine siting. Topographic maps, aerial photos, and online tools (such as Google Earth) can be useful in the preliminary identification of buildings. Since buildings may have been constructed or abandoned since such maps or photos were made, a field examination is recommended. In addition, since mobile homes or preconstructed buildings can be quickly moved into an area, it is recommended that additional field reconnaissance be performed immediately prior to the commencement of construction and periodically during the construction project.

8.7.2 Review of State and Local Ordinances

Check for state and local ordinances that will affect turbine siting (such as noise ordinances or setback requirements). Since enactment of such ordinances has been increasing as wind energy rapidly expands, it is always wise to check for new regulations even in areas where other wind projects currently exist.

8.7.3 Establish Setbacks

Establish internal recommended practices for noise limits and setbacks. Some of the computer models being used to design wind plants have capabilities to model turbine noise and plot setback areas.

8.7.4 Public Communication

Communicate with the public early and often. Town hall meetings can help to alleviate "fear of unknown" and to identify potential issues in the community.

8.7.5 Compare Noise Potential of Various Turbine Models

Include turbine noise as a consideration in the turbine procurement process. As discussed earlier, noise produced by turbines can differ by their general design characteristics and by turbine model.

8.7.6 Pre- and Postconstruction Noise Measurements

Take noise measurements for some period of time prior to beginning construction and operation of the wind project, as well as after project's commercial operation date. A long-term background noise level baseline should be established prior to operation of the wind project for the purposes of helping to resolve any potential questions about noise levels before and after the project.

8.8 SHADOW FLICKER

Shadow flicker occurs when the blades of a wind turbine rotate in sunny conditions, casting moving shadows on the ground or on windows of buildings resulting in alternating changes in light intensity, as depicted in Figure 8.2. This can be a nuisance to homeowners when the flicker occurs through the home's windows.

Some people have asserted that shadow flicker from wind turbines can trigger epileptic seizures in individuals suffering from photosensitivity. Shadow flicker has a frequency between 0.5 and 1.25 occurrences per second (0.5–1.25 Hz) and can easily be calculated based on the maximum rotational speed of wind turbines and the number of turbine blades. This is a much lower frequency than the strobe rate that is generally considered to be capable of triggering epileptic seizures. Small

FIGURE 8.2 Depiction of shadow flicker caused by wind turbines (Source: Reproduced by permission of Amy Marcotte).

wind turbines may turn at higher RPMs but are almost always on much shorter towers and have much shorter blades, meaning shadow flicker associated with such turbines is very minimal.

The book *Environmental Impacts of Wind Energy Projects* [1] produced by National Research Council includes the following information about shadow flicker:

- Shadow flicker is a function of several factors, including the location of people relative to the turbine, the wind speed and direction, the diurnal variation of sunlight, the geographic latitude of the location, the local topography, and the presence of any obstructions.
- Shadow flicker is not important at distant sites (e.g., >1000 ft from a turbine) except during the morning and evening when shadows are long.
- However, sunlight intensity is also lower during the morning and evening; this tends to reduce the effects of shadows and shadow flicker.
- Shadow flicker may be analytically modeled with software such as GH WindFarmer and WindPro.

AWEA's factsheet titled *Wind Turbines and Health* [8] includes the following information:

- Shadow flicker occurrence is easily calculated. Computer models in wind development software can determine the days and times during the year that specific buildings in close proximity to turbines may experience shadow flicker.
- Mitigation measures can be taken based on this knowledge and may include setbacks or vegetative buffers.
- Issues with shadow flicker are less common in the United States than in Europe due to the lower latitudes and the higher sun angles in the United States.
- Shadow flicker is not harmful to persons with epilepsy. The allegation is sometimes made that shadow flicker from wind turbines can cause epileptic seizures. This is not true—shadow flicker from wind turbines occurs much more slowly than the light "strobing" associated with seizures. The strobe rates generally necessary to cause seizures in people with photosensitive epilepsy are 5–30 flashes per second (based on information from the Epilepsy Foundation) and large wind turbine blades cannot rotate this quickly.

8.9 MITIGATION OF SHADOW FLICKER

Similar to addressing noise issues, the number one thing that wind developers can do to mitigate shadow flicker is to establish a "setback" distance between turbines and homes. Using setbacks of about 350 m between wind turbines and buildings should result in limiting shadow flicker occurrences in most cases to sunrise or sunset when turbine shadows would be longest. If projects are located in the northern US states or Canada, one may want to increase this distance in some directions from the turbine. In addition to establishing internal setback requirements, project designers should use one of the wind project design software packages that include the ability to model shadow flicker and produce an estimated number of minutes per year in which shadow flicker may affect a given residence.

8.10 ELECTRIC AND MAGNETIC FIELDS

Exposure to EMFs is very common. Electric distribution lines that run up and down streets and alleys are a source of EMFs, as well as electric substations and the high-voltage transmission lines that connect power plants to cities and towns. However, in most cases, the largest sources of such exposure are electric devices in one's own home (such as computers, televisions, hair dryers, and household appliances). Electrical fields fall off quickly with distance, which is why electrical appliances and home wiring are large contributors to most people's EMF exposure. EMFs are commonly measured in units of gauss (G) by an instrument known as a gaussmeter. A milligauss (mG) is 1000 times smaller than a gauss.

Components of wind power plants that can be potential sources of EMFs include the collection system (whether above- or belowground), the generators, the pad-mount transformers located either at the base of the turbine or in the nacelle, the transformers and other electrical equipment inside the project substation, and the transmission lines that connect the project to the electric grid.

An epidemiologic study by Wertheimer and Leeper [9] on childhood cancer undertaken in Denver and its subsequent report released in 1979 raised concerns about potential adverse health effects associated with exposure to EMFs. Since that time, volumes of research have been done on the topic with no consensus on whether or not there is a health risk associated with EMF exposure. Two national research organizations (the National Research Council and the National Institutes of Health) have examined many studies and concluded that there is no strong evidence that EMF exposures pose a health risk. The National Cancer Institute's factsheet titled *Magnetic Field Exposure and Cancer: Questions and Answers* [10] includes the following points:

- Overall, there is limited evidence that magnetic fields cause childhood leukemia, and there is inadequate evidence that these magnetic fields cause other cancers in children.
- Studies of magnetic field exposure from power lines and electric blankets in adults show little evidence of an association with leukemia, brain tumors, or breast cancer.
- Past studies of occupational magnetic field exposure in adults showed very small increases in leukemia and brain tumors. However, more recent, well-conducted studies have shown inconsistent associations with leukemia, brain tumors, and breast cancer.

8.11 MITIGATING EMF EXPOSURE FROM WIND PROJECTS

Once again, maintaining some established setback distance between wind plant facilities (including substations, transformers, and electric lines) is the best way to avoid issues related to EMF exposure. Since electrical fields fall off quickly with distance, reasonable setbacks will generally result in EMF levels produced by the wind plant being less than preexisting EMF levels in homes and other occupied buildings. The design of all wind plant electrical systems should adhere to applicable state guidelines and industry standards to minimize EMF exposure from any new overhead transmission lines, substations, or other equipment.

Employees of the wind plant who routinely operate near electric facilities (including in the turbine nacelle) should be briefed on EMFs as part of their overall electrical safety training, just as most electric utilities do with their employees. National and international organizations, such as the Institute of Electrical and Electronics Engineers, have established public and occupational EMF exposure guidelines on the basis of short-term stimulation effect, rather than long-term health effects.

8.12 SOLID AND HAZARDOUS WASTES

Proper handling of solid and hazardous wastes generated during the construction and operation of a wind energy project is very important, just as it is for any industry. Since wind plants are typically located on leased land, rather than land belonging to the project owner, it is important to handle even nonhazardous trash (such as empty soft drink cans and bottles or trash from fast food meals). Landowners generally have a high level of pride in their property and how it looks, so there are few things that make them as mad as seeing construction or maintenance personnel litter on their property. Some landowners insist on the inclusion of penalty provisions in lease documents in which they “fine” the project owner for litter attributable to the wind project. Another source of potential conflict between the landowner and the project owner is the project’s “boneyard” where retired or damaged turbine components and equipment are stored. If the components or equipment are never going to be used again, it is far better to dispose of them in a licensed disposal facility rather than leave them lying around.

During the construction process, there is a significant amount of solid waste that is generated (such as packing and crating materials). Another source of waste is the brush cleared for roads and turbine foundations and large rocks unearthed during construction. It is very likely that the landowner will consider this trash or solid waste, whereas a construction contractor may not see it this way and thus not include removal of brush and rocks in his or her contract price. Thus, it is very important that this particular item be addressed in both the property lease and the engineering/procurement/construction (EPC) contract. In addition, if an existing structure on the property is going to be removed, be aware of the potential for materials categorized as hazardous in the structure (such as asbestos or lead-based paints). These may require specialized practices for handling and removal.

During the construction and operation phases, the most likely sources of hazardous wastes are fluids including vehicle or equipment fuel, gearbox oil, hydraulic fluid, solvents, and cleaners. Storage facilities and handling procedures or protocols need to be developed for both new liquids and waste liquids prior to commencing project construction. It is likely that permits will be required in some areas for waste disposal/storage facilities, so these need to be obtained early in the process. Development of procedures for handing waste (the "Waste Management Plan") should consider the use of recycling where possible, which may help reduce waste disposal costs.

When turbines become operational, the project O&M team needs to be aware of any oil or hydraulic fluid leaks from the turbine. Often, the first sign of these may be streaking on the top of turbine tower just below the nacelle. This is not only an eyesore but it may mean that oil and/or hydraulic fluid is on the ground surrounding the turbine. This can necessitate soil removal or remediation and may also trigger requirements for reporting the spill to state environmental officials.

8.13 MITIGATION OF SOLID AND HAZARDOUS WASTE ISSUES

Following are some of the best practices for handling of solid and hazardous wastes associated with wind plant construction, operation, and retirement:

- Prior to beginning project construction, identify all potential wastes to be generated during construction and project operation; determine which ones will require permits and the regulations affecting disposal procedures and potential disposal locations.
- Verify the licensing of any disposal location that may be used.
- Address requirements for disposal of construction wastes, including removed brush and dislodged or unearthed rocks, in the EPC contract.
- Develop and implement a hazardous materials management plan establishing standard procedures for the reporting, handling, disposal, and cleanup of hazardous material spills and releases, including the availability and use of personal protective equipment.
- Locate storage areas for hazardous materials some distance away from occupied buildings and air-intake locations.
- If any hazardous materials will be handled with forklifts or other heavy equipment, ensure that all equipment operators are trained on the equipment.
- Use biodegradable lubricants and nonhazardous fluids wherever feasible; evaluate recycling and waste reduction as ways to minimize waste disposal and reduce disposal costs.
- Design containment or catch basins into turbine foundations or transformer pads in order to minimize impact of spills.
- Comply with all regulatory reporting requirements including reporting of any hazardous liquid spills.

- Promptly dispose of any failed turbine components or equipment once damage assessments have been completed and a determination has been made that the component or equipment is no longer usable or economically repairable.

8.14 CONCLUSIONS

As wind generation continues to expand rapidly in the United States and many other countries, existing issues or concerns about wind energy development will grow in importance and new issues will come up. If the wind energy industry desires to maintain its image as an environmentally benign source of electric generation with high levels of public support, it must continue to effectively respond to these questions or challenges rather than ignore them, and the development of projects must be done with a high degree of due diligence in order to mitigate or minimize potential environmental impacts of the projects.

As noted several times in this chapter, one of the best ways to minimize or mitigate impacts on humans is to ensure adequate separation between wind turbines and occupied buildings. That being said, a 10-fold expansion of wind energy capacity in the United States, as envisioned in the *20% Wind Energy by 2030* report [11], increases the likelihood that wind plants will be sited closer and closer to communities as sites with excellent wind speeds and proximity to transmission lines become scarce. This highlights the importance of addressing questions about or challenges to the technology with accuracy and in a way that respects the views of those with differing opinions.

REFERENCES

[1] National Research Council of the National Academies. *Environmental Impacts of Wind-Energy Project*. Washington, DC: The National Academies Press; 2007.

[2] American Wind Energy Association. Utility scale wind energy and sound. AWEA factsheet, Washington, DC; 2010.

[3] Colby D, Dobie R, Leventhall G, Lipscomb D, McCunney R, Seilo M, Sondergaard B. *Wind Turbine Sound and Health Effects: An Expert Panel Review*. Washington, DC/Ottawa: American Wind Energy Association and Canadian Wind Energy Association; 2009.

[4] The Scottish Office, Environment Department, Planning Advice Note, PAN 45, Wind Power, A.27, Renewable Energy Technologies, August 1994, The Scottish Office, Environment Department, Edinburgh, Scotland.

[5] GE Global Research and National Institute of Deafness and Other Communication Disorders. Available at http://www.gereports.com/how-loud-is-a-wind-turbine/. Accessed October 30, 2014.

[6] Pierpont, N. 2006. Wind turbine syndrome: noise, shadow flicker, and health. Accessed on website of Friends of the Highland Mountains. http://highlandmts.org/wp-content/uploads/2010/01/wind-turbine-syndrome-noise-shadow-flicker-and-health-pdf1.pdf. Accessed October 3, 2012.

[7] Oteri, F. An overview of existing wind energy ordinances, The National Renewable Energy Laboratory. Technical report NREL/TP-500-44439, Golden; 2008.

[8] American Wind Energy Association. *Wind turbines and health*, AWEA factsheet, Washington (DC); 2010.

[9] Wertheimer N, Leeper E. Electrical wiring configurations and childhood cancer. Am J Epidemiol 1979;109:273–284. Oxford University Press.

[10] The National Cancer Institute at the National Institutes of Health. Magnetic field exposure and cancer: questions and answers, factsheet, Research Triangle Park (NC); 2005.

[11] U.S. Department of Energy, Energy Efficiency and Renewable Energy. *20% Wind Energy by 2030: Increasing Wind Energy's Contribution to U.S. Electricity Supply*, DOE/GO-102008-2567. Washington, DC: U.S. Department of Energy, Energy Efficiency and Renewable Energy; 2008.

9

ENVIRONMENTAL AND ECOLOGICAL IMPACTS OF WIND ENERGY ON HUMANS: PUBLIC AND WORKFORCE SAFETY ISSUES

9.1 OVERVIEW OF WORKFORCE AND PUBLIC SAFETY ISSUES

Developers, construction contractors, owners, and operators of wind plants need to be aware of the issues and situations that can jeopardize the safety of their employees and the general public. Safety needs to be considered early in the project, beginning with the location of the site, which can impact aviation and telecommunications, and during turbine siting, which must consider proximity to publicly accessible roads and occupied buildings. Safety during the construction process is critical due to the immense size of the turbine components and the equipment used to erect the turbines and due to the impacts that the delivery of turbine components, concrete, and road materials can have on transportation in the area of the project. During operation of the project, employees will be working at substantial heights and around both rotating and high-voltage equipment. The issues and/or topics that will be discussed in this chapter include the following:

- Ice shedding
- Blade drop/throw
- Fire
- Lightning strikes
- Other weather emergencies (such as hurricanes, tornadoes, or hail)
- Impact on vehicular traffic

Wind Energy Essentials: Societal, Economic, and Environmental Impacts, First Edition.
Richard P. Walker and Andrew Swift.

- Electrical and pipeline safety
- Other wind energy industry workforce safety issues

9.2 THINGS CAN GO WRONG ANYTIME HUMANS ARE INVOLVED (IN ANY INDUSTRY)

As with pretty much any aspect of life involving humans, things can go wrong due to human error, and this is true of the wind energy industry. In addition, due to wind energy being intrinsically tied to nature's forces by trying to harness them, there will always be instances in which wind speeds attain levels so high that turbines cannot cost-effectively be designed to withstand them. The same is true for buildings or other structures being built in areas susceptible to hurricanes, tornadoes, flooding, earthquakes, or even straight-line winds exceeding 100 mph.

Engineers work with atmospheric scientists to estimate the probability of such events happening and affecting the structure they plan to build. Engineers will then base their structural design on withstanding forces that are likely to occur once or twice every 100 years. Differing models of utility-scale wind turbines will be designed to withstand differing levels of wind speed and turbulence that are classified by the International Electrotechnical Commission (IEC), a worldwide organization for standardization that promotes international cooperation on all questions concerning standardization in the electrical and electronic fields. For example, an IEC Class I-A turbine should be designed for installations at sites with a maximum 50-year return gust speed of 70 m/s (156 mph) and turbulence intensities of 0.21 or less. A 50-year return gust speed does not mean wind cannot exceed this level more than once in any 50-year period, just that the average expected recurrence of wind speeds this high is once in any given 50-year period.

Thus, designers of wind plants need to be aware of the potential for catastrophic failures of wind turbines, usually due to extreme weather conditions, and place turbines such that they are least likely to be hazardous to the public in the event of a turbine collapse or a blade breaking off.

9.3 ICE SHEDDING

The blades of wind turbines can accumulate ice in certain weather conditions (such as high humidity or misty conditions when temperatures near the hub height elevation are at or just below freezing) causing moisture to freeze upon contact with blades, forming rime ice. Ice can accumulate on the turbine blades with the turbine still in operation, although if icing gets bad enough to impede the wind speed and direction sensors used for turbine control, the control system should react by shutting the turbine down. In these situations, most turbines will restart only when ice has thawed and fallen from the stationary turbine and the project operator has reset the sensors. While icing events reduce the amount of electricity produced by the project, either due to turbine shutdowns or by reducing the aerodynamic properties of the blade, the bigger issue with turbine icing is its potential impact on the safety of the public and the wind plant's workforce when ice is dropped or thrown from the turbine blades.

Ice can break away from blades either due to increasing temperature or due to the aerodynamic and centrifugal forces of the turning blades. In some situations, the project's operator may contribute to ice shedding by thawing the control sensors and restarting the turbine while ice is still on the turbine blades. While wind plant operations and maintenance (O&M) staff and landowners need to be aware of the potential for chunks of ice falling straight down from stationary turbine blades due to rising temperatures, the bigger issue is the potential throw of ice when the blades are rotating, particularly when turbines are located near roads, parking lots, homes, or other occupied buildings.

A general rule that has been used for determining setback distances associated with ice throw, and one that is discussed in a paper by Tammelin et al. titled *Wind Energy in Cold Climate* [1], included the following simple formula for estimating the maximum ice throw distance:

$$d = (D + H) \times 1.5$$

where d = maximum ice throw distance in meters, D = rotor diameter of the turbine in meters, H = turbine hub height in meters.

This same formula is cited in a GE energy brochure titled *Ice Shedding and Ice Throw—Risk and Mitigation* [2], although it also notes that "the actual distance is dependent upon turbine dimensions, rotational speed and many other potential factors."

However, several studies have been performed to estimate the distance that ice can travel from the wind turbine. Garrad Hassan (GH) conducted such a study [3] in 2005 for the proposed East Haven Wind Farm (EHWF) that concluded that "based on our previous work and accounting for the terrain and machine size of the EHWF site, a very conservative estimate for the maximum achievable distance for ice to be thrown is considered to be 400 m (1315 ft)." Turbines for this project were 1.5-MW turbines with 65-m hub heights and 70.5-m rotor diameters. Note that GH's estimate of ice throw distance is almost twice as much as the general rule of 1.5 times the sum of hub height and rotor diameter, which results in 203.25 m. An interesting observation of this study was that roughly 15 people would have to be present on the wind site during an ice throw event for the risk of someone being struck by ice to be comparable with the risk of being struck by lightning in the United States, which they showed as having odds of approximately 1 in 750,000.

More recently, Dr. Pierre Héraud with Helimax Energy, a division of GL Garrad Hassan, testified in March 2011 about ice throw before the State of Connecticut's Siting Council regarding the proposed Wind Colebrook South project [4]. Dr. Héraud, whose PhD is in Physics, testified that he had calculated that a 1.6-MW turbine with a 100-m rotor diameter and 100-m hub height could throw ice a maximum distance of 285 m and that a 1.6-MW turbine with an 82.5-m rotor diameter and 100-m hub height could throw ice a maximum distance of 265 m. These values are fairly consistent with the general rule of 1.5 times the sum of hub height and rotor diameter.

9.4 MITIGATING ICE SHEDDING ISSUES

Like turbine noise and shadow flicker, use of some predetermined setback distances from homes, occupied buildings, and public roads is the primary means of mitigating issues with ice throw from wind turbines during the project design phase. As stated earlier, a general rule of thumb for such setbacks is 1.5 times the total of the turbine's hub height and its rotor diameter, but companies like GL Garrad Hassan will perform studies of ice throw for wind projects taking into account the variables specific to the project and turbine model that can affect this distance.

When a wind plant goes into operation, it is important to educate the operations staff about conditions likely to lead to ice accretion, about the risk of ice falling from the blades, and about operating protocols for stopping and restarting turbines during icing conditions. Within the wind project, use of warning signs alerting the wind farm personnel and the landowners about the potential for ice throw when turbines are operating during and after icing events can be beneficial. If it has been determined that ice throw may be a potential issue for one or more turbines that have already been erected, the operator may wish to implement an operating protocol calling for turbines to be shut down until it has been verified that all risk of ice throw has passed. Such protocols can be refined to minimize turbine curtailment by determining what conditions can actually lead to icing at the site and what range of wind speeds ice throw can occur in. In areas where icing is considered to be a significant issue, the operator may wish to install ice detectors on the nacelle of turbines considered to be of greatest concern. The operating protocol may call for turbines to be placed in Pause Mode pending inspection of the turbine(s) by O&M staff or with the use of remote video cameras.

9.5 BLADE THROW

Several wind turbine models now have rotor blades in excess of 50 m in length that are exposed to large centripetal, gravitational, and aerodynamic forces. Blade failures do periodically occur due to natural phenomena (such as hurricanes, tornadoes, or exceptionally high straight-line winds) or due to human error during the design, manufacturing, construction, or operations stages. This is just one more reason why it is very unwise to place turbines near occupied structures, even in situations when the homeowner wants a turbine there due to the royalties it would be able to generate. One of the authors of this book has been involved in the development of several large wind projects and has had landowners tell him "you can put a turbine in my backyard if you want." Even though the landowner may be willing, the potential liability is too great to justify such practices.

The Environmental Impact Statement for the Desert Claim Wind Power Project in Kittitas County, Washington [5], includes information showing the maximum blade throw distance is less than 150 m for a 1.5-MW turbine with potential hub heights of 65–85 m and potential rotor diameters of 70.5 or 77 m.

A study performed by Epsilon Associates, Inc. [6] in October 2010 for NextEra Energy's Montezuma II Wind Energy Center in Solano County, California, concluded that the maximum potential blade throw distance for a 2.3-MW turbine with 80-m hub height and 101-m rotor diameter was 172 m.

9.6 MITIGATING BLADE THROW ISSUES

Setback distances between wind turbines and occupied buildings established by wind developers to mitigate turbine noise and shadow flicker should exceed the maximum distance for blade throw. Thus, the main ramifications for blade throw are likely to be (i) siting of turbines near publicly accessible roads and (ii) site safety guidelines applicable to O&M staff, landowners, or visitors that should address weather conditions in which blade throw is more likely to occur and precautions that should be taken in such conditions.

Wind developers need to be aware of states or counties that may establish setback distances before they determine turbine locations. During the project construction process, procedures should be in place to identify any blades damaged during transportation or improperly manufactured. If blades are struck by lightning, inspections should be done to verify continued structural integrity of the blade, and if multiple blades begin to crack or fail during normal operations, it may be necessary to inspect all turbine blades to check for manufacturing flaws or serial defects.

9.7 RISK OF FIRE

Any activity that increases the number of people or vehicles on a property can increase the potential for fires, and the development of a wind plant on the property is no exception. During construction of a wind plant, there may be 100 or more personnel involved. Other possible contributors to increased risk of fire associated with wind energy development include the operation of vehicles and machinery, storage and handling of fuel, the addition of overhead electric lines, and the potential for turbine fires (Fig. 9.1).

The incidence of turbine fires is small, and suspected causes of such fires include electromechanical malfunctions (such as bearing failures), sparks or flames resulting from substandard machine maintenance, electrical shorts, equipment striking power lines, and lightning. Since the nacelle of wind turbines includes combustible materials (such as oil or hydraulic fluids) and much of the nacelle itself is made of fiberglass, it is possible that a fire could ignite the nacelle, allowing burning materials to fall to the ground. Failure of ground-level equipment such as padmount transformers could also start fires if weeds or brush are allowed to grow up around the equipment.

FIGURE 9.1 Remnants of a wind turbine nacelle following a fire (Source: R. Walker).

9.8 MITIGATING RISK OF FIRE

Actions that can be done to lower fire risk during the project design, construction, or operation phases include the following:

- Work with local emergency response organizations (fire and EMS) to develop and implement plans for fire and medical emergencies, including plans for obtaining access to the site from public roads. It is common for landowners to include provisions in lease documents requiring locks on all gates entering the property. Thus, emergency responders should be given keys or codes to locked gates that provide access to the property in emergency situations, although some landowners may resist providing these items to anyone (even emergency responders).
- Since discarded cigarette butts and vehicles driving over tall, dry grass can ignite fires, establish very clear policies for employees and visitors to the site with regard to locations where smoking is allowed, proper disposal of cigarette butts, and where vehicles can and cannot drive.
- Use underground wiring for the collection system connecting turbines to the project substation.
- Properly maintain and routinely inspect equipment; routinely maintain the site by ensuring that brush and weeds do not grow around equipment such as pad-mount transformers or along roads within the site.

9.9 LIGHTNING

Since wind turbines are usually the tallest structure in the vicinity in which they are located, lightning can be a significant issue for wind project owners, both from the standpoint of damage it can cause to turbines and other equipment at the site and from the standpoint of employee and public safety. The incidence of lightning varies significantly between regions in the United States or locations in the world. Data from the U.S. National Lightning Detection Network indicate that the frequency of lightning flashes may be as high as 8–16 flashes per square kilometer per year in parts of Florida, Alabama, and Mississippi. This compares to the west Texas area containing large concentrations of wind turbines, which only gets 2– 4 flashes per square kilometer per year, or those areas in California having much of the original wind development in the United States (Altamont Pass, San Gorgonio Pass, and Tehachapi), which get almost no lightning. The incidence of lightning can increase with the height of wind turbines, which continues to increase with some turbines now reaching almost 200 m in height.

Wind turbines are protected by lightning receptors and conductors embedded in the blades that are then grounded to the tower through the hub. The tower and its foundation are grounded to the earth, usually allowing lightning to dissipate harmlessly. Without such lightning protection systems, turbines might attract lightning and pose a threat to nearby persons or dwellings. However, an individual leaning against or standing next to a turbine when lightning strikes would be at some risk that ground potential rise could result in voltage between ground and tower or between two spots on ground. The project's operations staff are probably at the highest risk for such occurrences, but if property owners, hunters, or others are caught in a lightning storm, it may be that they seek shelter immediately downwind of the turbine. Therefore, the project's operating protocols or weather response plans need to address this risk by including safety procedures instructing staff and members of the public not to stand near turbines during lightning. Another consideration for O&M staff is the time it can require to descend from the nacelle of the wind turbine when a storm is approaching.

Since lightning strikes with varying force, no method of protection can guarantee absolute safety to individuals or equipment. However, effective lightning protection can significantly reduce the amount of risk. IEC's Code 61400-24 addresses lightning protection of wind turbines themselves, but the operating procedures of wind plant owners and the technologies that they may choose to install at the site will impact the safety of its workforce and the public.

Lightning detection systems can be installed at projects to monitor lightning strikes up to 60 miles (roughly 100 km) from the project site, allowing the site manager to implement various stages of lightning advisories to the project's employees or visitors to the site. The levels of such advisories can be dependent upon the height of turbines (which affects the amount of time to ascend and descend) and the overall size of the project (which affects how long it takes to drive from the furthest point of the project to the O&M building).

Examples of such advisories could include the following:

- "Lightning advisory," when lightning is first detected.
- "No-climb advisory," meaning no employee is to begin climbing a turbine.

- "Discontinue service advisory," meaning that employees in the turbine nacelle are to discontinue their work as quickly as possible and descend from the turbine.
- "Return to O&M building advisory," meaning all employees working outside or in turbines are to cease work and return to the shelter of the O&M building.
- "All clear advisory," meaning it is safe to commence normal operations.

9.10 OTHER WEATHER EMERGENCIES: HURRICANES, TORNADOES, AND HAIL

Safety procedures at wind plants need to include procedures for responding to other weather emergencies including hurricanes, tornadoes, and hail. Obviously, only those wind projects located offshore or in coastal areas have to worry much about hurricanes, but major hurricanes can also spawn multiple tornadoes and other severe weather many miles inland. With today's ability to forecast the landfall of hurricanes many days in advance (although the exact location of landfall may not be known), plant operators should have sufficient time to shut down turbines, to prepare the project for the storm, and to evacuate the site. This is not always the case for tornadoes, so immediate reaction to tornado warnings is vital to the safety of the workforce. Large hail can also be a source of danger to workers in the field and to the turbines themselves as hail may damage the turbine blades reducing their aerodynamic efficiency.

9.11 THE IMPACT OF WIND ENERGY ON VEHICULAR TRAFFIC

As wind turbines get larger and larger, transportation becomes a bigger issue or problem. Some turbine models have rotor diameters in excess of 120 m, meaning each blade can be 60 m or more in length. While the tower may be constructed and transported in more sections as turbine hub heights increase, the diameter and weight of the lowest section will continue to increase, with diameters being 4 m or more and weights reaching 50 tons or more. Special extended-trailer semitrucks are required to carry such blades and tower sections, and the size of turbine components can be a challenge even for rail carriers (Fig. 9.2).

Precise planning of equipment delivery routes is crucial for any wind project but can be especially complex when wind farms are located in rural areas with narrow roads and load-limited bridges or in areas with steep slopes (such as mesas, plateaus, or ridgelines). Additionally, when dealing with multiple states with differing regulations, obtaining hauling permits for transport of longer and heavier components is increasingly difficult and may sometimes require the use of circuitous routes for turbine delivery.

FIGURE 9.2 Wind turbine blade being delivered (Source: Reproduced by permission of Jon Goodman).

9.12 PLANNING FOR AND ADDRESSING TRANSPORTATION ISSUES

Following is a list of transportation considerations that need to be addressed during the development, engineering, materials procurement, and construction phases for a wind energy project:

- Impacts on the transportation system including roads, bridges, drainage systems, and road or highway turn-ins: County commissioners will expect all of these components of their local transportation system to be in the same or better condition after construction of a wind project than prior to the start of construction. It may be wise to document the preexisting condition of local roads through the use of video or sampling techniques so as to minimize future disagreements on road quality (Fig. 9.3).
- Impacts on traffic and public safety, including interstate and state highways and local roads: Heavy equipment making a slow, wide turn just below the crest of a hill could easily be a source of vehicle accidents.
- Adjustments to highways and other paved roadways should meet design criteria established by the American Association of State Highway and Transportation Officials in "A Policy on Geometric Design of Highways and Streets."

FIGURE 9.3 Wind turbine Nacelle and hub being delivered (Source: Reproduced by permission of Jon Goodman).

- Impacts of site roads on landowners should be considered. Construction traffic on unpaved roads can create lots of dust and noise. Savvy attorneys representing landowners frequently want to include provisions in lease agreements requiring the approval of road locations on the property in order to mitigate issues near homes, scenic areas of the property, or livestock-watering locations. Leases may also require that roads be narrowed to some specified width after project construction is complete.
- Speed limits should be established for all roads within the project that are used by employees of or visitors to the project. This is to ensure the safety of not only the driver and his or her passengers but also any livestock within the site. Ranchers, in particular, are likely to require some type of mandatory speed limit in the terms of their lease agreement.
- The quality of site roads or improvements made to county roads will impact the relationships with landowners and county officials. When wind projects are constructed in extremely rocky locations, landowners may greatly appreciate the new roads required for the wind farm, whereas roads in cultivated fields could be a concern for farmers.
- Workforce safety always needs to be given the highest consideration when planning delivery routes. For example, delivery of long turbine components

(such as turbine blades and tower segments) using a route that crosses a heavily traveled railroad track can be a significant safety concern, particularly if that crossing occurs in a rural area where crossing warning signals may not be used. Another issue may be the delivery of such components on winding dirt roads in steep terrain where one poor decision could be fatal.

- The potential cost of delays associated with transportation problems, which can be huge if a wind project is being constructed with a looming deadline for expiration of the Federal production tax credit.
- Transportation permitting or determining regulatory requirements for the delivery of large equipment is likely to involve interaction and planning across multiple states.
- The minimum turning radius for trucks delivering long components has to be considered, as well as the maximum slope that can be ascended by trucks delivering heavy components.
- Transportations plans have to include the identification of height restrictions of bridges and overpasses or weight restrictions of roads, bridges, and drainage culverts.
- Vendors of turbines and other large components (such as transformers) will have their own requirements for transporting large equipment, as well as their preferred shipping methods (truck, rail, barge, etc.). One very large cost component that this will affect is the construction of roads within the project.

In general, the larger the turbine is, the larger the delivery equipment will be as well as the size of crane needed to erect turbines. Site roads have to be constructed to bear the weight of very heavy equipment while also meeting requirements for small slopes and large turning radii (Fig. 9.4).

FIGURE 9.4 Wind turbine tower section being delivered (Source: Reproduced by permission of Jon Goodman).

9.13 ELECTRICAL AND PIPELINE SAFETY: HEIGHT, DEPTH, AND LOCATION OF ELECTRIC LINES

Large construction equipment accessing a wind energy plant under construction will have to deal with overhead electric lines on public roads and crossing the site. Great care must be taken to avoid contacting energized distribution or transmission voltage lines. One of the early steps in project construction may be working with the local electric service provider to put overhead electric lines crossing entrances to the project or roads within the project underground. Some electric service providers will work with the project's engineering, procurement, and construction (EPC) contractor, while others may be much harder to deal with, but in either case, it is almost certain that the project will have to bear the cost of any line relocations. Another good idea would be to have all vehicles entering the site pass underneath a bar or nonenergized wire indicating the lowest level of electric lines found on the site. One of the reasons that the majority of the wind project's electric collection system is placed underground is to minimize the potential for large construction or delivery equipment hitting an energized electric line, in addition to minimizing aesthetic impacts and limiting locations for birds to roost.

Identifying the location and depth of any existing underground electric lines and oil or gas pipelines is also critically important prior to undertaking any trenching, soil boring, or excavation activities. While many underground lines will be owned by the local electric service provider, construction personnel need to be aware of the potential for privately owned underground electric circuits, particularly in areas of irrigated farmland or oil and gas production. Pipelines also may have limits on the weight of vehicles that can safely cross them, so wind developers and/or EPC contractors need to find out if permits for crossing pipelines are required and what weight limitations may apply. Another pipeline-related issue to be aware of is the potential use of blasting during foundation construction. Care must be taken to ensure that such blasting does not fracture or otherwise harm any nearby pipelines.

The depth and location of underground electric lines being installed for the wind project is also important to landowners and for public safety reasons. Anyone else working on the project site including the landowner, construction personnel, oil field crews, or utility companies needs to be able to clearly identify where the wind project's underground electric lines are located, so adequate signage identifying their location is needed. Such lines also need to be well below "plow depth" in cultivated fields.

9.14 OTHER WIND ENERGY INDUSTRY WORKFORCE SAFETY ISSUES

Wind plant construction, maintenance, and operation can involve working with very heavy components and equipment, working at heights, working around rotating equipment, working around high-voltage equipment, and working outside in extreme heat or cold. This requires a heavy emphasis on safety both during the construction

FIGURE 9.5 Turbine Nacelle being lifted into place (Source: Reproduced by permission of Jon Goodman).

process and after the project goes into operation. Previous sections within this chapter touched on several of these issues.

Vocational, trade, or technical schools specializing in the education and training of wind energy technicians (such as Texas State Technical College, Iowa Lakes Community College, South Plains College, Mesalands Community College, and Columbia Gorge Community College) stress safety throughout their curriculum (Fig. 9.5).

One of the largest challenges in finding wind technicians stems from the height of modern utility-scale wind turbines. Typical tower heights have steadily increased, and it is likely that 100-m hub heights will soon be commonplace. Acrophobia, or the fear of heights, can eliminate many individuals from even considering work in this field, while others may think they can deal with this until having to face their first turbine climb. Others may make it to the top of the tower but then be overwhelmed with fear due to a swaying sensation in windy conditions.

Required use of fall protection equipment is standard throughout the wind energy industry, helping some individuals to overcome their fear of heights. Another change that may help with the acrophobia issue is the use of tubular steel towers with ladders inside the tower for almost all utility-scale turbines instead of lattice steel towers where technicians are outside and can easily see the ground below them.

Another issue related to the increasing height of turbines is the physical fitness level required to ascend and descend from three or four turbines a day in extreme

FIGURE 9.6 Rotor blade being installed (Source: R. Walker).

heat, thus eliminating another large sector of the population. Dehydration and exhaustion can impair one's thought processes, so technicians need to constantly be aware of the condition of themselves and their coworkers. As turbine heights have increased, the use of climb-assist devices is becoming more common, and some turbines even have manlifts or small elevators inside them, which eases the physical stress of climbing the towers. Preemployment physicals and climb-testing are useful in early identification of individuals who may not be suited for climbing towers.

Electrical safety in the wind energy industry is similar to that required in the electric utility industry, following many of the same principles such as lockout/tagout procedures when one is working on high-voltage equipment to help ensure that another staff member does not inadvertently reenergize equipment while one of his or her coworkers is working on it.

The size and strength of cranes needed to erect state-of-the-art wind turbines has had to increase just as rapidly as the size and height of wind turbines. Thus, stability of cranes, both while lifting components and while moving from turbine location to turbine location, is a large issue. Site roads and crane pads have to be carefully designed and built to ensure that these massive cranes remain as level as possible in all situations (Fig. 9.6).

Erection of towers, nacelles, and rotor sets is complicated and dangerous even in dry, low wind conditions. The addition of weather forces (such as high wind, rain, and snow) can further complicate the situation. And since wind plants are built in places with high winds, excessive wind speed is a very common issue that must be effectively dealt with to ensure a high degree of workforce safety. "Wind-out" days are built into almost every wind plant construction schedule, reflecting those days when it is unsafe to lift and set in place the extremely heavy components of a wind turbine.

9.15 CONCLUSIONS

The combination of public accessibility to projects, the use of huge components and equipment, the proximity to high electrical voltages, and working at heights and outside in weather conditions (including extreme heat, extreme cold, high wind, ice, rain, and snow) requires that developers, constructors, owners, and operators of wind plants constantly emphasize safety in all aspects of the project from the very beginning of project design to the time when turbines are retired from service. There are entire courses and seminars offered on the subject of public and workforce safety in wind farms, so this chapter barely skims the surface of many of the issues. Those individuals contemplating careers in either construction or operation of wind plants should take advantage of any opportunity presented to learn more on this subject.

REFERENCES

[1] Tammelin B, Cavaliere M, Holtinnen H, Morgan C, Seifert H, Säntti K. *Wind Energy in Cold Climate: Final Report*, WECO (JOR3-CT95-0014). Helsinki: Finnish Meteorological Institute; 2000.

[2] Wahl, D., Giguere, P. Ice shedding and ice throw—risk and mitigation; brochure, GE Energy, Wind Application Engineering, Greenville; 2006.

[3] LeBlanc, M. Assessment of ice throw risk for the proposed East Haven Wind Farm report, Garrad Hassan America, Inc; 2005. Proposal Document No. 9606/AP/01 ISSUE B, December 1, 2004.

[4] Héraud, P. Pre-filed Testimony before State of Connecticut's Siting Council. Petition No. 983, New Britain; 2011.

[5] State of Washington Energy Facility Site Evaluation Council. Final EIS, Desert Claim Wind Power Project, Kittitas County, Washington, Olympia; 2008.

[6] O'Neal, R. Wind turbine blade throw analysis; report, Epsilon Associates, Maynard; 2010.

10

WIND ENERGY'S IMPACT ON AVIATION, RADAR, AND TELECOMMUNICATIONS

10.1 IMPACT OF WIND ENERGY ON MILITARY AND CIVILIAN AVIATION

Wind energy development can affect aviation in a number of ways, and one of the objectives of a wind developer is to work with governmental officials to minimize any potential impacts, obtain all required permits, and do so as economically as possible. One does not wish to have spent large amounts of money and time securing property rights and developing a wind project only to find out that the required permits from the Federal Aviation Administration (FAA) cannot be obtained due to the likelihood of impacts on aviation. The aviation-related topics that will be discussed in this chapter include the following:

- Regulatory requirements
- FAA permitting process
- Location of turbines near airports, air force bases, private runways, and flight paths
- Preliminary screening process available on FAA website
- FAA requirements for marking or lighting of obstructions
- Marking and/or lighting of meteorological towers
- Potential impact on navigation radar

Wind Energy Essentials: Societal, Economic, and Environmental Impacts, First Edition.
Richard P. Walker and Andrew Swift.

10.2 REGULATORY REQUIREMENTS AND FAA PERMITTING PROCESS

In the United States, the FAA oversees and manages navigable airspace, develops plans and policies for its use, and is charged with ensuring the safety of aircraft and the efficient use of the airspace. Navigable airspace is a limited national resource and the FAA's primary mission, in this context, is to preserve that resource for aviation and to negotiate equitable solutions to conflicts over the use of the airspace for nonaviation purposes.

The FAA requires that persons or entities wishing to build structures of certain heights and/or proximity to military or civilian airports or heliports provide public notice of the construction by filing FAA Form 7460-1, *Notice of Proposed Construction or Alteration*. Generally, public notice is required if the structure is more than 200 ft in height above ground level or near or on an airport or heliport (either military or civilian). Therefore, construction of almost all utility-scale wind turbines, some small wind turbines, and even some meteorological towers will require review and approval from the FAA. Knowingly and willingly violating FAA notice criteria for a structure requiring notice can result in civil penalties of $1000/day until the notice is received. In recent years, the volume of proposed wind turbines submitted to the FAA for review has increased dramatically. The number of wind turbine cases handled by the FAA increased from 3,030 in 2004 to 25,618 in 2009.

The Notice of Proposed Construction or Alteration provides the FAA the opportunity to identify potential hazards to aviation. The initial FAA study normally takes 30 days, but the cumulative implications of a wind plant can require more extensive evaluation within the FAA, the Department of Defense (DoD), and the Department of Homeland Security (DHS). The initial evaluation includes review by FAA's Offices of Airports, Flight Standards, Frequency Management, and the appropriate military organizations. Therefore, the FAA recommends that wind developers file notice with the FAA 8–12 months before the planned construction date due to the extensive studies that wind turbines require, allowing sufficient time for the FAA, DoD, and DHS to review the proposals and to engage in any negotiations. Notices can be filed electronically, and there are no fees associated with filing the notice or obtaining a determination from the FAA, although it may not be wise to arbitrarily submit notices on highly speculative projects. Information needed to file the notice includes proposed coordinates of each wind turbine, the ground elevation at each of these locations, and the total height of the wind turbine at the very tip of blade pointing straight up (i.e., hub height plus one-half of the rotor diameter). This would infer that the turbine model should be known when notice is filed, otherwise information about turbine spacing requirements and total height of the turbine would not be known.

If a wind turbine or meteorological tower is identified as being a potential hazard, it is the responsibility of the person or entity wishing to erect the structure to propose a plan for mitigating the potential hazard, which of course could be to not place the wind turbine or meteorological tower at the proposed location. Mitigation plans can include recommending appropriate markings and lighting to make the structure visible to pilots, depicting structures on aeronautical charts to inform pilots, or actions taken by the FAA such as revising published data or issuing a notice to airmen (NOTAM) to alert pilots to airspace or procedural changes made because of a structure, although the cost of mitigation, including actions taken by the FAA, would have to be borne by the wind project or wind developer.

10.3 FAA SCREENING TOOLS

The public has access to the FAA's website, which includes several useful tools or programs for gathering information about other companies' proposed wind project locations, determining whether notice is required for their project, or identifying potential issues for their proposed project. The Obstruction Evaluation/Airport Airspace Analysis page can be found at https://oeaaa.faa.gov/oeaaa/. Useful tools or applications on this website are listed as follows.

10.3.1 Circle Search for Airports

This application allows one to enter a given coordinate, such as the location of a planned wind project, and to determine the location of all airports located within 50 miles of that coordinate.

10.3.2 Circle Search for Cases

This application allows one to enter a given coordinate, such as the location of a planned wind project, and determine the location of other previously filed FAA notices located within 50 miles of that coordinate. Sorting the results by structure height can help one identify existing and potential locations of other wind projects, since the presence of several sequential permit requests for structures having identical structure heights (usually within a range of 300–500 ft aboveground, typical of today's utility-scale turbine heights) is a good indicator of a wind project.

10.3.3 DoD Preliminary Screening Tool/Long-Range Radar

This application allows one to enter either a single coordinate or four coordinates of a polygon (with a perimeter of <100 miles) to obtain a preliminary review of potential impacts to Long-Range Radar prior to making official FAA filings. This tool will produce a map relating the structure to any Long-Range Radar. The use of this tool does not replace the need to make official filings with the FAA but can give wind developers an early indication of potential problems. Figure 10.1 shows an example of results from entering a single point, while Figure 10.2 shows an example of results from entering a polygon.

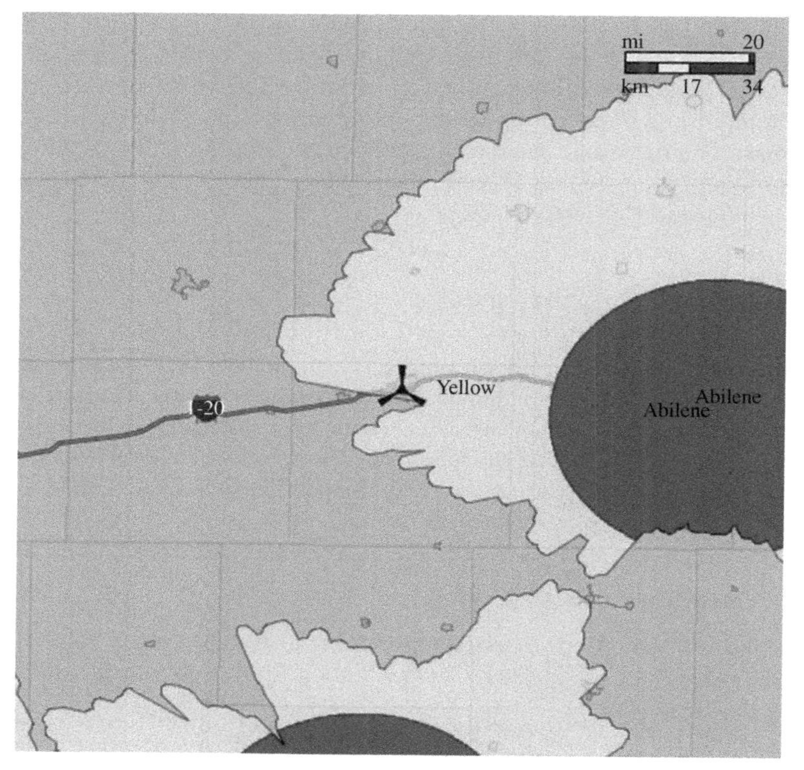

FIGURE 10.1 Example of results from FAA/DoD preliminary screen tool for long-range radar around a single point (Source: R. Walker produced using FAA Screening Tool).

- Grey or green areas represent no anticipated impact to Air Defense and Homeland Security radars.
- The yellow area indicates that impact is likely to Air Defense and Homeland Security radars.
- The red area indicates that impact is highly likely to Air Defense and Homeland Security radars.
- In all cases, an aeronautical study is required.

10.3.4 DoD Preliminary Screening Tool/NEXRAD Radar

This application allows one to enter either a single coordinate or four coordinates of a polygon (with a perimeter of <100 miles) to obtain a preliminary review of potential impacts to Next-Generation Radar (NEXRAD; weather) prior to making official

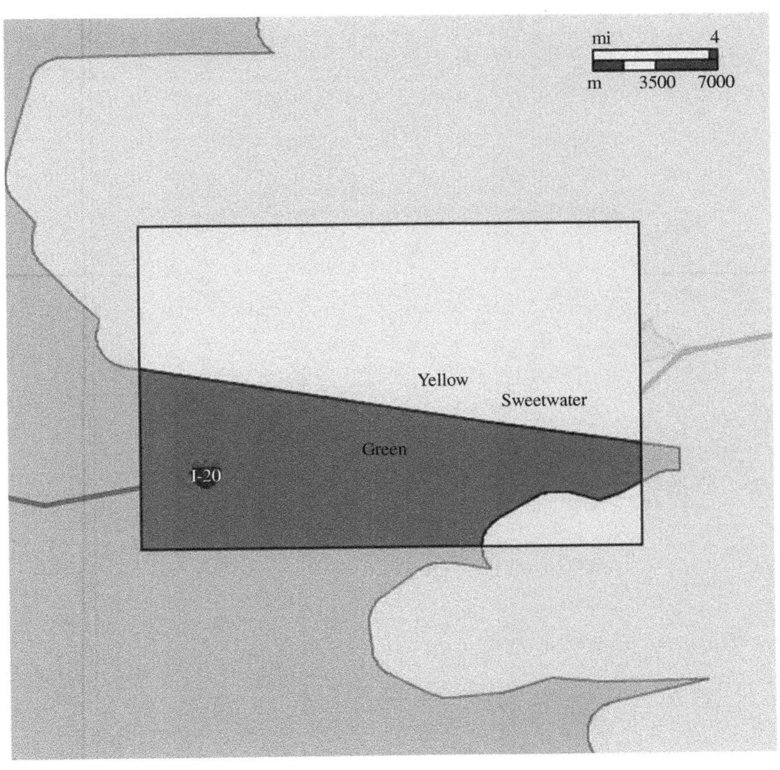

FIGURE 10.2 Example of results from FAA/DoD preliminary screen tool for long-range radar within a polygon (Source: R. Walker produced using FAA Screening Tool).

FAA Obstruction Evaluation/Airport Airspace Analysis (OE/AAA) filings. This tool will produce a map relating the structure to any NEXRAD Radar. The use of this tool does not replace the need to make official filings with the FAA but can give wind developers an early indication of potential problems. Figure 10.3 shows an example of results from entering a single point, while Figure 10.4 shows an example of results from entering a polygon.

10.3.5 DoD Preliminary Screening Tool/Military Operations

This application allows one to enter a single coordinate to obtain a preliminary review of potential impacts to military training and special airspace prior to making official FAA filings. This tool will produce a map such as the example in Figure 10.5.

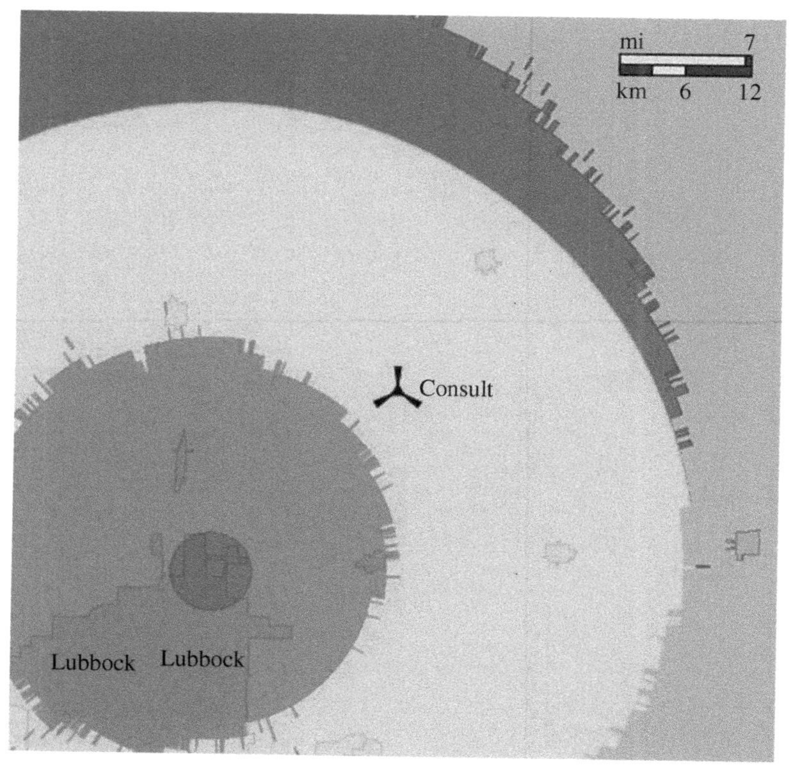

FIGURE 10.3 Example of results from FAA/DoD preliminary screen tool for NEXRAD radar around a single point (Source: R. Walker produced using FAA Screening Tool).

- Light green or grey areas represent no impact zones where impacts are not likely. NOAA will not perform a detailed analysis but would still like to know about the project.
- Dark green represents notification zones where some impacts are possible. Consultation with NOAA is optional, but NOAA would like to know about the project.
- Yellow represents consultation zones where significant impacts are possible. NOAA requests consultation to discuss project details and to perform a detailed impact analysis and may request mitigation of significant impacts.
- Orange represents mitigation zones where significant impacts are likely. NOAA will likely request mitigation if a detailed analysis indicates that the project will cause significant impacts.
- Red represents no-build zones where severe impacts are likely. NOAA requests developers not to build wind turbines within 3 km of the NEXRAD. Detailed impact analysis is required.

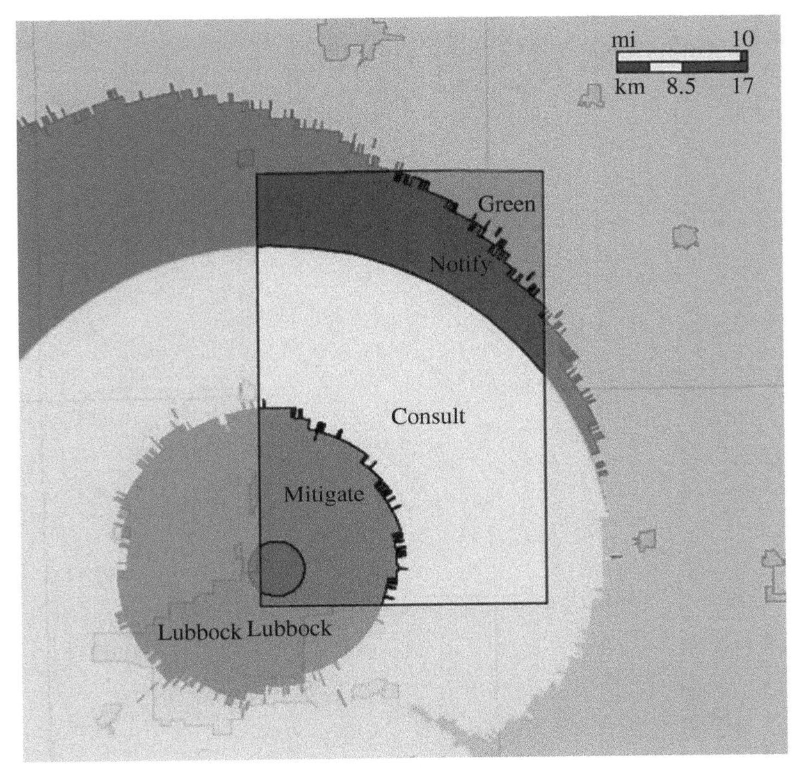

FIGURE 10.4 Example of results from FAA/DoD preliminary screen tool for NEXRAD radar within a polygon (Source: R. Walker produced using FAA Screening Tool).

The use of this tool does not replace the need to make official filings with the FAA or to contact military officials about the proposed project but can give wind developers an early indication of potential problems.

10.4 FAA REQUIREMENTS FOR MARKING OR LIGHTING OF OBSTRUCTIONS

Marking and/or lighting of wind turbine structures are intended to provide day and night conspicuity and to assist pilots in identifying and avoiding these obstacles. An FAA Advisory Circular AC 70/7460-1K titled *Obstruction Marking and Lighting* [1] includes requirements for marking and lighting of "wind turbine farms," defined as a wind turbine development that contains more than three (3) turbines of heights greater than 200 ft above ground level.

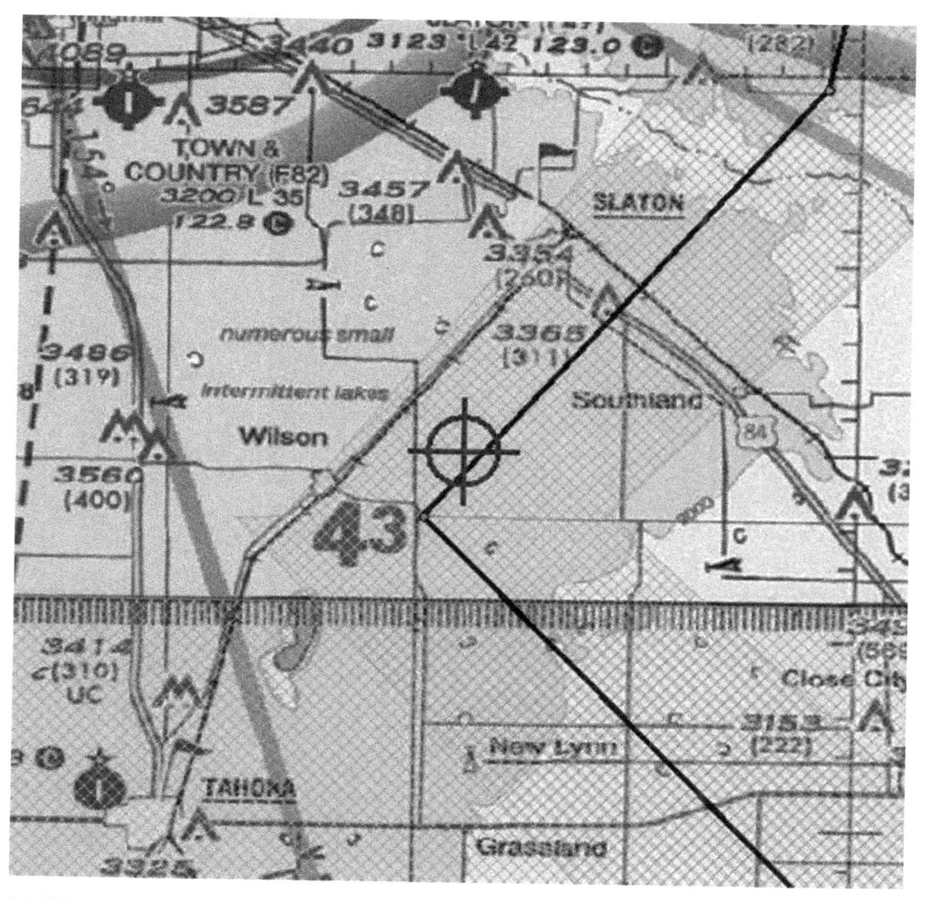

FIGURE 10.5 Example of results from FAA/DoD preliminary screening tool for military operations (Source: R. Walker produced using FAA Screening Tool).

Your structure falls within the confines of SR274 and may have an impact on military operations. For a more detailed review, please contact Joe Smith at (233) 273-5490. This Point of Contact (POC) will review the analysis and identify any additional areas of concern. Upon completion of this process, the POC will provide you a letter stating the results of the review.

Your structure falls within the confines of SR275 and may have an impact on military operations. For a more detailed review, please contact Joe Smith at (233) 273-5490. This POC will review the analysis and identify any additional areas of concern. Upon completion of this process, the POC will provide you a letter stating the results of the review.

Following are excerpts of some of the more useful information from the FAA Advisory Circular:

- The development of wind turbine farms is a very dynamic process, which constantly changes based on the differing terrain they are built on. Each wind turbine farm is unique; therefore it is important to work closely with the sponsor to determine a lighting scheme that provides for the safety of air traffic. The following are guidelines that are recommended for wind turbine farms. Consider the proximity to airports and Visual Flight Rules (VFR) routes, extreme terrain where heights may widely vary, and local flight activity when making the recommendation.
- Not all wind turbine units within an installation or farm need to be lighted. Definition of the periphery of the installation is essential; however, lighting of interior wind turbines is of lesser importance unless they are taller than the peripheral units.
- Obstruction lights within a group of wind turbines should have unlighted separations or gaps of no more than one-half statute mile if the integrity of the group appearance is to be maintained. This is especially critical if the arrangement of objects is essentially linear.
- Any array of flashing or pulsed obstruction lighting should be synchronized or flash simultaneously.
- Nighttime wind turbine obstruction lighting should consist of the preferred FAA L-864 aviation red-colored flashing lights.
- White strobe fixtures (FAA L-865) may be used in lieu of the preferred L-864 red flashing lights but must be used alone without any red lights, and must be positioned in the same manner as the red flashing lights.
- The white paint most often found on wind turbine units is the most effective daytime early warning device. Other colors (such as light gray or blue) appear to be significantly less effective in providing daytime warning. Daytime lighting of wind turbine farms is not required as long as the turbine structures are painted in a bright white color or light off-white color most often found on wind turbines.
- If darker paint is used, wind turbine marking should be supplemented with daytime lighting, as required.
- Flashing red (L-864) or white (L-865) lights may be used to light wind turbines. Studies have shown that red lights are most effective and should be the first consideration for lighting recommendations of wind turbines.
- Should the synchronization of the lighting system fail, a lighting outage report should be made.
- Light fixtures should be placed as high as possible on the turbine nacelle so as to be visible from 360°.
- On occasion, one or two turbines may be located apart from the main grouping of turbines. If one or two turbines protrude from the general limits of the turbine farm, these turbines should be lit.

10.5 MARKING AND/OR LIGHTING OF METEOROLOGICAL TOWERS

When the average hub height of new wind turbines being installed was only 60 or 65 m, meteorological tower heights of 60 m were usually sufficient. However, with hub heights of 80 and 100 m (or greater) now becoming commonplace, meteorological towers are also getting taller in order to improve the accuracy of measurements. A 60-m meteorological tower is 198 ft tall, falling just below the FAA requirement for lighting unless the meteorological tower is going to be located near an airport or a heliport. But the use of towers in excess of 60 m will require the installation of FAA warning lights, increasing the cost and complexity of such meteorological towers.

There have been reported cases of small airplanes (such as crop dusters) hitting meteorological towers. Thus, it is fairly common to place aircraft marker balls on the guy wires of meteorological towers, such as shown in Figure 10.6.

However, some states like South Dakota have enacted requirements specifying marking of meteorological towers, and in late 2010, the FAA issued a notice that it was considering establishing guidance for the voluntary marking of meteorological

FIGURE 10.6 Sixty-meter tall meteorological tower with aviation market balls (Source: R. Walker).

towers that are less than 200 ft tall. The FAA guidance recommends that the towers be painted in alternate bands of aviation orange and white and recommends that spherical and/or flag markers on guy wires be used when additional conspicuity is necessary. The FAA notice also indicates that it is considering recommending high-visibility sleeves on the outer guy wires of the towers.

Wind developers installing new meteorological towers or lowering existing towers for maintenance should determine if new state regulations or FAA regulations require specific markings of the towers or if such regulations are in the process of being phased in. Even if not mandated, developers should consider the use of markings similar to that proposed by the FAA as an additional safety measure.

10.6 POTENTIAL IMPACT OF WIND TURBINES ON AVIATION OR NAVIGATION RADAR

With the rapid growth of wind energy in the United States, the potential for turbines to impact aviation, navigation, or weather radar facilities has also increased, adding greater significance to the issue. Good sources of information on the issue of wind energy development's potential impact on airspace and radar include American Wind Energy Association (AWEA's) *Wind Energy Siting Handbook* [2] (Section 4.1.6), the National Research Council's book *Environmental Impacts of Wind-Energy Projects* [3], and AWEA's factsheet titled *Airspace, Radar and Wind Energy* [4].

As discussed earlier in this chapter, the FAA oversees and manages navigable airspace in the United States and is charged with ensuring the safety of aircraft and the efficient use of the airspace. When a person or an entity files a *Notice of Proposed Construction*, one of the issues that the FAA reviews is the potential impact on aviation radar systems.

The US military in particular has expressed concern about the potential impact of wind energy development on radars necessary for military and Homeland Security use. In 2006, the US DoD issued a report titled *Report to the Congressional Defense Committees: The Effect of Windmill Farms On Military Readiness* [5] reviewing the issue of wind plants potentially impacting their radar systems. Figure 10.7

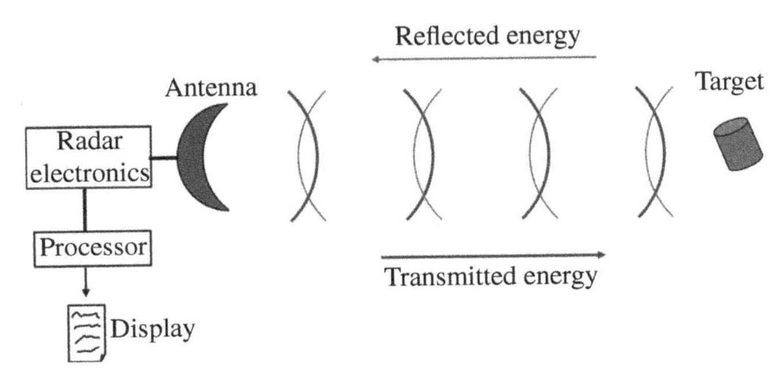

FIGURE 10.7 Illustration of a basic radar system (Source: *Report to the Congressional Defense Committees: The Effect of Windmill Farms On Military Readiness* [5]).

(from that report) illustrates how a radar system works to detect distant objects. If wind turbine blades reflect energy from the radar instead of intended targets such as airplanes, this may confuse the radar operator.

10.6.1 Conclusions and Recommendations from *Report to the Congressional Defense Committees: The Effect of Windmill Farms on Military Readiness* [5]

The Department of Defense's report evaluating the potential impact of wind energy development on the military's radar system includes the following conclusions and recommendations:

- "Although wind turbines located in radar line of sight of air defense radars can adversely impact the ability of those units to detect and track, by primary radar return, any aircraft or other aerial object, the magnitude of the impact will depend upon the number and locations of the wind turbines. Should the impact prove sufficient to degrade the ability of the radar to unambiguously detect and track objects of interest by primary radar alone this will negatively impact the readiness of U.S. forces to perform the air defense mission.
- The mitigations that exist at present to completely preclude any adverse impacts on air defense radars are limited to those methods that avoid locating the wind turbines in radar line of sight of such radars. These mitigations may be achieved by distance, terrain masking, or terrain relief and requires case-by-case analysis.
- The Department has initiated efforts to develop additional mitigation approaches. These require further development and validation before they can be employed.
- Wind turbines in close proximity to military training, testing, and development sites and ranges can adversely impact the 'train and equip' mission of the Department. Existing processes to include engagement with local and regional planning boards and development approval authorities should be employed to mitigate such potential impacts.
- Wind turbines located in close proximity to Comprehensive Test Ban Treaty monitoring sites can adversely impact their ability to perform this mission by increasing ambient seismic noise levels. Appropriate offset distance criteria should be developed to mitigate such potential impacts.
- The Federal Aviation Administration (FAA) has the responsibility to promote and maintain the safe and efficient use of U.S. airspace for all users. The Department defers to the FAA regarding possible impacts wind farms may have on the Air Traffic Control (ATC) radars employed for management of the U.S. air traffic control system. The Department stands prepared to assist and support the FAA in any efforts the FAA may decide to undertake in that regard.
- The National Weather Service (NWS) has the primary responsibility to provide accurate weather forecasting services for the nation. The Department defers to the NWS regarding identification of impacts wind farms may have on weather

radars and development of appropriate mitigation measures. The Department stands prepared to work with the NWS in this area on NWS identified mitigation measures that have the potential to benefit Department systems."

10.6.2 Insights from AWEA's *Airspace, Radar and Wind Energy Factsheet*

AWEA produces "factsheets" on a number of issues surrounding wind energy development, and since the potential impact of wind generation on radar systems is one of the bigger issues that the wind industry is currently dealing with, they have produced one on this issue. Some of the more interesting insights included in AWEA's *Airspace, Radar*, and *Wind Energy Factsheet* [4] include the following:

- "The wind industry strongly supports responsible, effective actions designed to identify and address any potential conflicts with airspace and radar due to proposed wind farms.
- Wind turbines, radar, and military and civilian airspace needs can co-exist. Experience shows that technical case-by-case analysis of potential airspace and radar interactions can resolve some concerns, and mitigation measures should be able to resolve others.
- Decades of experience in developing wind turbine facilities in the U.S. and around the world, often near civilian or military radar installations, have demonstrated that wind turbines and radar can coexist.
- Depending on location, wind turbines may interfere with some types of civilian and military radar, causing 'clutter' or other interference. Wind energy developers have to be cognizant of military and civilian airspace needs as well."

10.6.3 Insights from *Wind Farms and Radar* Produced by the MITRE Corporation [6]

The US DHS contracted with JASON, a division of the MITRE Corporation of Mclean, Virginia, to review the current status of the conflict between the ever-growing number of wind turbine farms and the air-security radars that are located within some tens of miles of a turbine farm. The resulting report is another excellent source of information on this subject and includes the following findings:

- Wind farms interfere with the radar tracking of airplanes and weather.
- The velocity of the blade tips can reach 170 mph, causing significant Doppler clutter. This creates problems and issues for several stake holders, including DHS, DOD, FAA, and NOAA.
- Examples of issues include:
  - a wind farm located close to a border might create a dead zone for detecting intruding aircraft;
  - current weather radar software could misinterpret the high apparent shear between blade tips as a tornado;

- current air traffic control software could temporarily lose the tracks of aircraft flying over wind farms.
- Despite these difficulties, there is no fundamental physical constraint preventing detection and mitigation of windmill clutter. The technologies of wind turbines and radar can coexist.
- On the other hand, the nation's aging long range radar infrastructure increases the challenge of distinguishing wind farm signatures from airplanes or weather; this is especially so since many promising mitigation measures (discussed below) are based on digital processing capabilities.
- The challenge is to evolve the current system, and to design future systems to effectively distinguish and mitigate a source of clutter that was not anticipated in the original design specifications for either radar or wind farms.
- Progress forward requires the development of not only mitigation measures, but also of quantitative evaluation tools and metrics to determine when a wind farm poses a sufficient threat to a radar installation for corrective action to be taken.

This same report also included several suggested methods or technologies for mitigating the potential impact of wind plants on radar systems, including the following:

- Reducing the radar signature of wind farms by
  - Modifying the turbine blades to modify or reduce their radar signature by shifting the Doppler frequency spectrum from the blades to lie outside the range of frequencies processed by the radar.
  - Modifying the inside of the blades with layers of circuits and reflectors that would reduce the strength of the radar return from the blades.
- Using telemetry from turbines to radars to blank out turbine radar returns while preserving returns from objects of interest, such as aircraft.
- Modifying radar system software to process data for long-range radar to better distinguish between the radar signatures of aircraft and wind farms.
- Modifying radar designs to have shorter pulses, a higher-pulse repetition frequency (PRF), local oscillators coherent over a turbine blade period, or multiple elevation beams to avoid ground scraping. The higher PRF allows for painting a given turbine blade with more pulses before the blade rotates significantly.
- Replacing older radar systems with newer ones that incorporate multidimensional detection, with greatly enhanced processing and pulse shapes designed to optimally distinguish between aircraft and wind farms. Note that the cost of a single radar installation is in the range of $3,000,000 to $8,000,000.
- Using gap fillers in cases when a wind farm causes an unacceptable loss of coverage. Gap fillers allow a second view of the wind farm radar interference, making it considerably easier to process this interference out through data fusion.

10.7 POTENTIAL IMPACT OF WIND TURBINES ON WEATHER RADAR

In addition to impacting aviation radar systems, wind turbines may also have an effect on weather radar systems or NEXRAD . The NEXRAD network uses Doppler radar technology, which obtains information about the weather such as precipitation or movements of the atmosphere (wind) by emitting a burst of energy and then evaluating the returned energy.

If the energy emitted strikes an object (such as a raindrop or hailstone), the energy is "scattered" in many directions, with some small fraction of the energy being returned to the radar (Fig. 10.8).

Returned signals are processed into a map, and iterations of the maps can be used to track the movement of storm cells. But if the emitted energy strikes a wind turbine blade instead, energy reflected from the blade may be detected and processed by the NEXRAD system as an area of precipitation or storm cell. The most obvious problems this might create include the following:

- Cluttering of radar imagery that may confuse or distract radar operators
- Mistaking a wind farm for a shower or thunderstorm
- Mistaking a tornado or thunderstorm for a wind farm

FIGURE 10.8 Representation of NEXRAD radar signal being reflected by a wind turbine (Source: Reproduced by permission of Amy Marcotte).

10.7.1 Mitigation of Wind Plant Impacts on Weather Radar

To reduce the potential impact of wind energy development on weather radar systems, the wind energy industry and the NOAA must work closely together. Wind developers need to identify locations of weather radar systems early in the development process to avoid wasting time and money on a project site that cannot obtain FAA approval and to avoid setting up confrontational situations. Some manufacturers of wind turbines are already working on stealth blade technology that can reduce the radar image of turbines, so research in this area or other potential technology-related solutions will continue.

Where wind projects already exist within the boundaries of the weather radar's scanning ability, radar operators can scan at higher altitudes to give forecasters a view above the rotating turbine blades, establish exclusion zones in precipitation processing, use adjacent radar coverage to see over or beyond wind projects, and be trained in how to better to recognize wind farm signatures and better understand the impacts wind projects may have on the data being received and processed.

In addition, NOAA is either currently using or considering the following initiatives:

- Using limited operational curtailment of wind farms during some severe weather events under certain conditions:
  - Voluntary agreement between wind farm owners and local Weather Forecast Offices
  - Stopping turbines for 30 min to a couple of hours during select, preagreed severe weather events over or near wind farms
  - Involving only wind farms located within about 10 miles from radar
- Sharing wind farm meteorological tower data with the local forecast office to compensate/correct for contamination of precipitation estimates by wind farms
- Developing signal processing identification that can remove wind turbine clutter
- Using phased-array radar transmitters

10.8 POTENTIAL IMPACT OF WIND TURBINES ON TELECOMMUNICATIONS

Wind plant facilities including turbines, transmission lines, distribution-voltage collection systems, and substations may impact the following types of telecommunications systems:

- Microwave or fixed-link signals
- TV broadcast signals
- AM/FM broadcast signals
- Mobile telephone service

Impacts can be caused by the following:

- Electromagnetic noise from the wind plant facilities that can interfere with telecommunications signals, in much the same way that facilities of an electric utility can impact telecommunications. Think of what the AM radio in your car sounds like when driving under or by some electric lines.
- Physical obstructions that distort communications signals in the same way that large buildings or other types of structures can be an obstruction to communications signals.

The types of communications systems that may be affected include microwave systems, off-air TV broadcast signals, land mobile radio (LMR) operations, and mobile telephone services.

10.8.1 Microwave Telecommunications Issues

Microwave telecommunication systems are wireless point-to-point (or line-of-sight) links that communicate between two sites (antennas) and require clear line-of-sight conditions between each antenna. Any large obstructions between the transmitters can reduce the reliability of the transmission. Wind developers need to perform a Licensed Microwave Search and Worst Case Fresnel Zone Analysis to identify microwave paths within a project area. In some cases, unlicensed microwave paths may exist, so it may be useful to have surveyors and consultants perform field reconnaissance to be on the alert for microwave towers not plotted on project maps. The length of the turbine blade also needs to be considered when evaluating potential impacts on microwave paths, so it is wise to always allow some additional "buffer" when performing turbine layout. Performing this analysis during the development stage is highly preferable to discovering the problem after construction, which could result in very expensive mitigation actions such as relocating a turbine or a microwave facility.

Consultation with federal agencies such as the National Telecommunications Information Administration and the National Weather Service is also necessary to identify federal government microwave communication systems.

10.8.2 TV Broadcast Signal Issues

Over-the-air, analog television broadcast signals are subject to scattering, distortion, or signal reflection from the turbine blades and by the attenuation of the signal passing through the wind turbines, which may result in multipath distortion or "ghosting." Since 2009, digital television signals are required in the United States, which are not affected in the same way, but wind developers active in other nations still using analog signals need to be aware of the potential for such issues.

With regard to digital television signals, a white paper by Lester Polisky of Comsearch [7] indicates that wind turbines located in fringe areas of digital television service could cause some minor signal attenuation from obstructions, which could result in going from a high-quality video picture to "no signal found" due to the nature of digital modulation. Residential television viewers who utilize "rabbit ears" type antennas and live within a few miles of a wind project are probably more susceptible to potential issues. Other factors that may affect the extent of any issues include pre-existing signal strength and the extent of cable and satellite television use in the area.

Wind developers should identify television broadcast facilities within 100 miles of the project and determine the potential for any issues arising. A TV Broadcast Reception Analysis can be done prior to project construction to characterize the baseline signal strength and reception conditions in the area of the planned project. After the project goes into service, a similar analysis can be performed to determine if there are any impacts resulting from the wind project. If issues arise after a wind plant has been constructed, possible mitigation methods include the following:

- Relocation of the homeowner's antenna to a location that receives a better signal
- Installation of a higher-quality or directional outside antenna, with amplifier if necessary
- Purchase of a new HDTV set or converter box
- Installation of satellite or cable TV
- Setting up a wireless television distribution system for a cluster of homes affected by a wind project
- Construction of a repeater station if a large area is impacted

10.8.3 AM/FM Broadcast Signal Issues

Since most radio transmitters are omnidirectional, sending signals out in several directions, wind energy projects will not affect AM/FM radio broadcast signals in the same way that they affect microwave signals. However, the nearer that the wind project is located to the broadcast antenna, the larger the potential there is for some impact on signal quality. Therefore, wind developers should attempt to identify all directional broadcast antennas within 2 miles of a planned wind project and omnidirectional antennas within 1 mile of a planned wind project.

10.8.4 Other Potential Issues

It is possible that wind projects could have minor effects on LMR systems, cellular systems, or personal communications services. If issues are identified, they may be mitigated by installing repeater antennas for LMR systems or base station antennas for cellular and personal communications services. Wind developers should attempt to identify all such radio towers within 1 mile of a planned wind project.

REFERENCES

[1] FAA Advisory Circular AC 70/7460-1K (2007) *Obstruction Marking and Lighting*, U.S. Department of Transportation, Federal Aviation Administration, Washington, DC.

[2] Tetra Tech EC, Inc. and Nixon Peabody LP (2008) *Wind Energy Siting Handbook*, American Wind Energy Association, Washington, DC.

[3] National Research Council of the National Academies (2007) *Environmental Impacts of Wind-Energy Projects*, National Academies Press, Washington, DC.

[4] American Wind Energy Association. Airspace, radar and wind energy, AWEA factsheet, Washington, DC: American Wind Energy Association; 2012.

[5] U.S. Department of Defense, Office of the Director of Defense Research and Engineering. *Report to the Congressional Defense Committees: The Effect of Windmill Farms On Military Readiness*. Washington, DC: U.S. Department of Defense, Office of the Director of Defense Research and Engineering; 2006.

[6] Brenner, M., Cazares, S., Cornwall, M.J. *et al.* (2008) *Wind Farms and Radar*, JASON, the MITRE Corporation for the U.S. Department of Homeland Security, Washington (DC).

[7] Polisky L. *Post Digital Television Transition—The Evaluation and Mitigation Methods for Off-Air Digital Television Reception In-and-Around Wind Energy Facilities*. Asburn: Comsearch; 2009.

11

OTHER ENVIRONMENTAL ISSUES OF WIND ENERGY DEVELOPMENT: AESTHETICS, CULTURAL RESOURCES, LAND USE COMPATIBILITY, WATER RESOURCES, AND SITE RESTORATION

11.1 AESTHETIC ISSUES: WIND ENERGY'S IMPACT ON SCENERY OR VIEWSHED

As the saying goes, "Beauty is in the eye of the beholder." People's viewpoints on what is attractive or ugly are highly subjective in nature. There are those who say things like "wind turbines are a blight on the landscape" or call wind turbines "eyesores," while others think they are quite graceful and beautiful, even including them in posters or paintings used for decoration.

Apparently, the advertising industry thinks wind turbines are reflective of something good since one cannot watch very many television commercials these days without noticing wind turbines in the background, even when the product being advertised may have absolutely nothing to do with wind energy. While many of them are obviously trying to portray that the company producing the product being advertised is environmentally responsible or "green," it is unlikely that the use of wind

Wind Energy Essentials: Societal, Economic, and Environmental Impacts, First Edition.
Richard P. Walker and Andrew Swift.

turbines in advertisements would be as widespread if a large percentage of the public perceived them as being "ugly" (Fig. 11.1).

In addition to advertising companies, many architects are interested in wind energy, particularly in regard to how smaller turbines can be integrated into buildings for decorative purposes while also helping produce some of the energy used by the building and its occupants. Landscape architects may be interested in the use of turbines for landscaping, with small vertical-axis turbines in particular being designed for use on buildings or in homeowners' backyards. For example, two small wind turbines are located on the grounds of the United States Botanic Garden adjacent to the U.S. Capitol.

Since wind turbines are usually sited in locations with a good wind resource, they also tend to be highly visible since better wind speeds are typically found on ridgelines, mesas, or plains rather than in more visually secluded areas (such as in valleys or forests). The real question becomes whether or not the turbines are intrusive. While the question of the impact of wind turbines on scenery can generate much debate, there are other aspects of wind projects that will also impact an area including transmission lines, substations, roads, and the operations buildings. All of these facilities should be considered when designing a wind energy project and evaluating the project's impact on local scenery. Proper siting and design of wind projects can help

FIGURE 11.1 Wind turbines used in an advertisement on an electronic billboard in New York City's Times Square (Source: R. Walker).

mitigate some of the opposition to such projects. Factors that will affect the impact of a wind energy project on an area's scenic quality are as follows.

11.1.1 Size of Turbines

The height of typical utility-scale turbines being installed in 2010–2011 is somewhere in the range of 350–400 ft, but the largest commercially produced turbines reach nearly 650 ft in height, and the trend continues to be larger and larger turbines. The greater the height of the turbine, the greater the size of the viewshed in which the turbines can be seen.

11.1.2 Spacing of Turbines

As wind turbines get larger, the spacing between turbines within rows and between turbine rows increases in order to reduce the effects of turbulence and wake effect. Within rows, turbines may be 800–1200 ft apart, while the distance between rows may be in the range of 2000–3000 ft. Increasing the spacing generally reduces how "cluttered" a project may look. Figures 11.2 and 11.3 contrast the look of older projects to modern projects.

FIGURE 11.2 An older wind project with many small turbines on lattice towers (Photo credit: Stan Shebs, http://en.wikipedia.org/wiki/File:Tehachapi_wind_farm_3.jpg).

FIGURE 11.3 Megawatt-scale turbines on tubular steel towers (Source: Reproduced by permission of K. Jay).

11.1.3 Turbine Design Factors: Horizontal Versus Vertical Axis, 2-Bladed Versus 3-Bladed, Tubular Steel or Lattice Towers, Color, Turbine Lighting

The vast majority of state-of-the-art, utility-scale turbines are 3-bladed horizontal-axis turbines, with tubular steel towers painted white, with blades also colored white. Some people believe lattice steel towers present a cluttered look, while others may argue that they are less visually intrusive from a distance than tubular steel towers.

There is a general belief that 3-bladed turbines are more visually pleasing than 2-bladed turbines, but there are certain to be others that disagree with this. Black blades are occasionally used for wind turbines, sometimes to minimize ice buildup on the blades, but most people seem to prefer the look of white blades.

The primary reasons for the dominance of horizontal-axis turbines are efficiency and economics, but many people have a preference for vertical-axis turbines from an aesthetic perspective.

With regard to the color, the Federal Aviation Administration (FAA) preference for aviation safety reasons is that the turbines be bright white. Other colors (such as light gray or blue) are believed to be significantly less effective in providing daytime warning to aircraft. However, from an aesthetic perspective, people might prefer turbines that are painted to match the landscape. FAA regulations also require warning lights on some portion of the turbines (assuming the turbines are >200 ft in height). The synchronized red flashing lights visible at night may be annoying to some people, although the red lights now used are generally less annoying than the white strobe lights that were used in the past.

11.1.4 Overall Size of Project/Number of Turbines per "Cluster"/Number of Turbine Rows

Some of the wind projects in Texas contain several hundred wind turbines spread out over many miles of land. Obviously, to someone driving down a highway near such projects, these are visible for a much longer period of time than a much smaller project would be. These large projects may also have multiple rows of turbines more or less parallel to each other (Fig. 11.4).

Parallel rows of turbines are more common in projects located in relatively flat areas of the country (such as in Kansas or the Texas panhandle), whereas in other areas projects may meander along ridgelines or the edge of mesas. Again, there are some people who will prefer the "order" of parallel turbine rows, while others may prefer the meandering look.

11.1.5 Shielding by Trees, Topography, or Other Structures; Distance from Roads and Communities

Wind projects in the Great Plains region may be visible from many miles away, whereas ones located in West Virginia, Pennsylvania, or New York may only be visible from a few vantage points due to shielding by trees or topographic features.

FIGURE 11.4 Wind project in Texas with many turbine rows (Photo source: R. Walker).

In addition, the distance that projects are located from highly traveled roads and from communities will influence the amount of opposition due to visual impacts. Designers of wind projects may wish to consider these factors when performing turbine layouts if they believe that the aesthetics of the project may become a significant issue.

11.1.6 Scenic Quality of the View Prior to Installation of Turbines

In general, wind projects located in agricultural areas of the Midwest or desert areas of west Texas and New Mexico have been favorably received by the community, even though some of these projects are among the largest wind projects in the world. Perhaps the widespread use of water-pumping windmills by settlers in the region is part of the reason why. On the other hand, projects containing only one or two turbines in scenic areas like Cape Cod may encounter large amounts of opposition. The proposed Cape Wind project, which would be located offshore in Nantucket Sound, Massachusetts, has probably received the largest amount of opposition of any wind project. Siting a project near state or national parks, national forests, hiking trails, or wilderness areas is likely to result in opposition to the project.

11.1.7 Electric Lines, Roads, Substations, and Operations Buildings

In recent years, the electric utility industry has experienced greater resistance to electric transmission lines. While transmission lines are essential to the reliability and affordability of electric energy, most landowners would rather have the lines located on someone else's property, preferably far enough away to be completely out of their own view. On the other hand, most landowners would very much appreciate the opportunity to have one or more wind turbines located on their property due to the potential for thousands of dollars in royalties over a two-decade period.

However, the transmission and distribution lines, roads, substations, and buildings that come with the wind turbines are usually not quite so welcome. Most landowners understand the need for these facilities and will tolerate them as long as the project developer works with them to minimize the visual and logistical impact on their property. But to other people living in the immediate area who are not receiving wind energy royalties, these facilities may represent eyesores (Figs. 11.5 and 11.6).

Wind project designers need to be aware that these facilities will also have aesthetic impacts on the region and, if possible, locate them so as to minimize their visibility. For example, primary access roads constructed to project sites can stand out from many miles away if no consideration is given to their visual impact. And like turbine design choices, the design of transmission towers will affect their visibility and the likelihood of opposition. In general, the public seems to prefer single-pole, tubular structures (wood, steel, or concrete) to lattice towers.

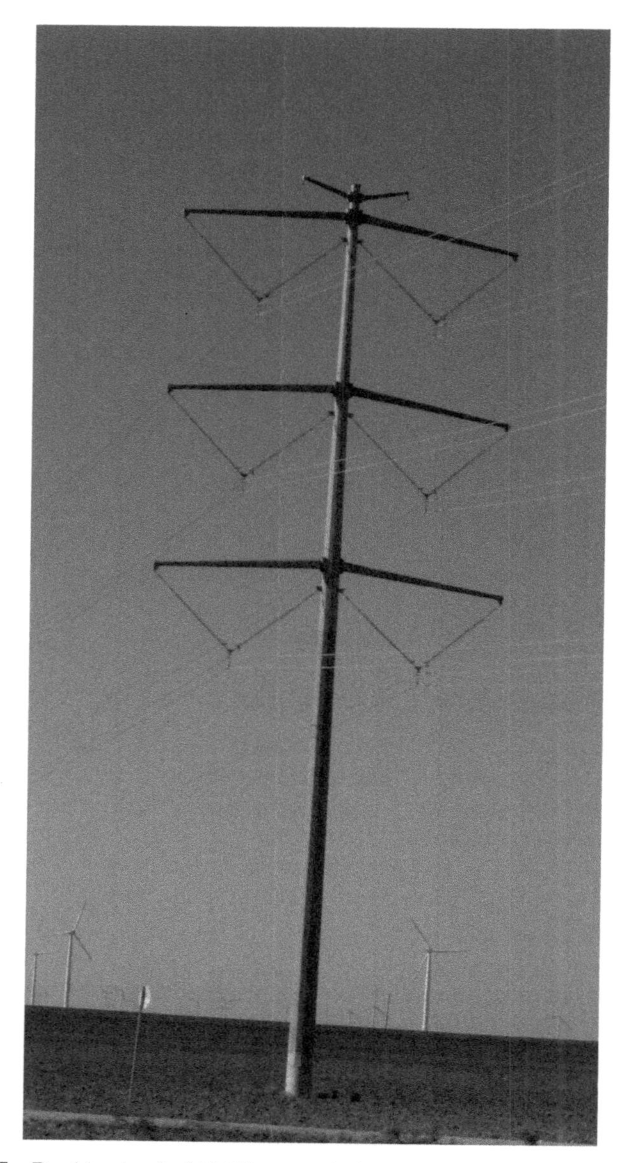

FIGURE 11.5 Double-circuit 345-kV transmission line with tubular steel tower (Photo source: R. Walker).

Substations and operations facilities can be quite large and can also have a substantial amount of parked vehicles and traffic around them. Therefore, project designers need to consider their aesthetic impact when siting them as well as when designing the facilities. Landscaping of the operations building can play an important role in determining the public's perception of the project (Fig. 11.7).

FIGURE 11.6 Double-circuit 345-kV transmission line with lattice steel tower (Photo source: R. Walker).

FIGURE 11.7 Invenergy's Camp Springs Energy Center (Source: Charles Norland, Norland Photographic Art, St. Louis, MO).

11.2 MITIGATING AESTHETIC IMPACTS AND ISSUES

Some of the ways that developers of wind energy projects can reduce the likelihood of controversy and opposition to the project based on aesthetic concerns include the following:

- Avoid siting projects near well-known scenic or cultural resources.
- Perform an inventory or assessment of scenic resources in the region being considered for wind development.
- Prepare photo simulations of the project showing several views of the area before and after the installation of wind turbines. Display these simulations at community or town hall meetings held to discuss the project.
- Keep aesthetic impacts in mind during all aspects of project design including turbine selection and siting, transmission line structural design and routing, substation location, and operations building location and design.
- Take advantage of trees or terrain to shield views of the wind project facility from heavily traveled roads or nearby communities if reasonable.

11.3 IMPACT OF WIND ENERGY ON CULTURAL RESOURCES

In addition to impacting the aesthetics of an area or the traditional usage of land, wind energy development may affect certain cultural resources in an area. Cultural resources can include archaeological sites, prehistoric artifacts or areas, historic sites or buildings, recreational areas, and community meeting places.

Examples of recreational areas or facilities that can be affected by wind development include national parks, national forests, state parks, lakes, seashores, rivers, streams, hiking trails (such as the Appalachian Trail), scenic drives (such as the Blue Ridge Parkway or Skyline Drive), snow skiing slopes, or golf courses. Aesthetic impacts, as discussed in the previous section, are probably the largest potential impact for most of these. Of course, while many ski resorts proudly announce that their ski lifts are powered by wind energy, they may not be too receptive to a wind developer placing a wind turbine in the middle of a ski slope.

After the potential aesthetic impacts of wind energy on cultural resources, probably the next largest issue for wind development with regard to cultural resources is the potential for damage to archaeological sites, historical structures, or sites sacred to native or indigenous cultures. Development of wind projects in the United States necessitates that developers be aware of several national laws addressing the preservation of archeological or historical sites or artifacts. In addition, it is very likely that states have statutes that may also come into play, although a review of state laws will not be included in the scope of this book.

Desktop and in-the-field archeological surveys should be performed by qualified experts for all areas within the proposed wind project where construction activities may disturb the land including turbine locations, roads, electric line routes, substation locations, building locations, and storage yards.

Some of the laws, regulations, and resources that wind project developers and construction companies need to be aware of are mentioned as follows:

11.3.1 National Historic Preservation Act

The National Historic Preservation Act (NHPA) is intended to promote the preservation of historic properties significant to the nation's heritage in recognition of the growth of urban centers; highways; and residential, commercial, and industrial development. This can include the protection, rehabilitation, restoration, and construction of districts, historic sites, buildings, structures, and objects significant in American history, architecture, archaeology, or culture.

11.3.2 National Register of Historic Places

The NHPA authorized the Secretary of the Interior to maintain the National Register of Historic Places, which is the official list of historic places worthy of preservation in the United States. It is part of a national program to coordinate and support public and private efforts to identify, evaluate, and protect our historic and archeological resources. The National Register of Historic Places is administered by the National Park Service, and its searchable database can be accessed at: http://nrhp.focus.nps.gov.

Properties listed in the National Register include districts, sites, buildings, structures, and objects that are significant in American history, architecture, archeology, engineering, and culture. It includes the following:

- All historic areas in the National Park System
- National Historic Landmarks that have been designated by the Secretary of the Interior for their significance to all Americans
- Properties significant to the US states or communities that have been nominated by State Historic Preservation Offices (SHPOs), Federal agencies, and tribal preservation offices and have been approved by the National Park Service

Therefore, this is a good starting point for the identification of such sites within the boundaries of a planned wind project.

11.3.3 Advisory Council on Historic Preservation

The NHPA also directs that the Advisory Council on Historic Preservation (ACHP) serve as the primary federal policy advisor with regard to national preservation policies. The ACHP is an independent federal agency that promotes the preservation, enhancement, and productive use of our nation's historic resources and advises the President and Congress on national historic preservation policy.

11.3.4 State Historic Preservation Offices

State Historic Preservation Offices (SHPOs) carry out many of the responsibilities for historic preservation dictated by federal laws such as surveying, evaluating, and nominating significant historic buildings, sites, structures, districts, and objects to the National Register. The SHPO should be contacted soon after the boundaries of a potential wind project are identified.

11.3.5 Archaeological Resources Protection Act

The Archaeological Resources Protection Act (ARPA) is intended to preserve archeological resources on public lands and intended lands. If a project is being considered on federal, state, or tribal lands, it may need to obtain an ARPA permit before any construction activities begin.

11.3.6 Native American Graves Protection and Repatriation Act

The purpose of this act is to protect Native American graves that may be discovered during construction activities taking place on federal or tribal land.

11.3.7 Federal Land Policy and Management Act and National Environmental Policy Act

Both of these acts protect paleontological resources (such as fossils, imprints, or bones) that are found on federal land.

11.4 COMPATIBILITY OF WIND ENERGY DEVELOPMENT WITH EXISTING LAND USE

Wind turbines can be very compatible with existing land use in many areas of the United States, but there are definitely some areas that it would be wise to avoid. Where the primary use of the land is farming, ranching, oil and gas development, and/or hunting, it is likely that wind development can also take place as long as careful and thoughtful consideration is given to ensuring compatibility with these existing uses and that appropriate setbacks are used to mitigate issues with noise, ice shedding, or shadow flicker.

11.4.1 Compatibility with Farming

One of the largest wind energy projects in the United States is the 782-MW Roscoe Wind Farm, which was developed by one of this book's authors (Fig. 11.8). The majority of the land covered by this project is used for raising cotton. The project originated when one of the landowners, himself a cotton farmer and entrepreneur, began talking to his neighbors about trying to organize a group of landowners to attract a wind developer to their area. Their primary reasons for doing this were to diversify the sources of income to themselves and the community, to increase the tax base needed to support county services and schools, and to bring jobs and economic development opportunities to the area. Since that time, many similar groups of landowners have formed around the country in an attempt to entice wind projects to their properties and communities.

The turbine pad sites, roads, substations, electric lines, and operations building needed for a wind energy project will generally require only about 2–5 % of the surface area within the entire project site. The remaining 95–98% can continue to be used for its original purpose in most cases.

FIGURE 11.8 Roscoe Wind Farm amid fields of cotton and grain sorghum (Source: R. Walker).

Following are the primary issues that need to be considered when developing, constructing, and operating wind energy projects in areas of cultivation:

- **Irrigation:** Crops can be irrigated in a variety of ways including underground drip irrigation systems, aboveground circular irrigation pivots, large rotating spray nozzles, or irrigation channels. Wind turbines are not very compatible with underground drip irrigation systems since the heavy equipment used to construct the project may crush the underground pipes. So only in rare cases where the landowner is in full agreement should wind turbines be placed in fields with drip irrigation systems. Wind turbines should not be placed within the area traversed by circular irrigation pivots since the tower would prevent the full rotation of the pivot. However, turbines can be placed in the corners of the tract where the circular irrigation system does not reach. Even in these situations, great caution must be taken to avoid underground water lines and electric lines associated with the water wells and pumps needed to provide water to the irrigation pivot.

- **Road locations:** During the construction process, roads will need to be constructed allowing the delivery of the huge turbine components and the construction equipment necessary to erect the turbines. These roads will usually need to be hard-packed caliche or crushed rock and capable of supporting cranes and cement trucks weighing several tons. Roads may need to be as wide as 40 ft during the construction process, with large turn radii to facilitate delivery of blades and towers. After construction is complete, the roads are often narrowed down to 16–20 ft wide for use by service vehicles. Rarely are the roads removed until the entire project is retired since the turbines will require routine maintenance and inspection, as well as unplanned maintenance caused by weather events or equipment failure. These roads will be an obstacle for farmers plowing their fields and harvesting their crops, sometimes necessitating a change in the patterns used for plowing. However, for the majority of landowners, the economic benefits from wind royalties will far exceed the nuisance factor of these roads (Fig. 11.9).
- **Control of weeds, dust, rocks, and trash:** Most farmers take great pride in their land and thus look unfavorably upon anyone littering on their property or introducing weeds into their crops. Dust from roads needed to access wind turbines can also have an impact on crop productivity, and farmers do not want

FIGURE 11.9 Road accessing wind turbine locations (Source: Reproduced by permission of K. Jay).

large rocks in their fields as a result of wind plant construction since this can damage their farm equipment. Therefore, it is customary to address each of these issues in property lease agreements.

The owner of the wind project should be responsible for the control of weeds along the roads it constructs or uses as well as around any other facilities including the turbine pads, substation sites, and electric line easements. However, farmers also do not want someone inexperienced with agricultural practices spraying weed killer in their fields if it is also likely to kill their crop, so care must be taken in addressing this issue to ensure that only individuals or companies well qualified in agricultural weed control practices are employed for this purpose. Methods to control rocks and dust must be utilized during construction, and roads must be well maintained by the wind project owner during operation of the project (Fig. 11.10).

- **Storm water and erosion control:** Engineers designing roads and turbine pad sites need to consider impacts on storm water runoff during the design process and determine the best practices needed to avoid creating undue soil erosion or pooling of water in fields. Roads crossing streams, creeks, or runoff ditches should use culverts where necessary to prevent altering water flow.
- **Crop dusting:** Many farmers contract with crop dusters to apply chemicals to their fields. Crop dusters use small, highly maneuverable aircraft. Even if crop dusting occurs when wind turbines are shut down, there is still some potential for collisions with the turbines. If a crop duster attempted to fly between turbines while they are turning, the turbulence caused by rotation of the massive blades would most likely make it very difficult for the crop duster to effectively maneuver the small plane. Therefore, it is highly recommended that crop dusting be avoided in and around wind energy projects and that only ground-based spray rigs be used for applying chemicals. Another related issue that can arise during the construction of the project is the potential for construction crews to be exposed to crop dusting chemicals if crop dusters come too close to the crews and winds cause the chemicals to drift toward the construction area. Landowner lease documents should address this by requiring landowners to notify the project owner before initiating any crop dusting.

FIGURE 11.10 Many farmers see wind energy as a great way to diversify their sources of income (Source: Charles Norland, Norland Photographic Art, St. Louis, MO).

- **Tenant farmers:** Sometimes, the person farming a tract of land is not the owner of the property. Such farmers are sometimes referred to as "tenant farmers." This can cause issues when the landowner leases the land for wind development particularly if he or she does not inform the tenant farmer prior to signing a wind lease. It is customary that wind royalties go to the property owner, and not to tenants or lessors. While tenant farmers should definitely be reimbursed for any damage to crops occurring during the construction process, it would also be reasonable for the landowner and tenant farmer to make some adjustment to the farm rental agreement to account for any loss of cultivated land and the additional nuisance created by the placement of roads and turbine pad sites on the property.

11.4.2 Compatibility with Ranching

Ranching, or the raising of livestock, is very compatible with the development of wind energy. Some of the largest wind projects in the United States are entirely on ranch land. Farming and ranching can be a tough way to make a living in tight economic times or extended periods of drought. Therefore, royalties from wind energy add diversity to a rancher's income and may be the only means that allow him or her to hold onto the family farm or ranch.

Following are the issues to be aware of when developing a wind project on ranch land.

- **Gates and fencing:** In most cases, there is no need to place fences around the wind turbines or padmount transformers since it would be very difficult for livestock to damage the equipment or be harmed by the equipment, although perhaps ice shedding events could result in livestock being struck by a piece of ice large enough to do harm. Some livestock will even seek shelter from the sun in the shadow cast by the wind turbine tower. The project substation should be fenced for safety reasons, and they almost always are. It is also wise to fence a small area around each of the guy wires to any guyed meteorological towers since cattle and other livestock seem to like to scratch their backsides on the guy wires, which can shake the tower enough to disrupt accurate data collection.

  Gates and/or cattle guards may need to be inserted in existing fences on the property in order to allow the efficient movement of construction and maintenance equipment within the wind project. This is an issue that needs to be addressed within the property lease document. Most ranchers are very particular about how such gates and/or cattle guards are constructed. It is also very likely that they will be concerned about precautions taken by construction and operations and maintenance (O&M) crews to prevent livestock from escaping as well as preventing entrance to the property by trespassers. It is customary to include in wind leases provisions requiring locks on all external gates and provisions requiring that gates be closed immediately after vehicles pass through them so that livestock do not escape from pastures or wander into public roadways where they may be struck by vehicles.

- **Speed limits on the property/off-limit areas:** Wind developers and landowners should agree upon speed limits applicable to the property. This should be done to ensure the safety of wind plant personnel, the landowner, visitors, and livestock. In addition, the landowner should specify areas of the property where construction traffic is not allowed. This may include the landowner's home, where noise and dust can be a problem, or livestock-watering areas, where construction equipment may pose an added danger due to the higher concentration of livestock, including young or newborn animals. The wind project owner needs to inform its staff members and contractors about speed limits and off-limit areas and be very specific about the consequences of violating these provisions.
- **Weed control:** Good ranchers work very hard to ensure the absence of noxious weeds on their property. Weeds (such as bitterweed or hemlock) can be poisonous to grazing livestock. Vehicles entering a property free of noxious weeds can introduce seeds onto the property on their tires or frame if proper precautions are not taken, and some lease agreements may require vehicle washing prior to entering the property and treatment of any noxious weeds adjacent to areas traversed by vehicles involved in construction or maintenance of the wind project.
- **Restrictions on firearms, hunting, or fishing:** Many wind leases prohibit the wind project's staff and contractors from bringing firearms onto the property and prohibit hunting or fishing on the property. Again, wind project staff members and contractors need to be informed about such provisions and the consequences of violating them.
- **Livestock:** Just as there are people contending that wind turbine noise and shadow flicker can impact the health of humans, there are those who assert that wind turbines can have similar impacts on livestock. Some animals may be more sensitive to noise than humans, and the shadow cast by a rotating turbine blade may "spook" animals on occasion. However, there have been a few anecdotal stories or claims of significant impacts on livestock, including goats, dairy cattle, horses, and chickens. For example, a 2009 news story carried by BBC [1] discusses a Taiwanese farmer's assertion that many of his goats began to die after the installation of a small wind project nearby, attributing their death to either refusal to eat or lack of sleep.

  Stray voltage from wind farm facilities has also been alleged to adversely affect the health of livestock. Stray voltage is defined by the Institute of Electronics and Electrical Engineers as "a voltage resulting from the normal delivery and/or use of electricity (usually smaller than 10V) that may be present between two conductive surfaces that can be simultaneously contacted by members of the general public and/or their animals. Stray voltage is caused by primary and/or secondary return current, and power-system induced currents, as these currents flow through the impedance of the intended return pathway, its parallel conductive pathways, and conductive loops in close proximity to the power system. Stray voltage is not related to power system faults, and is generally not considered hazardous."

11.4.3 Compatibility with Wild Game and Hunting

Game hunting has become a big business and income source for many landowners. Hunters may pay several thousand dollars each year for the right to hunt deer, elk, dove, quail, or other wild game on someone's property. Therefore, landowners receiving revenue from hunting activities do not want it impacted by the presence of wind turbines on their property. In fact, the appraised value of some properties may be primarily tied to hunting revenues in some parts of the country (Fig. 11.11).

During the project construction period, which may last from just a few weeks to several months depending upon the size of the project, it is likely that the amount of noise, human activity, and vehicular traffic will cause some wild game to move away from the area of activity. After the construction crews and equipment go away, the impact of wind turbines on the property will not be much different than other activities requiring humans to enter the property, such as oil and gas production or ranching.

Landowners who allow hunting on their property need to address both monetary and safety issues within their lease agreements. Some wind developers may want landowners to discontinue hunting activities on the property altogether, and if the wind royalties far outweigh revenues from hunting, the landowner may agree to this. On the other hand, some leases allow hunting other than during the construction period. These leases typically require the project owner to reimburse the landowner

FIGURE 11.11 Antelope near wind turbines (Source: David Young, West Inc.).

for any lost hunting revenues during the construction period and specify communications channels and procedures that can ensure the safety of workforce and visitors to the wind project. For example, the lease may require that the landowner notify the wind project site manager when hunters will be present, and it may also require that the site manager notify the landowner when major turbine repairs are being undertaken during hunting season. Requiring that deer blinds be oriented away from any wind turbines is another common condition in wind leases that allows hunting to continue on the property.

11.4.4 Compatibility with Oil and Gas Development

Many of the best wind resource areas in the United States also have active oil and gas exploration and production. Again, wind energy development is generally going to be compatible with oil and gas development. Most wind lease agreements have provisions in them addressing oil and gas exploration and production, and hopefully, wind leases and oil or gas leases will each have reasonable terms allowing such activities to continue as long these activities do not unreasonably interfere with wind generation or oil/gas production.

For example, once the wind project is constructed, it would be unreasonable to require relocation of a turbine in order to allow for drilling of an oil well exactly where the wind turbine is located, unless the oil company is willing to pay for relocation of the turbine (Fig. 11.12). On the other hand, it would be unreasonable for the wind developer to locate a turbine so close to an existing oil or gas well that maintenance could no longer be performed on the oil or gas well. Slant drilling, or directional drilling, allows oil and gas developers to hit pockets of oil or gas without having to be located directly on top of the formation.

Assuming all parties involved including the landowner, the wind developer, the oil/gas producer, and their attorneys are reasonable, it is very likely that legal agreements can be prepared that allow both wind development and oil/gas exploration to occur on the tract with very little impact on the other activity. The joint use and maintenance of roads is one of the issues that should be addressed within such agreements.

Many states, including Texas, allow the severance of mineral rights from the surface rights. Wind energy is a surface right, so it can frequently occur that the owner of the surface rights, who stands to profit from wind development on the property, is different from the owner of the mineral rights, who stands to profit from the development of oil and gas underneath the property. Some states may have laws based on the Accommodation Doctrine, which generally states that the mineral owner can use the portions of the surface reasonably necessary for the exploration and production of minerals as long as it is done in a nonnegligent manner complying with statutory limitations and conducts these activities with due regard to the owner of the surface rights and his or her activities. Since slant drilling is available to oil and gas producers, requiring relocation of a turbine would not be reasonable.

Pipelines used for transportation of oil and gas present another issue for wind development related to oil and gas production and development. In very large wind

FIGURE 11.12 Oil well near wind turbine at West Texas wind farm (Photograph by Stephen W. Williams).

projects, it is likely that one or more such pipelines are located within the boundaries of the project. Wind developers need to identify any pipelines very early in the development process and determine who owns the pipeline and what the requirements will be to safely cross the pipeline with heavy construction equipment or perform construction in the vicinity of the pipeline. Pipeline-crossing agreements signed by the pipeline owner are likely to be required by attorneys representing lenders or investors of the wind project.

11.4.5 Compatibility with Residential Land Development

Chapters 8 and 9 of this book discuss issues such as noise, shadow flicker, and ice shedding and their potential impact on humans. Following the guidelines suggested in those chapters will reduce but not eliminate the potential issues that the proximity of turbines to occupied structures creates. The authors of this book have been approached in the past by developers of residential subdivisions asking our opinions on allowing the use of homeowner-owned or community-owned turbines in a planned subdivision. Some small turbines may be quite noisy and thus should not be installed in subdivisions unless the lots are fairly large, such as 5-acre tracts. Even in those cases, the turbine should usually be sited in the middle of the tract rather than on the exterior of the tract in order to minimize the potential for problems with neighbors. Some of the small, vertical-axis turbines available to landowners today have minimal noise levels and can be quite attractive and thus would be better candidates for use in subdivisions, but even these should be selected and sited very thoughtfully with due consideration given to one's neighbors.

Many community wind projects have been constructed in the United States as well as in European nations. These typically may include one or two medium- to large-sized turbines. Hopefully, these turbines would be sited at locations that minimize the potential for issues with property owners, but a quick Internet search can lead one to several stories about complaints associated with such projects.

With regard to the development of large wind projects, the best way to avoid significant opposition to proposed wind projects is to find large areas of land with very sparse housing. The closer one gets to communities or to clusters of homes with a wind project, the more likely they are to encounter opposition and potential lawsuits. Some of the situations that wind developers need to be aware of include the following:

- The presence of "ranchettes" in the area in which wealthy individuals whose primary residence is in the city have purchased some property in the country for the purpose of getting away from the hustle and bustle of the city. These people are unlikely to need money from turbine royalties and may look upon a wind plant as urbanization, from which they are trying to escape.
- Property purchased by a land developer for future development of residential or commercial purposes.
- Small, unleased tracts of land within the project area that may suddenly have a mobile home or ready-built home located on it. This can happen so quickly that the wind developer is unaware of the new home location until construction crews arrive on adjoining properties ready to build a road or foundation.

11.5 IMPACT OF WIND ENERGY DEVELOPMENT ON WATER RESOURCES

The needs or uses for water by wind energy projects is very small compared with traditional sources of electric generation (such as nuclear power plants, coal-fired power plants, and natural gas–fired power plants) and even some of the renewable energy sources (such as biomass-fueled power plants). Even though water usage is low, meeting these needs still needs to be considered during the project development and construction phases. Some wind energy leases include provisions allowing the project owner to drill one or more water wells on the site to obtain a source of water. Other projects may be near enough to water system pipelines of a municipality or a rural water district to connect to the pipelines and purchase water from its owner. A final option that may have to be resorted to if water cannot be obtained from wells or water systems is to have water hauled in by truck and stored in large containers.

A potential greater impact that wind energy projects can have on water resources is related to waste water and storm water runoff from the project site.

11.5.1 Water Uses During Construction of a Wind Project

During construction of a wind project, uses of water can include the following:

- Mixing concrete for turbine foundations and transformer pads
- Watering of roads for purposes of compaction or dust control
- Water supply for temporary construction offices

11.5.2 Water Uses During Operation of a Wind Project

After a wind project begins commercial operation, uses of water can include the following:

- Watering of roads for dust control
- Periodic blade washing to remove bugs or dirt
- Water supply for the operations and maintenance building(s)
- Washing maintenance equipment

11.5.3 Potential Impacts of Wind Energy Projects on Water Resources and Methods to Mitigate Such Impacts

Care must be taken during the project design and construction processes in order to minimize potential impacts to the existing water resources in an area, particularly since many US wind projects are sited in arid regions of the county or in fields used for farming or ranching. Water resource issues that should be addressed include the following:

- The preservation of wetlands and associated wildlife habitat. Wetlands, including playa lakes, which may be dry for large parts of the year, need to be identified prior to siting turbines and roads and avoided whenever feasible. Erosion control must also be used to ensure that wetlands do not fill up with silt any faster than they would normally do so.
- The potential for roads or turbine foundations to change storm water runoff. Construction of a wind energy project will require the development of a Stormwater Pollution Prevention Plan (SWPPP). The U.S. Environmental Protection Agency has a helpful guide [2] available online for the development of an SWPPP. Construction of bridges or installation of culverts is likely to be

required in some locations to avoid altering runoff patterns or creating excess erosion.

- Potential effects may include changes in runoff patterns to terraces in cultivated fields. Terraces are used by farmers to prevent erosion and rapid runoff from rainfall in cultivated fields. Changes in runoff patterns may cause terraces to be breached, resulting in rapid erosion and loss of fertile topsoil during periods of heavy rainfall.
- Control of soil erosion throughout the project site that may be caused by wind project facilities including turbine foundations, transformer pads, roads, underground electric lines, substations, and creek or stream crossings. Reseeding using native plants and grasses should be done during or shortly after project construction to prevent undue erosion.
- The potential effect of hazardous chemical spills (oil from turbines, vehicles, transformers, etc.) on water resources, particularly if it can contaminate livestock-watering sites.
- The potential for blasting during project construction to stop or to reduce the flow of water wells or natural springs. Property lease provisions should address this issue, and it is wise to test water-pumping capabilities or spring outflow before and after construction of the project.
- The potential contamination of groundwater resources by blasting agents used during construction. Perchlorate is a chemical compound used for some but not all blasting agents and explosives. If perchlorate gets into water supply, it can affect the function of the thyroid gland, which regulates the body's metabolism. Landowners may wish to include provisions in lease agreements prohibiting the use of blasting agents containing perchlorate on their property.
- Disruption of livestock by construction traffic around watering holes. Landowners may wish to include a provision in lease agreements requiring construction traffic to avoid watering holes by some specified distance.

11.6 REMOVAL OF RETIRED EQUIPMENT AND RESTORATION OF PROPERTY

When wind energy development was in its infancy in the United States during the 1980s, federal and California state tax subsidies made proposed wind projects in California an attractive investment opportunity even though many of the turbines being used were unproven. A combination of changes in tax policies, state energy regulations, and mechanical failure of turbines led owners of several projects to just walk away from them. Most of these were located in California's three large areas of early wind development—Altamont Pass (east of San Francisco Bay area), San Gorgonio Pass (near Palm Springs), and Tehachapi (north of Los Angeles, near Bakersfield)—although some projects in Hawaii also suffered from this fate.

Opponents of wind energy still frequently drag out these ghosts of the past, and it is very important that current developers of wind projects have the financial backing and mechanisms to ensure that wind project facilities will be removed following the end of a project's life and that the property will be restored as near as feasible to its original condition. Thus, it is common that wind leases include provisions requiring the project owner to post security at the point in time that the retirement cost of the project exceeds its salvage value and that the amount of security increases over time as the salvage value of turbines and other equipment declines.

11.7 BEST MANAGEMENT PRACTICES FOR DECOMMISSIONING OF WIND ENERGY PROJECTS (FROM USFWS *LAND-BASED WIND ENERGY GUIDELINES*)

The U.S. Fish and Wildlife Service's *Land-Based Wind Energy Guidelines* [3] released in March 2012 included a list of best practices for decommissioning of wind energy projects, which is shown as follows:

- Decommissioning methods should minimize new site disturbance and removal of native vegetation, to the greatest extent practicable.
- Foundations should be removed to a minimum of 3 ft below surrounding grade, and covered with soil to allow adequate root penetration for native plants, and so that subsurface structures do not substantially disrupt ground water movements. Three feet is typically adequate for agricultural lands.
- If topsoils are removed during decommissioning, they should be stockpiled and used as topsoil when restoring plant communities. Once decommissioning activity is complete, topsoils should be restored to assist in establishing and maintaining pre-construction native plant communities to the extent possible, consistent with landowner objectives.
- Soil should be stabilized and re-vegetated with native plants appropriate for the soil conditions and adjacent habitat, and of local seed sources where feasible, consistent with landowner objectives.
- Surface water flows should be restored to pre-disturbance conditions, including removal of stream crossings, roads, and pads, consistent with storm water management objectives and requirements.
- Surveys should be conducted by qualified experts to detect populations of invasive species, and comprehensive approaches to preventing and controlling invasive species should be implemented and maintained as long as necessary.
- Overhead pole lines that are no longer needed should be removed.
- After decommissioning, erosion control measures should be installed in all disturbance areas where potential for erosion exists, consistent with storm water management objectives and requirements.
- Fencing should be removed unless the landowner will be utilizing the fence.
- Petroleum product leaks and chemical releases should be remediated prior to completion of decommissioning.

REFERENCES

[1] BBC News. 2009. "Wind farm kills Taiwanese goats" news story. http://news.bbc.co.uk/2/hi/asia-pacific/8060969.stm, posted May 21, 2009.

[2] U.S. Environmental Protection Agency. *Developing Your Stormwater Pollution Prevention Plan: A Guide for Construction Sites*. Washington, DC: U.S. Environmental Protection Agency; 2007.

[3] U.S. Fish and Wildlife Service. *Land-Based Wind Energy Guidelines*. Washington, DC: U.S. Fish and Wildlife Service; 2012.

12

IMPACT OF WIND ENERGY ON THE ELECTRIC GRID

12.1 OVERVIEW OF WIND INTEGRATION ISSUES

One of the biggest challenges for greater reliance on wind energy to meet the growing demand for electricity in the United States and most other nations is the intermittent nature of wind generation. Storage of large amounts of electricity is not cost-effective in most locations or circumstances, so for the most part, electricity is consumed at the same time that is produced, whether produced from coal plants, nuclear power plants, natural gas plants, or wind turbines.

It takes wind speeds of about 3–4 m/s, or 7–9 miles/h, for most utility-scale wind turbines to begin producing electricity. Even in the windiest parts of the United States, there are times when the wind speeds are below these levels and levels of wind generation are low. In addition, the locations of the most cost-effective wind farms do not always match up with the locations in which power is needed by the electric grid. Therefore, a diversity of generation resources, and not just wind energy alone, is needed to meet the electricity needs of the United States. While there are more than enough windy land and offshore sites to produce far more electricity than the nation consumes, the timing and the location of wind generation would prevent it from meeting all of the nation's demand for electricity. Just as diversification of investments in the stock market is highly recommended by almost all credible financial advisors, diversification of electric-generation resources is a wise idea, including increasing the use of renewable resources (such as wind, solar, and

Wind Energy Essentials: Societal, Economic, and Environmental Impacts, First Edition.
Richard P. Walker and Andrew Swift.

biomass energy). A nation cannot rely solely on nuclear power, as demonstrated by the recent accident at the Fukushima nuclear power plant in Japan. Natural gas prices are highly volatile, and supplies can become limited fairly quickly, so reliance on natural gas alone is not reasonable. While the United States has large reserves of coal, the emissions issues associated with its use pose formidable barriers to permitting of new plants.

In 2006, President Bush highlighted the need for greater energy efficiency and a more diversified energy portfolio. He noted that wind resources in the United States could provide up to 20% of America's electricity needs, and this subsequently led to an effort by the Department of Energy, the National Renewable Energy Laboratory, other governmental agencies, and industry representatives to study the feasibility of reaching such a high level of wind penetration and the benefits, potential problems, and barriers to do so. In 2008, the Department of Energy released a report titled *20% Wind Energy by 2030: Increasing Wind Energy's Contribution to U.S. Electric Supply* [1].

While the report identified many challenges to reaching a level of 20%, it also identified large benefits of doing so, including reduced emissions from electric generation, reduced water use associated with electric generation, and lower natural gas use, which can translate to lower natural gas prices.

The rapid growth of wind power has already created challenges for the nation's electric grid and its system planners and operators, although these challenges are not altogether different than ones they have dealt with for many decades. For example, the amount of electricity used by customers of electric utilities has always fluctuated, and system operators have effectively dealt with such fluctuation for the vast majority of the time. Cold fronts may blow into an area causing large spikes in electricity demand as electric heaters are turned up. As discussed in previous chapters, natural gas–fired generators are the primary source used to meet fluctuations in customer load since many of them are capable of ramping-up or ramping-down very quickly. Thus, it does not require a large leap in technology or operating protocols for utilities to ramp-up or ramp-down gas-fired generators in order to react to fluctuations in wind energy production.

However, large-scale integration of wind generation into the electric grid involves much more than ensuring that there are enough generation resources online to meet customer demand for electricity. Issues such as flicker, harmonics, reactive power requirements, and low-voltage ride-through (LVRT) have to be addressed in order to ensure the continued high reliability of our nation's electric grid.

12.2 THE ELECTRIC GRID

The first commercial power plant in the United States was constructed in 1882. This and other early plants generally only provided electricity to the community immediately surrounding it. When the plant was shut down due to mechanical problems or for scheduled maintenance, the community's lights went out. This led to communities and/or owners of the power plants agreeing to interconnect the power plants

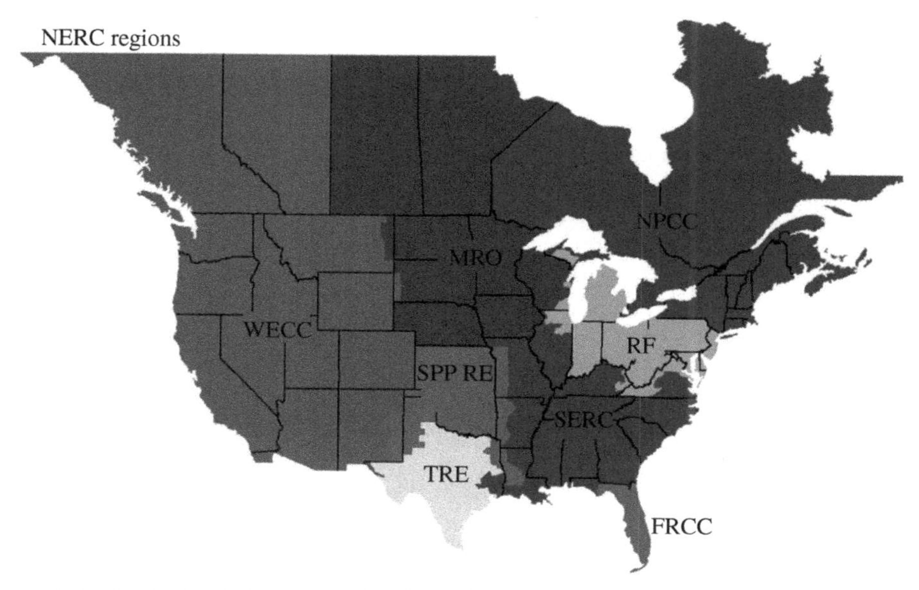

FIGURE 12.1 North American Electric Reliability Corporation (NERC) regions in the United States and Canada (Source: This image from the North American Electric Reliability Corporation's website is the property of the North American Electric Reliability Corporation and is available at http://www.nerc.com/AboutNERC/keyplayers/Documents/NERC_Regions_Color.jpg. This content may not be reproduced in whole or any part without the prior express written permission of the North American Electric Reliability Corporation [2]).

and the communities they served with electric lines, leading to the first "electric grids," thus providing backup sources of power when one of the plants had to be shut down. Initially, such grids may have only included two or three plants or communities, but over time, the grids began to grow larger and larger. With the advent of the Rural Electrification Administration as one of the New Deal agencies in 1935, the electric grid was extended to rural areas that private or investor-owned companies had largely ignored due to low population density.

Figure 12.1 shows the various electricity regions covering the majority of the United States and Canada, or NERC regions. NERC stands for the North American Electric Reliability Corporation, which is a not-for-profit international regulatory authority whose mission is to ensure the reliability of the bulk power system in North America. These regions may also be depicted as three separate "interconnections"—the Western interconnection (shown as WECC in green), the Electric Reliability Council of Texas (ERCOT) or Texas Reliability Entity (shown as TRE in yellow), and the Eastern interconnection, which includes all of the remaining areas. These regional interconnections, or Wide-Area Synchronous Grids, all operate at an average frequency of 60 Hz. Due to geographic reasons, Hawaii and Alaska both still operate as multiple smaller grids. Within a regional interconnection, all generators operate synchronously (i.e., the same electric frequency and phase), but across interconnections, the generators do not operate on the same frequency or phase. High-voltage direct current facilities are used to connect the regional interconnections.

Notice that within the Eastern Interconnection, there are several smaller regional reliability organizations (such as the Southwest Power Pool or the Midwest Reliability Organization). System planners, system operators, and management of electric companies within these regions work closely together to ensure the continued reliability of the electric grid. While reliability (i.e., keeping the lights on) and system stability are their highest priorities, they also work together to try to meet the demand for electricity as economically as possible using a process referred to as "economic dispatch." Economic dispatch refers to the operation of generation facilities to produce energy at the lowest cost while also ensuring reliability of the electric grid, thus recognizing any operational limits of generation and transmission facilities. Concerns about emissions from electric generation have increased over time and will probably continue to do so, resulting in new and more stringent regulations being enacted by state or federal governments. As a result, regional organizations are already placing greater emphasis on managing such emissions and evaluating different mixes of electric generation that would comply with or exceed any future regulations being considered or those that are already being phased-in.

12.3 WIND PENETRATION

Wind penetration refers to the percentage of total power or energy being produced that comes from wind generation. It may be expressed as the wind penetration level for a nation, state, or electricity region. There are also differing types of wind penetration levels that can be discussed, so it is important to be able to distinguish between them. Such references to wind penetration levels may include the following.

12.3.1 Energy Penetration

Energy penetration refers to the ratio of energy (as measured in kWh, MWh, or GWh) produced from wind energy to total energy produced for a given period of time (such as a month or year), usually expressed as a percentage. The *20% Wind Energy by 2030* report refers to energy penetration, or 20% of all electric energy (as measured in kilowatt-hours or megawatt-hours) consumed in the United States being produced by wind energy.

12.3.2 Capacity Penetration

This usually refers to the cumulative nameplate capacity of wind plants (as measured in megawatts or gigawatts) compared with the peak demand for electricity. Alternatively, some may refer to capacity penetration as the ratio of cumulative nameplate wind plant capacity to the cumulative generation capacity of all generating resources.

12.3.3 Instantaneous Penetration

This refers to the ratio of wind power generation (as measured in megawatts or gigawatts) at a specific point in time to the total electric load (also measured in megawatts or gigawatts) at that same point in time, again usually expressed as a percentage.

12.3.4 Existing Wind Penetration Levels

Statistics from the American Wind Energy Association indicate that wind energy provided 2.3% of the kWh of electricity consumed in the United States during 2010, thus the energy penetration level of the United States was 2.3%. While Texas leads the nation in installed capacity of wind generation, Iowa leads the nation in terms of energy penetration levels, producing greater than 15% of its electricity from wind energy. North Dakota and Minnesota also have energy penetration levels of 10% or greater, and eleven states have energy penetration levels of 5 % or greater.

During 2010, several European nations had much higher levels of wind energy penetration than the United States did, including the following:

- Denmark: 21%
- Spain: 12%
- Portugal: 9%
- Ireland: 8%
- Germany: 7%

With regard to instantaneous penetration levels, the ERCOT region has had several instances in which wind energy accounted for more than 20% of electricity production, while in regions of Europe, there have been times in which wind generation was meeting almost all load requirements.

12.4 ISSUES THAT ARISE DUE TO INTEGRATION OF WIND ENERGY

Following are some of the issues that arise from large-scale integration of wind energy into the electric grid. These issues need to be thoroughly understood and addressed by both the electric utility industry and wind energy industry if wind energy is going to continue growing as rapidly as it has in recent years.

12.4.1 Wind Energy's Intermittent Nature

As depicted in Figure 12.2, wind energy generation is highly variable, even when many diversely located projects in a region are included in an analysis. Utility system operators and planners have traditionally dealt with dispatchable generation resources (such as coal-fired plants, nuclear power plants, or natural gas–fired

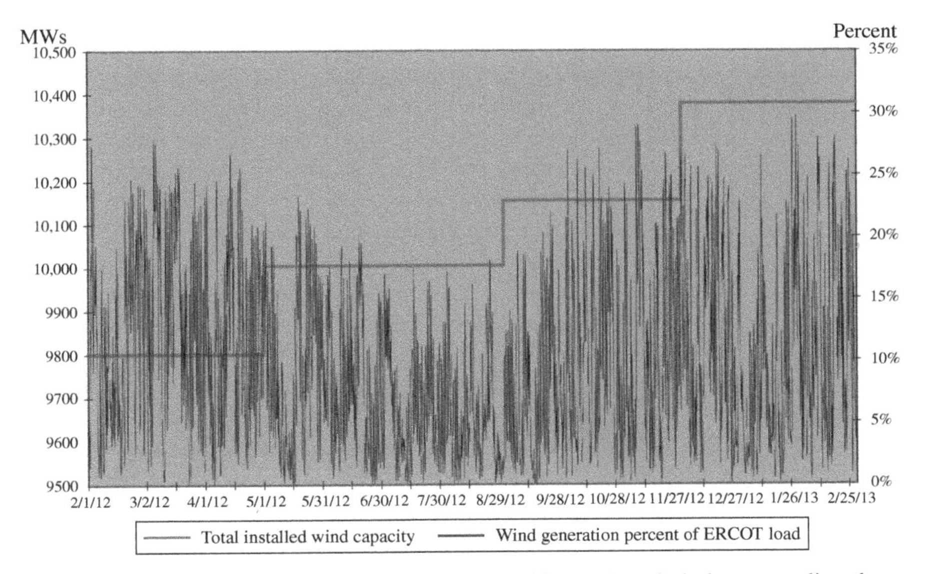

FIGURE 12.2 Wind Generation in ERCOT over a 12-month period; the orange line shows total installed wind capacity in MW, while the purple line shows the percentage of ERCOT's total electric load that was being met with wind energy (Source: ERCOT [3]).

plants). Even hydroelectric plants can be dispatched by controlling the amount of water released through the plant's turbines. So getting comfortable with a resource over which utilities have less control does take some time and experience, and they typically want to ease their way into this rather than suddenly having several hundred megawatts of wind generation added to their system in a very short time. One can see in Figure 12.2 that there are times in which wind generation provides in excess of 25% of the electricity being consumed in ERCOT, while at other times, it may be as low as 1 or 2%.

As previously mentioned, utility system operators have been dealing with fluctuating load for many decades, so a fluctuating resource is not altogether different. Since the variability of wind generation and the demand for electricity by the utility's customers (or electric load) are statistically uncorrelated, the combination of electric load less wind generation (sometimes referred to as "net load") will only have slightly more variability than if no wind generation existed on the system.

Predictability is the key to managing wind power's variability, making wind forecasting a valuable tool for utility system operators as they deal with large-scale wind integration on their systems. Wind forecasts let system operators know the estimated amount of wind generation that is expected to occur the following day, within 12 h, within 6 h, the following hour, etc., allowing them to then improve plans for meeting the upcoming demand for electricity. Several studies have been performed showing the value of wind forecasting, and many such studies can be found at the website of the Utility Variable-Generation Integration Group (UVIG) http://variablegen.org.

12.4.2 Remote Locations of Wind Resources/Transmission Congestion

Many of the best wind resources in the United States lie in rural areas where there is little demand for electricity. As shown in Figure 12.3, much of the nation's prime wind resource is located from west Texas all the way to North Dakota and Montana, in areas where there are very few large cities (i.e., 500,000 or greater in population). Figure 12.4 shows the US population density by county, and if one were to overlay the two maps, many of the areas with the highest wind also have some of the lowest population density. The electric lines in such regions have historically been sized to deliver energy to small communities in these rural areas and in most cases were not designed to deliver several hundred MW of wind power from the rural area to an urban area.

Wind developers tend to look for areas having high average annual wind speeds that are also near to existing transmission lines. Since most transmission lines in these rural yet windy areas were not designed to move large volumes of wind energy, many of the transmission lines in the windiest areas quickly reached their available capacity as wind energy began to grow rapidly beginning in 2005. This has led to transmission congestion, which can prevent the use of the most economic or most environmentally friendly generation resources and which may also jeopardize system reliability.

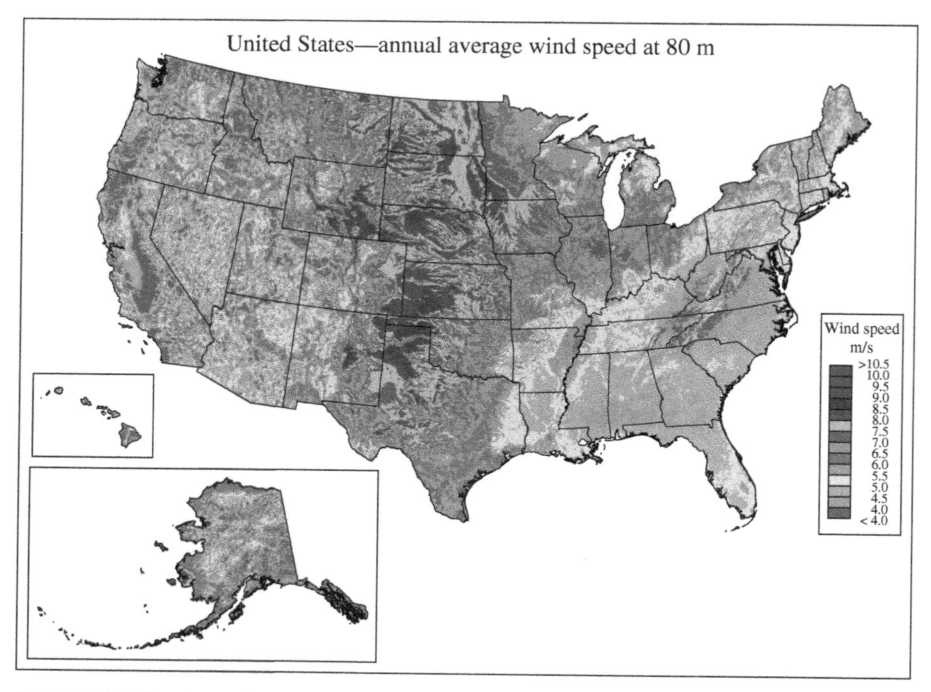

FIGURE 12.3 Annual average wind speed at 80 M aboveground (Source: NREL and AWS Truepower [4]).

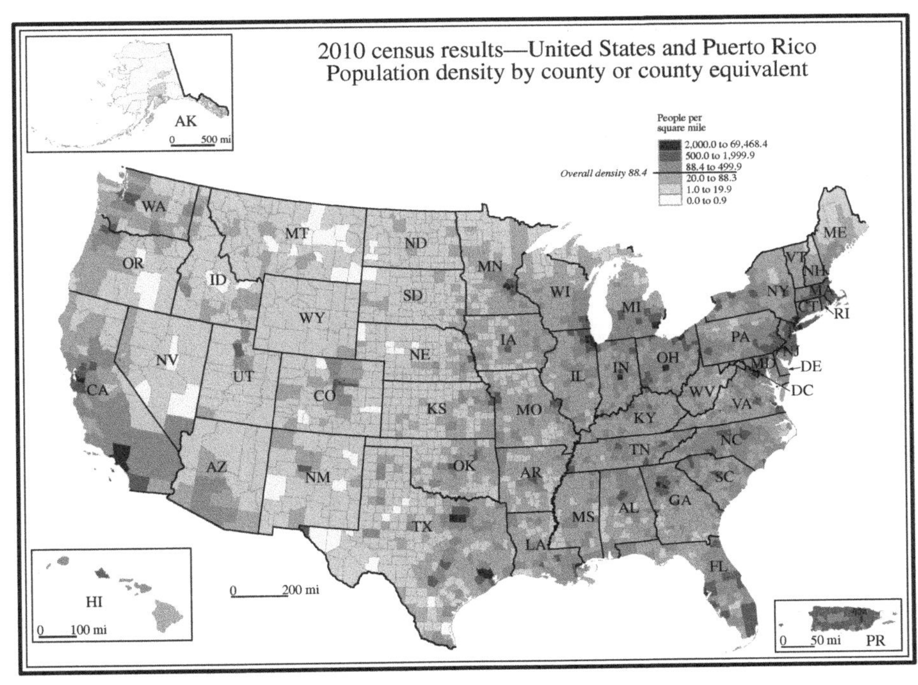

FIGURE 12.4 US population density by county (Source: U.S. Census Bureau [5]).

In areas or times in which transmission congestion occurs, grid operators have had to curtail production from wind plants, or in other words, they have required that output from the wind plants be reduced in order to prevent transmission lines from becoming overloaded. Electric current running through a transmission line causes the conductors (typically made of aluminum and steel) to heat up and to sag closer to the ground. The National Electric Safety Code specifies minimum clearances between the ground and the transmission lines, depending upon their voltage, so grid operators must pay close attention to this criterion and take the necessary precautions to prevent the lines from overloading and overheating.

When a wind plant is curtailed, its owner will usually suffer some financial harm. This can be in the form of the following:

- Lost revenues from the sale of electricity
- Lost revenues from the sale of renewable energy credits
- Lost Federal production tax credits
- Penalties for underperformance included in some power purchase agreements

The portion of west Texas lying within the ERCOT reliability region includes an area around Sweetwater, Texas, that has more than 6000 MW of wind generation within 75 miles of Sweetwater. Unfortunately, this led to significant curtailment of wind plants in

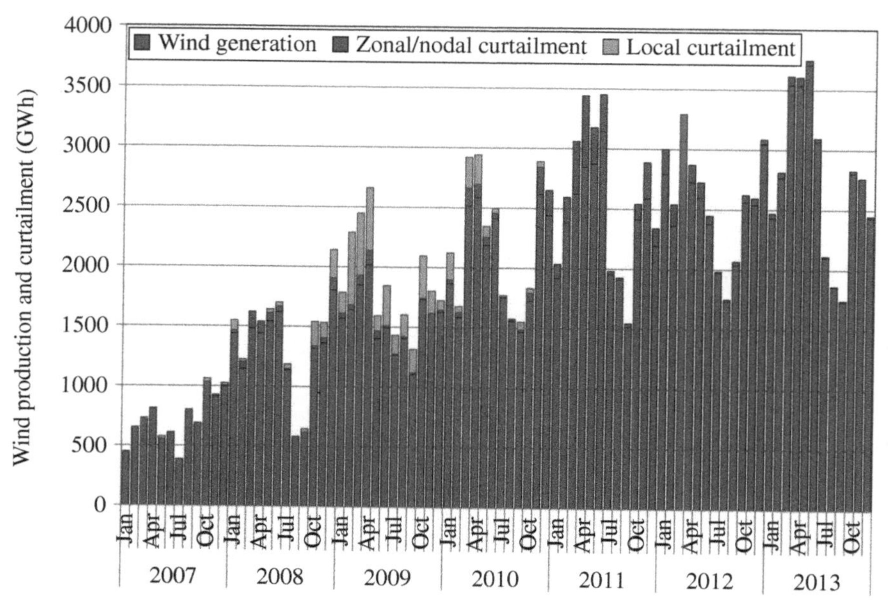

FIGURE 12.5 Curtailment of wind generation in ERCOT 2007–2013 (Source: Developed by Potomac Economics, the ERCOT Independent Market Monitor [6]).

that area beginning in 2007 (see Fig. 12.5). This reached a peak in 2009, when more than 17% of wind generation that could have been produced was not produced due to curtailment necessitated by transmission congestion. The good news is that such curtailments resulted in Texas moving forward with construction of about \$6 billion of new 345-kV transmission lines to areas known as Competitive Renewable Energy Zones, greatly reducing the amount of curtailments of wind generation in 2013 and beyond.

12.4.3 Need for Balancing Area Cooperation or Consolidation

The NERC defines a balancing authority as "the responsible entity that integrates resource plans ahead of time, maintains load-interchange-generation balance within a Balancing Authority Area, and supports Interconnection frequency in real time." Within the reliability regions, one or more entities serve as balancing authorities. Usually, these are large, vertically integrated utility companies, but in some cases (such as the ERCOT region), the regional reliability organization serves as the balancing authority.

Consolidation of balancing authorities, such as has been done in ERCOT, tends to benefit the addition of large amounts of wind generation. In lieu of consolidation, cooperation of adjoining balancing authorities is essential for large-scale wind development. For example, a balancing authority covering a large area of rural but very windy land may not be able to provide balancing for large volumes of wind energy, whereas consolidation of that area with other areas containing higher volumes of electric load and generation will more easily accommodate the large volume of variable wind energy.

Benefits of balancing area consolidation or cooperation include the following:

- Aggregation of several diverse renewable resources over larger geographic areas reduces the overall variability caused by the individual projects.
- Load aggregation reduces the overall variability of the load.
- Aggregation of nonrenewable generation resources provides access to more balancing resources, such as natural gas–fired plants.

12.4.4 Effect on Operating Reserves, Regulation Service, Ramping, and Load Following

Electric companies or electric reliability regions must keep some level of generation in reserve in order to meet changes in customer demand or to ensure the stability of system frequency when a generating unit unexpectedly trips off-line. The addition of large-scale wind generation in a region can increase the amount of reserves needed and may increase the utility's costs for regulation, ramping, load following, and unit commitment. Not only is it important to understand physical requirements such as these for wind integration, it is also important to determine the increased costs of these items when caused by the addition of wind generation.

Some of the key concepts to remember in this regard include the following:

- **Operating reserve** is the amount of generating capability above firm system demand that is required to provide for regulation, load forecasting error, forced outages, scheduled outages, and local area protection. Reserves are analogous to an insurance policy to help when the unexpected occurs. Operating reserves can consist of spinning reserve and nonspinning reserve.
- **Spinning reserve** refers to online reserve capacity that is synchronized to the electric grid and that can rapidly meet electric demand within seconds or minutes.
- **Nonspinning reserve** refers to off-line generation capacity that can be ramped-up and synchronized to the grid within some specified time period but may also include interruptible load that can be removed from the system in a specified time. Interruptible load refers to electricity normally being consumed by an industrial customer of the utility who has agreed to reduce some or all of its electricity consumption during emergency situations in exchange for a lower electricity price during periods of normal operation. Removing or reducing the electric load of industrial customers served under interruptible or curtailable service tariffs means that the utility will have greater reserves to meet the needs of its other customers.

- **Regulation or regulation service** is an ancillary transmission service one balancing authority may provide to another balancing authority to correct deviations in the frequency of the electric grid and to balance out transactions between neighboring systems.
- **Ramping** refers to changing the output of an electric generator, either up or down, to accommodate fluctuations in load or in wind generation output.
- **Ramp rate** refers to how quickly the generator can change output and is normally expressed in megawatts per minute. Some generators (such as state-of-the-art natural gas–fired generators) have high ramp rates, whereas coal-fired plants and nuclear power plants may have low ramp rates. Therefore, regions with higher percentages of natural gas–fired generation may be able to better accommodate higher levels of wind penetration than other regions with greater reliance on coal or nuclear power.
- **Load following** refers to the constant adjustment of generation levels as demand for electricity fluctuates throughout the day. Natural gas generators predominately perform this function.
- **Unit commitment** is a process used by system operators while planning to meet the upcoming day's projected electricity requirements. Since some generators (such as coal-fired plants) may take from several hours to a day or more to bring them back into service after being shut down, the system operator has to determine when to recommit such units well in advance of when they are finally producing electricity. Using unit commitment, they identify a mix of generation that can reliably meet all needs while also trying to minimize costs.

Since the addition of large-scale wind generation can increase the variability of net load (i.e., total load minus wind generation), this may necessitate an increase in the amount of operating reserves needed for regulation, ramping, and load following. This increase represents an added cost, although several studies have indicated that these added costs are not unreasonable when compared with the average cost of electric generation. In fact, such costs may be far outweighed by the benefits of wind energy (such as emissions reductions, stabilization of utility fuel costs, easing price pressures on natural gas, and conservation of water).

In one study evaluating the significance of such impacts, it was determined that the addition of 1500 MW of wind generation, representing about 15% of the system's peak load, only increased regulation requirements by 8 MW in order to maintain the same level of NERC control performance standards. A second similar study of another region estimated that the addition of 3300 MW of wind, or about 10% of the regional peak load, increased regulation requirements by 36 MW.

According to the U.S. Department of Energy's (DOE's) *2010 Wind Technologies Market Report* [7], "Integrating wind energy into power systems is manageable, but not free of costs, and system operators are implementing methods to accommodate increased penetration. Recent studies show that wind energy integration costs are below $10/MWh – and often below $5/MWh – for wind power capacity penetrations up to about 40% of the peak load of the system in which the wind power is delivered." Note that power prices typically range from $30 to $100/MWh, so these types of wind integration costs are not disproportionate and, in most cases, will not significantly discourage the purchase of wind energy.

12.4.5 Capacity Value of Wind Energy

The likelihood or probability of a generating resource being able to help meet part of a utility's peak load multiplied by its nameplate capacity rating determines the capacity value of that resource. Since nuclear power plants, coal-fired plants, and natural gas–fired plants do periodically break down, even they would not have capacity values of 100% times their nameplate capacity ratings.

The variable nature of wind energy makes the determination of a wind plant's capacity value a much more difficult and uncertain process. While the primary value of wind plants is customarily viewed as its use as an energy resource based on its ability to displace the use of fossil fuels, there is some capacity value that should also be assigned to wind plants. System operators or planners unfamiliar with wind energy may say things like "since the wind does not always blow, wind energy has no capacity value." Advocates of wind energy might then point out that nuclear power plants and coal-fired power plants do not always work, so perhaps they too would have no capacity value if 100% certainty is a requirement to be assigned capacity value. Another common misconception about wind energy is that the addition of wind plants to a utility's system requires the addition of a comparable amount of conventional generation as a backup to the wind. This is not the case since wind generation is primarily used as an energy resource.

While determining the capacity value of wind generation can be complicated, there are several studies that have already been performed demonstrating methods for doing so. Such studies have shown wind energy capacity values as high as 40% of nameplate capacity and as little as 5% of nameplate capacity, although the majority of studies come in somewhere between 10% and 20% of nameplate capacity. The value is highly dependent upon the correlation between the electric system's load profile and the output of a wind plant. This can vary significantly by region and even within regions. Increased diversity of wind plant locations within a region typically will increase the average capacity value of wind generation within that region. As an example, winds in west Texas may be greatest during early morning or evening periods, while winds along the Texas Gulf coast or offshore may be greater during the middle part of the day. Just as diversification of generation resources is a good idea, so is diversification of wind plant locations.

In regions of the country with very hot summertime temperatures and significant electric load from air-conditioning, many utilities experience their annual peak loads during July or August and somewhere between the hours of 4 P.M. and 7 P.M. In some of these same areas, wind speeds tend to be lower during these months and tends to die down during these hours, meaning that wind may not contribute significantly to meeting the utilities' peak loads, as depicted in Figure 12.6.

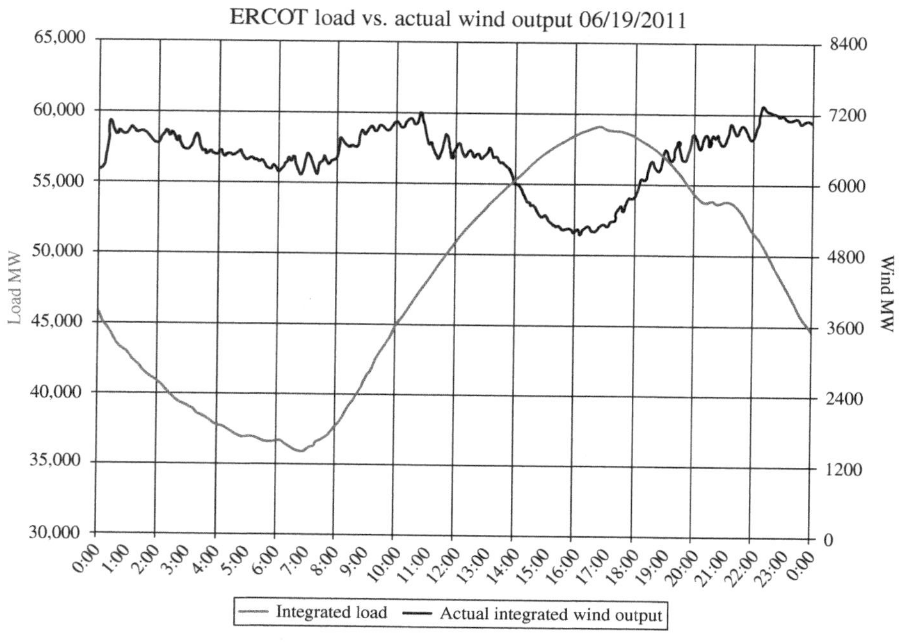

FIGURE 12.6 Wind output versus electric load in ERCOT, June 19, 2011 (Source: ERCOT [8]).

12.5 CAPABILITY OF WIND TURBINES TO MEET APPLICABLE STANDARDS FOR GRID INTERCONNECTION

The electrical and mechanical systems making up today's utility-scale wind turbines are significantly different from the traditional generators utilities have used for decades, but acceptance of wind generation as a routine part of the generation mix continues to improve through utilities' actual experience in dealing with wind plants and better communication between the utility industry, grid operators, wind turbine manufacturers, and wind plant owners. When wind plants were small and wind generation was a negligible percentage of generation on an electric grid, wind generators did not always have to meet the same types of standards for interconnection to the grid that a large, fossil-fueled or nuclear power plant would have to meet. Now that many wind plants exceed several hundred megawatts and since wind generation is becoming a substantial percentage of all new electric generation added to the grid, utilities and grid operators are paying much more attention to the electrical characteristics of wind turbines. A discussion of some of the more important criteria relative to interconnection of wind turbines and wind plants follows in the next section of this chapter.

Following are some of the most important characteristics of wind turbines and wind plants that will impact their ability to be effectively integrated into the electrical grid on a large scale without significantly affecting the grid's reliability.

12.5.1 Power Quality

Electrical equipment used by utilities and their customers are designed to operate within a range of voltages, usually within ±5% of the nominal voltage. Deviation from this may cause light bulbs to dim, motors to overheat or be damaged, or electronic equipment not to work. Therefore, the connection of large wind plants to the electric grid needs to be done without degrading the existing power quality of the grid. Utilities strive to keep voltage and frequency levels as stable as possible. Thus, it is very important that wind turbines and wind plants be designed to provide "clean" power to the grid, minimizing voltage and current distortions created by harmonics, and to prevent wind plant self-excitation, which may occur due to the loss of transmission service to the plant. While older turbines still in service may not meet current standards, state-of-the-art variable-speed turbines with power electronics allow for much more flexible operation and smoother regulation.

12.5.2 Low-Voltage Ride-Through

Disruptions or faults do periodically occur on the transmission grid that can cause a dip in voltage for some period of time. If wind turbines are not "electrically robust," then they may also trip off when a fault occurs, thus compounding the problem on the grid. Low-Voltage Ride-Through (LVRT) capabilities of most wind turbines have improved significantly in recent years, and capabilities of state-of-the-art wind turbines are now comparable with that of conventional generators.

12.5.3 Reactive Power Control

Electric power can be divided into real and reactive components. While reactive power is needed in alternating current electric systems to support the transfer of real power, too much reactive power can be problematic. Reactive power consumes transmission and generation resources and increases electrical losses. The relationship between real power and reactive power can be expressed as a term called "power factor." Wind plants are required by grid operators or transmission owners to maintain a certain level of power factor, usually as determined from the results of an interconnection study. Some wind plants may be interconnected at remote locations where the transmission system is relatively weak, thus creating an even greater need for the wind plant to meet criteria determined in the interconnection study. Since various turbine models perform differently, requirements related to reactive power control and power factor can impact turbine selection as well as design of the overall wind plant. Some turbine models have the ability to dynamically control reactive power, while others do not. In cases where the turbines themselves cannot meet the necessary requirements, other reactive power–controlling devices (such as capacitors, reactors, or static VAR compensators) can be added to the wind plant. Note that VAR is an acronym for "volt-ampere reactive," an electrical unit that is a counterpart to a watt.

12.5.4 Communication with Grid Operators

Grid operators need to have the ability to communicate with large wind plants at all hours, just as they do with large fossil-fueled or nuclear plants. As the number of wind plant owners/operators has increased, and as their technical sophistication or capabilities have improved, many wind plants are no longer routinely staffed 24×7; many of them are instead remotely monitored and controlled by the project owner or turbine vendor from a centralized location allowing site personnel to go home in evenings or on weekends. As transmission congestion has become a significant issue for grid operators and wind project owners/operators, wind plants have to be able to respond quickly to instructions from the grid operator.

12.5.5 Wind Forecasting

As discussed earlier in this chapter while discussing wind energy's intermittent nature, predictability is critical to managing wind power's variability. Wind forecasting is one of the most valuable tools a utility system operator can have, and some reliability areas are now requiring wind plant owners to provide forecasts of their expected energy production based on increasingly sophisticated weather and prediction models. Without such forecasts, system operators have to plan conservatively, leading to higher reserve margins and higher costs.

In addition, some dramatic weather events (such as severe cold fronts entering a region) may cause large increases in customer load and large decreases in wind production. This situation may be further complicated by forced outages at fossil-fueled plants where the severe cold may cause pipes in cooling systems to freeze and burst or coal piles to freeze up in conditions that may also include snow, sleet, or freezing rain. Transmission outages may also increase in such conditions. The ERCOT system, which includes a large amount of wind generation, has experienced Emergency Electric Curtailment Plan events in the past due to the combination of such occurrences, leading to curtailment or interruption of customer load. Accurate forecasting of both electric load and wind generation can help to mitigate these types of events.

Of the many wind integration studies that have now been performed, virtually all of them point to the need for accurate forecasts of wind generation as a means to minimize the costs of wind integration and to mitigate the potential for wind generation to reduce the reliability and stability of the electric grid.

12.6 CONCLUSIONS

Addition of one or more large wind plants in a region does create new challenges for grid operators and can result in additional costs to regional balancing authorities. However, these complications of high levels of wind penetration can be effectively dealt with through moderate changes to the operating protocols of electric utilities and grid operators and through the use of state-of-the-art wind turbine technology.

Keys to ensuring successful integration of large-scale wind generation include use of wind forecasting and continued dialogue between electric utilities, grid operators, turbine vendors, and wind plant owners/operators so that all parties understand each other's issues and can work together to reach the most cost-effective solutions.

It is only fair that utilities or balancing authorities incurring new costs associated with wind integration be reimbursed for such costs, either from the wind project owner or the purchaser of the wind energy, but it is important that utilities or balancing authorities have studies showing the true costs of wind integration and not arbitrarily impose fees or tariffs with no basis behind them.

REFERENCES

[1] U.S. Department of Energy, Energy Efficiency and Renewable Energy. 20% Wind Energy by 2030: Increasing Wind Energy's Contribution to U.S. Electricity Supply, DOE/GO-102008-2567. Washington, DC: U.S. Department of Energy, Energy Efficiency and Renewable Energy; 2008.

[2] North American Electric Reliability Corporation (NERC). Available at http://www.nerc.com/AboutNERC/keyplayers/Documents/NERC_Regions_Color.jpg.Accessed October 31, 2014.

[3] Electric Reliability Council of Texas. ERCOT's February 2013 Monthly Status Report from the ERCOT System Planning Department, Austin TX; 2013.

[4] AWS Truepower. Study for the National Renewable Energy Laboratory's Wind Powering America program, Golden CO: 2010.

[5] U.S. Census Bureau. Available at http://census.gov/2010census/data/. Accessed August 13, 2012.

[6] Potomac Economics. Potomac Economics the ERCOT Independent Market Monitor, Austin; 2014.

[7] Wiser R, Bolinger M. *2010 Wind Technologies Market Report*. Berkeley: U.S. Department of Energy, Environmental Energy Technologies Division; 2011.

[8] Electric Reliability Council of Texas. Report of the ERCOT Grid Operations and Planning Committee to the ERCOT Board of Directors, ERCOT, Austin; July 19, 2011.

13

NON-WIND RENEWABLE ENERGY SOURCES USED TO MEET ELECTRICITY NEEDS

13.1 COMPARING SOURCES OF ELECTRIC GENERATION: ATTRIBUTES THAT WILL BE CONSIDERED

All sources of electric generation have positive attributes and negative attributes, and they all seem to have both supporters and detractors. In the following two chapters, we will discuss renewable energy sources (other than wind energy) used for electricity generation and nonrenewable sources of electric generation. Factors relevant to environmental and economic impacts will be considered in hopes of presenting an "apples-to-apples" comparison. The various facts, factors, impacts, and/or issues that will be evaluated include the following:

- Installed capacity and net generation in the United States/average capacity factors
- Basics of the technology/efficiency
- Primary operational mode within the electric grid
- Capital cost/variable costs including fuel/levelized cost of energy
- Availability of fuel source and price variability
- Emissions/effect on human health
- Production of solid and/or hazardous wastes and disposal options
- Cooling water requirements/other water requirements

Wind Energy Essentials: Societal, Economic, and Environmental Impacts, First Edition.
Richard P. Walker and Andrew Swift.

- Impacts to surface water and groundwater quality
- Land use requirements/compatibility with other uses
- Workforce health and safety

13.2 RENEWABLE ENERGY GENERATION RESOURCES IN THE UNITED STATES

The U.S. Energy Information Administration compiles data regarding electricity generation, as well as other types of energy use such as transportation fuels, in their *Annual Energy Review* [1]. Table 13.1 shows electric generation capacity and annual production data from renewable and nonrenewable resources for calendar year 2012. Renewable energy accounted for only 12.2% of energy produced, with hydroelectric making up the majority of that.

13.3 HYDROELECTRIC GENERATION

13.3.1 Overview

In the 1940s, hydroelectric power accounted for about a third of electric generation in the United States. Since that time, the percentage contribution of hydroelectric power has progressively declined as electric usage has risen dramatically, yet very few additions of hydroelectric generation have been made. Hydroelectric generation is considered a renewable resource, although many environmentalists may oppose the construction of new hydroelectric facilities due to the large area of the lake impounded by the dam, the resulting disruption of animal habitat, and the large volume of water lost to evaporation (Fig. 13.1).

Some hydroelectric facilities can provide a dispatchable resource for electric system operators, meaning they have the ability to control the production of electricity by controlling the amount of water released to flow through the turbines. Droughts or floods may disrupt this ability to some extent. In other situations, regulatory authorities may also place restrictions on the electric system operator's ability to dispatch the resource due to other uses of the water, such as minimum flow requirements in rivers or the desire to keep lakes at a certain level. Run-of-river hydroelectric facilities are favored by environmentalists, as they do not require damming of a river or stream.

- Installed capacity: 78.7 GW (end of 2012).
- Electric generation: 276,240 GWh (calendar year 2012).
- Capacity factor: 40%.
- Percent of US electric production: 6.8%.
- Land requirements: Highly variable, 50–1000 acres/MW.
- Water use: Evaporation of water in reservoirs at US hydroelectric plants has been estimated to average 18,000 gallons/MWh produced.

TABLE 13.1 Electric Generation Capacity and Annual Production During Year 2012[a]

	Coal	Natural gas	Nuclear	Wind	Solar thermal and PV	Wood and wood-derived fuels	Other biomass	Geothermal	Hydro-electric	Total
2011 year-end capacity (MW)	317,640	415,191	101,419	45,676	1,524	7,077	4,536	2,409	78,652	1,051,251
2012 year-end capacity (MW)	309,680	422,364	101,885	59,075	3,170	7,508	4,811	2,592	78,738	1,063,033
Average 2012 capacity (MW)	313,660	418,778	101,652	52,376	2,347	7,293	4,674	2,501	78,695	1,057,142
2012 generation (GWh)	1,514,043	1,225,894	769,331	140,822	4,327	37,799	19,823	15,562	276,240	4,047,765
Average annual capacity factor (%)	55.0	33.3	86.2	30.6	21.0	59.0	48.3	70.9	40.0	43.6

[a]Source: R. Walker using data from USEIA Annual Energy Review; public domain information [1].

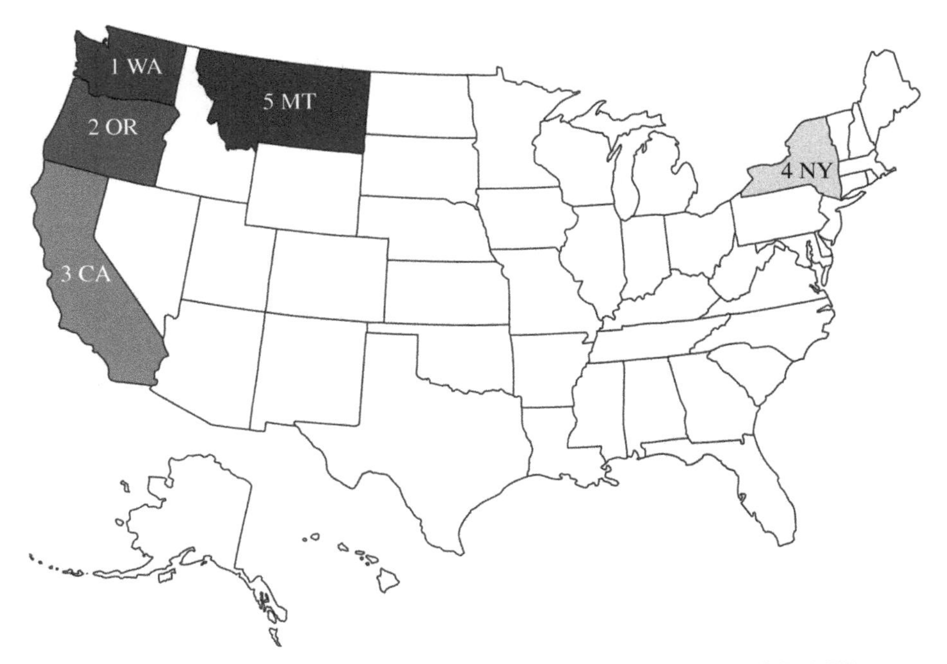

FIGURE 13.1 States with most hydroelectric generation (Source: US-EIA [2]).

13.3.2 Positive Attributes of Hydroelectric Power Generation

- Renewable resource
- Can be dispatchable by electric system operator.
- Low life cycle carbon emissions (including construction).
- No greenhouse gas emissions during operation.
- Long lifetime of projects (over 100 years).
- Additional benefits of dams including regulation of river flow, flood prevention, municipal water source and recreation.
- Run-of-river hydro has very little impact on wildlife habitat.
- Very low operating cost after initial construction.

13.3.3 Potentially Negative Attributes of Hydroelectric Power Generation

- Limited locations for new, large-scale facilities, therefore unlikely to be significant contributor to future solution to global climate change
- May inundate large areas of land, impacting wildlife habitat
- Not compatible with existing land use (homes, agriculture, etc.)
- Barrier to fish migration during spawning
- Huge amount of water evaporation per MWh of electricity produced

13.4 SOLAR ENERGY

13.4.1 Overview

Solar energy is one of the most environmentally benign sources of electric generation and is thus a favorite of many environmentalists and the public, or at least until the cost of energy comes into the equation. Solar energy can be cost-effective for direct thermal applications, such as home water heating or heating swimming pools but when used to produce electricity can be several times as expensive on a cost-per-kWh basis as the use of natural gas, coal, or wind. Yet, the production of solar modules is growing very fast. In the United States, the capacity (in kW or MW) of modules produced increased 30% during 2009 and 90% in 2008.

The two primary methods used to convert solar energy into electricity are (i) solar thermal generation, which operates much like a traditional steam (or Rankine cycle) power plant in that heat from the sun is used to make steam, which is then run through a turbine generator, and (ii) solar photovoltaics, which directly converts photons from the sun into electricity using panels containing semiconductors (such as silicon) that collect and absorb solar energy and produce DC electricity.

Figure 13.2 shows a solar photovoltaic panel, while 13.3 and 13.4 show two types of solar thermal generation technologies. Figure 13.3 shows a solar trough collector that reflects direct sunlight onto a receiver tube to make steam directly. Figure 13.4 shows a solar central receiver, sometimes called a solar power tower, concept where a field of heliostat mirrors are used to reflect sunlight to a central receiver located in the tower. The receiver contains the boiler for steam generation.

Figure 13.5 shows the location of the best solar resources in the United States. The cost of solar energy has declined significantly due to (i) improved efficiency of photovoltaic modules, (ii) increased economies of scale as the number of modules being produced has increased, (iii) larger projects being constructed again leading to economies of scale, and (iv) advancements in the material used to produce photovoltaic modules, such as the use of multijunction, thin-film materials.

- Installed capacity: 2347 MW (end of 2012)
- Electric generation: 4327 GWh (calendar year 2012)
- Capacity factor: 21.0%
- Percent of US electric production: 0.1%
- Land requirements: 5–15 acres/MW
- Water requirements: very little for photovoltaic systems (module washing); 100–1000 gallons/MWh generated by concentrating solar thermal power systems, depending on system design

FIGURE 13.2 Solar photovoltaic system located at the Central & Southwest (CSW) Solar Park near Fort Davis, Texas (Photo Credit: R. Walker).

FIGURE 13.3 Parabolic trough solar thermal electric power plant located at Kramer Junction, California (Photo Credit: kjkolb, http://commons.wikimedia.org/wiki/File:Parabolic_trough_solar_thermal_electric_power_plant_1.jpg).

FIGURE 13.4 Solar thermal electric power plant commonly known as a "power tower" (Photo Credit: Torresol Energy, http://commons.wikimedia.org/wiki/File:Gemasolar.jpg, http://commons.wikimedia.org/wiki/Commons:Free_Art_License_1.3).

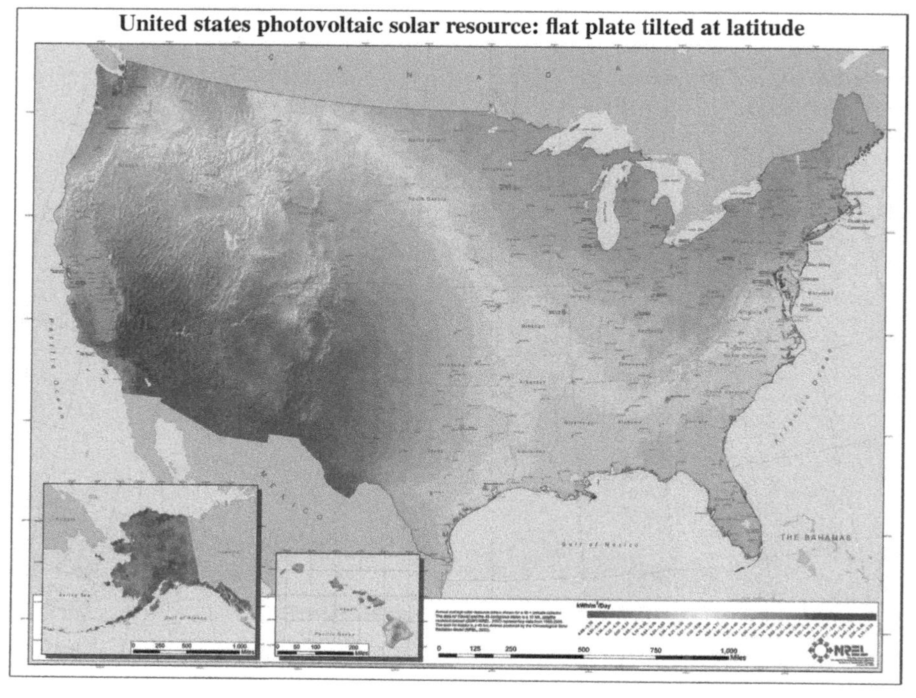

FIGURE 13.5 Solar energy resources in the United States (Source: NREL [3]).

13.4.2 Positive Attributes of Solar Energy for Electric Power Generation

- High levels of public support.
- Renewable resource.
- Use of energy-storage mediums such as molten salt with solar thermal systems can enable dispatchability by electric system operators.
- Low life cycle carbon emissions (including manufacturing and construction).
- No air emissions during operation.
- Able to locate systems in areas remote from electric grid.
- Potential use as building materials—roof, siding, and windows.
- Very little water use for photovoltaic system generation.

13.4.3 Potentially Negative Attributes of Solar Energy for Electric Power Generation

- High cost of energy
- Low capacity factors
- Most photovoltaic systems not dispatchable—daylight required
- Limited economic locations for large applications, principally desert southwest
- Dust accumulation can reduce efficiency without regular cleaning
- High water use for concentrating solar power projects, which are often located in desert areas already struggling with water supplies

13.4.4 Salinity Gradient Solar Pond Technology

Although most solar energy technologies provide intermittent power, the Salinity Gradient Solar Pond (SGSP), developed and tested for power generation in both Israel and the United States, uses stored solar energy to provide on-demand electrical power. Figure 13.6 is a schematic cross-section of an SGSP—typically about 3–5 m (10–15 ft) in depth. The salinity gradient acts as a transparent insulator allowing solar energy to be transmitted and then absorbed in the bottom storage zone. Due to the increasing density of the gradient zone, there is no convection of heat from the lower zone to the surface, as would occur in a usual pond, and it therefore continues to heat—storing high temperature salt water. This hot water at near-boiling temperatures can be extracted and used in an organic Rankine cycle engine to generate electrical power on demand. Figure 13.7 shows the first SGSP electrical generation demonstration project in the United States [4].

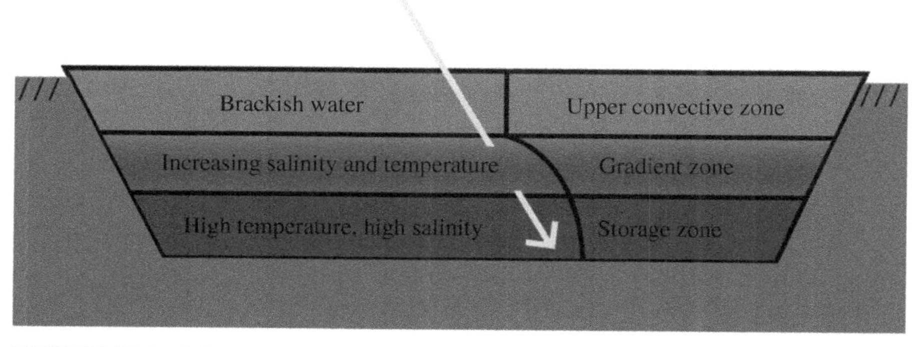

FIGURE 13.6 Schematic cross-section of a salinity gradient solar pond showing the temperature and salinity profile. Temperatures in the storage zone can exceed boiling if energy is not extracted on bright sunny days. The stored energy can be delivered on demand at night or during periods of cloudy days (Source: A. Swift and K. Jay).

FIGURE 13.7 The 3355 m^2 El Paso Solar Pond Project; 1983–2003. The first SGSP to generate electrical power in the United States. Rated at 100 kW the organic Rankine cycle engine was provided by Ormat Turbines of Israel and provided both heat and on-demand electrical power for Bruce Foods, a food canning operation (Source: A. Swift).

13.5 BIOMASS ENERGY

13.5.1 Overview

Biomass has been used as an energy source for many centuries, essentially since humans first began burning wood to cook with or keep warm. In 2010, biomass provided over 4% of US energy as measured in Btu but only accounted for a little over 1% of electric generation. Significant amounts of biomass are used for transportation purposes (such as ethanol or biodiesel) or for industrial heat processes. Figure 13.8 shows an estimate of the amount of biomass resources available throughout the United States. Industrial wastes such as scrap wood (from sawmills or furniture-manufacturing plants) or agricultural wastes are often used as fuel for industrial heat processes.

A key concept for understanding why biomass is usually considered as a renewable resource is the time scale of the "carbon cycle." When coal or natural gas is burned, emissions including carbon dioxide are released into the atmosphere; but the same can be said of biomass. However, the key difference is that fossil fuels release carbon dioxide captured by photosynthesis millions of years ago, whereas biomass releases carbon dioxide that is offset or balanced by the carbon dioxide captured during its own growth.

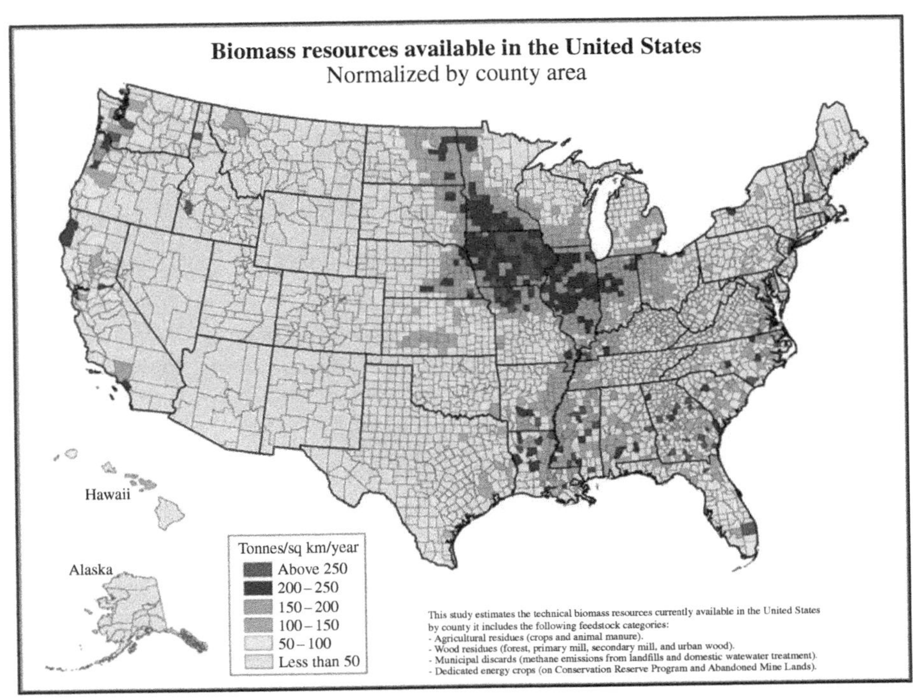

FIGURE 13.8 Biomass resources of the United States (Source: NREL [3]).

Biomass can be used for electricity production in several different ways, some of which are more environmentally preferable than others. These include (i) growing crops such as switchgrass or fast-growing willow trees specifically for the purpose of providing fuel to an electric generator, either by burning the fuel to make steam or by "gasifying" the biomass, (ii) using agricultural wastes such as sugarcane bagasse or rice hulls as fuel for an electric generator, (iii) burning or gasifying municipal solid waste (MSW; discussed separately in Section 13.5), (iv) capturing methane from decomposing organic matter in landfills, which can then be used to fuel an electric generator (discussed separately in Section 13.6), and (v) cofiring, or mixing, biomass matter such as pulverized urban wood waste with coal as fuel for a traditional coal-fired power plant.

- Installed capacity: 12,319 MW (end of 2012).
  - Wood and wood products: 7508 MW
  - MSW, landfill gas, sludge waste, agricultural byproducts, other biomass solids or liquids, and other biomass gases: 4811 MW
- Electric generation: 57,622 GWh (calendar year 2012).
- Capacity factor: 53.4%.
- Percent of US electric production: 1.4%.
- Land requirements: Can be substantial, with estimates ranging from 500 to 1500 acres/MW of electric-generating capacity, depending on assumed fuel source and other assumptions. Several hundred acres of switchgrass must be grown to provide fuel for 1 MW of generation.
- Water requirements: 300–500 gallons of water/MWh produced is consumed for cooling purposes, similar to fossil-fueled power plants; crops grown for use in biomass facilities may also consume large volumes of water.

13.5.2 Positive Attributes of Biomass Energy for Electric Power Generation

- Considered "carbon neutral" other than fuel used for growing and transportation
- Renewable resource
- Creates new market for agricultural crops, helping farming remain viable
- Effective use of something that might otherwise be considered waste and fill up landfills
- Can be dispatchable by electric system operator

13.5.3 Potentially Negative Attributes of Biomass Energy for Electric Power Generation

- Produces emissions such as particulate matter, SO_2, NO_X, CO, and volatile organic compounds (VOCs).
- Harvesting and transportation costs for fuel source can be significant for large generation facility.

- Lower energy intensity than in fossil fuels can mean higher transportation costs if local supply is not sufficient.
- Use of biomass crops for fuel can drive up prices for other uses (food, livestock feed, etc.).
- Some environmentalists oppose "cofiring" of biomass in existing coal plants as emissions may slightly increase.
- May contribute to deforestation in developing countries.
- Uses conservable amounts of water, both for crop irrigation and power plant cooling.

13.6 MUNICIPAL SOLID WASTE

13.6.1 Overview

The U.S. Environmental Protection Agency (EPA) estimates that each person in the United States generates an average of 1130lb of waste/year [5] and that only about 30% of the waste Americans generate is recycled, with most of the remainder disposed in landfills. In some places, though, rather than landfilling trash, this "municipal solid waste" is used to generate electricity. On its face, this does seem to be a much better use for our nation's MSW than filling up landfills, but there are significant issues with regard to emissions from incineration and with disposal of the residue or ash remaining after incineration.

Power plants utilizing MSW are often referred to as waste-to-energy plants. Some states consider MSW to be a source of renewable energy since no new fuel sources are used other than the waste that would otherwise be sent to landfills. But other states, such as Texas, specifically exclude MSW as a renewable resource due to the emissions and ash discussed earlier and since MSW may include nonrenewable materials derived from fossil fuels, such as tires and plastics. EPA statistics indicate that there are 87 MSW-fired power generation plants in the United States with a combined capacity of approximately 2500MW or only about 0.22% of total generating capacity in the United States.

13.6.2 Positive Attributes of Using MSW to Fuel Electric Power Generation

- Considered as a renewable resource in some states.
- While CO_2 is released due to incineration of MSW, it is still considered carbon neutral.
- Preserves fossil fuels such as coal and natural gas for other uses.
- Effective use of something that would otherwise fill up landfills.
- MSW is dispatchable by electric system operators.

13.6.3 Potentially Negative Attributes of Using MSW to Fuel Electric Power Generation

- Produces emissions such as particulate matter, SO_2, NO_X, CO, VOCs, and trace amounts of toxic pollutants, such as mercury compounds and dioxins
- Variation in content of MSW will affect emissions rates, as it may contain small amounts of toxic pollutants such as lead, mercury, or cadmium
- Due to the aforementioned emissions issues, it may be difficult to obtain the necessary permits for construction of MSW plants near urban areas where most of the potential fuel source is produced, which can result in increased transportation costs if MSW plants have to be located in rural areas.
- MSW plants also require water in about the same amount as conventional fossil-fueled plants, about 300–500 gallons/MWh generated.
- MSW plants also discharge much of the water that has been used, and the discharges may include pollutants from the power plant boiler and cooling systems.
- Ash and other residues remaining after incineration of MSW may contain toxic materials that must be handled and disposed of as hazardous waste.

13.7 LANDFILL GAS

13.7.1 Overview

Landfill gas is generated by the anaerobic decomposition of organic refuse deposited in landfills and, like natural gas, landfill gas is primarily methane. This then begs the question: "If natural gas is not considered a renewable source of energy, why is landfill gas considered renewable?"

Although CO_2 emissions are released when using gas collected from landfills to fuel electric generators, the net effect is judged to be positive for the atmosphere because CO_2 emissions are significantly less radiative (i.e., the "greenhouse effect" is less) than methane emissions. If landfill gas is not collected, it will eventually escape into the atmosphere. The EPA requires that landfills above a certain size install gas collection systems to minimize the release of methane. Some landfills simply "flare" the landfill gas converting methane into CO_2, whereas others clean up the landfill gas and use it to fuel electric generators or possibly to sell as pipeline-quality gas.

Landfill gas collection systems are installed during the active life of a landfill and then completed at closure of the landfill or an area of the landfill. Vertical wells are drilled to the bottom of the waste, typically about one well per acre. Horizontal collector pipes are placed in active fill areas, and a compressor is used to place a vacuum on the gas wells to draw out the gas that is produced as the organic material decomposes. Gas collected from landfills can be quite corrosive, so a "scrubber" may be required to clean up the gas before it goes into the generator in order to reduce the corrosive characteristics of the gas. Landfill gas consists primarily of methane

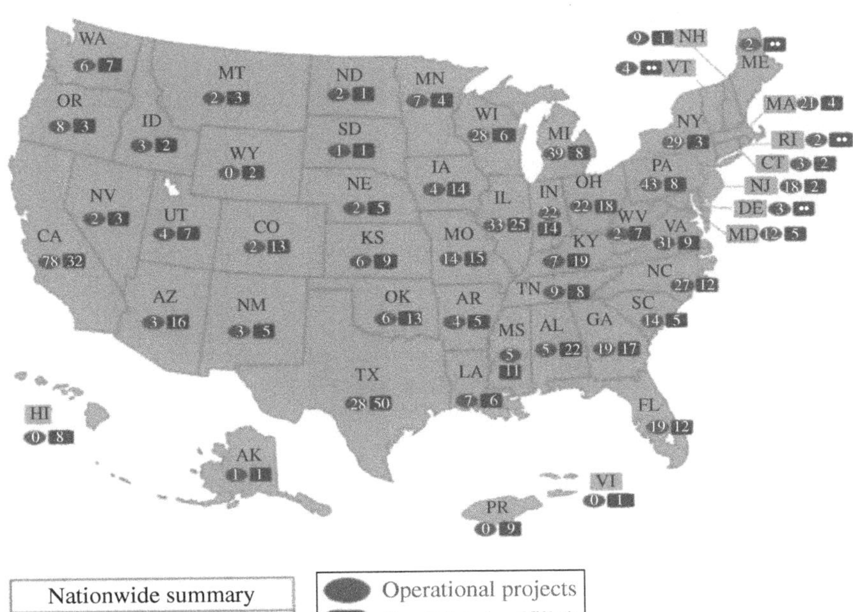

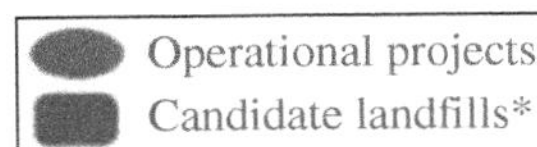

Nationwide summary

621 Operational projects (1978 MW and 311 mmscfd)

~ 450 candidate landfills (850 MW or 470 mmscfd, 36 MMTCO2e/yr potential)

* Landfill is accepting waste or has been closed 5 years or less, has at least 1 mm tons of waste, and does not have an operational, under-construction, or planned project; can also be designated based on actual interest by the site.

These data are from LMOP's database as July 16, 2013.

** LMOP does not have any information on candidate landfills in this state.

FIGURE 13.9 Existing and potential landfill gas projects (Source U.S. EPA [4]).

(35–60%) and carbon dioxide (35–55%), with small amounts of oxygen (0–2.5%), nitrogen (0–20%), and non-methane organic compounds (Fig. 13.9).

US EPA statistics indicate about 558 operational landfill gas projects are in operation in the United States, totaling about 1727 MW of generation capacity. Thus, the average-size facility is only about 3 MW. The facilities produce about 8,000,000 MWh of electricity with net capacity factors of around 50%. Some landfill gas collection facilities produce pipeline-quality gas, and others produce thermal energy for industrial processes.

13.7.2 Positive Attributes of Using Landfill Gas for Electric Power Generation

- Considered as a renewable resource in some states.
- While CO_2 is released due to incineration of landfill gas, the net effect is to reduce the total impact of greenhouse gas releases.
- Decreased emissions of methane; non-methane organic compounds; and toxics such as benzene, carbon tetrachloride, and chloroform.

- Preserves fossil fuels such as coal and natural gas for other uses.
- Landfill gas projects are dispatchable by electric system operators.
- Each MW of electricity generated by landfill gas projects is estimated to provide electricity for about 600 homes.

13.7.3 Potentially Negative Attributes of Using Landfill Gas for Electric Power Generation

- Relatively small project size makes it hard for projects to be economically competitive.
- Gas production builds as organic decomposition occurs after a section of the landfill is closed and covered, but production may slowly begin decreasing after only a few years.
- Obtaining required air permits can be difficult and time-consuming, especially for projects located in ozone, nitrogen oxide, and carbon monoxide nonattainment areas.
- Depending upon the type of generator used to produce electricity from landfill gas, water may be required for cooling purposes.

13.8 GEOTHERMAL ENERGY

13.8.1 Overview

A geothermal power plant is very similar to a traditional Rankine cycle power plant in that steam is forced through a turbine connected to a generator, but instead of using coal, natural gas, or uranium to heat water, heat from the Earth is used by drilling hot water or steam wells. The key drivers behind the economics of using geothermal energy for electric power generation are (i) depth of the geothermal resources, which can range from shallow ground to hot water and rock several miles below the Earth's surface and (ii) temperature of the resource, which can range from room temperature to over 300 °F, with the higher temperature resources generally being more cost-effective. In the United States, most economically viable geothermal reservoirs are located in the western states, Alaska, and Hawaii. The first US facility for electricity generation was constructed in 1904.

- Installed capacity: 2592 MW (end of 2012).
- Electric generation: 15,562 GWh (calendar year 2010).
- Capacity factor: 70.9%.
- Percent of US electric production: 0.4%.
- Land requirements: 1–8 acres/MW.
- Water requirements: Small amounts of water may be needed for cooling purposes.

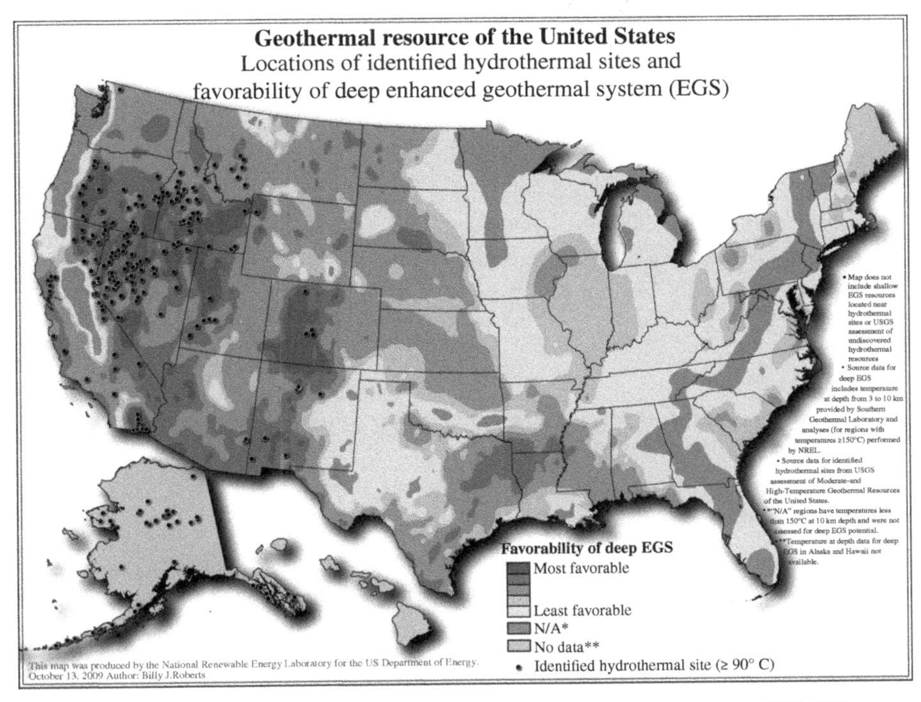

FIGURE 13.10 Geothermal resources in the United States (Source: NREL [3]).

Figure 13.10 shows the location of geothermal resources throughout the United States.

13.8.2 Positive Attributes of Geothermal Energy

- Renewable resource that helps conserve fossil fuels.
- Dispatchable, base-load generation resource.
- Few emissions; some projects reduce sulfur emissions that would have occurred from natural venting.
- Land use requirements among the lowest of any generation resource.
- Among the most reliable of generation resources due to their simplicity.
- No fuel mining or transportation.
- Very little waste production.

13.8.3 Potentially Negative Attributes of Geothermal Energy

- Limited availability of economic resources.
- Fluids drawn from deep earth may carry carbon dioxide (global climate change) and hydrogen sulfide (acid rain, noxious smell).

- Emission-control systems may be needed to reduce exhaust of acids and volatile chemicals.
- Hot water from geothermal sources may contain trace amounts of dangerous elements such as mercury, arsenic, and antimony, which must be handled and disposed properly.
- Plant construction can adversely affect land stability potential causing subsidence and even small seismic events.
- Scaling and fouling of the well casing is often a problem due to the high mineral content of the geothermal water.

13.9 ENERGY FROM THE OCEAN: TIDAL POWER AND WAVE POWER

13.9.1 Overview

Energy from the ocean is just beginning to be harnessed. Coastal areas experience two high and two low tides over the course of a day, but for tides to be harnessed for economically viable electricity generation, the difference between high and low tides must be fairly large (perhaps 5 m or greater), and there are few sites with tidal differences of this magnitude. Currently, there is only one major tidal generation station in operation: the 240-MW La Rance plant located on the northern coast of France, which has been in operation since 1966.

Waves are caused by the wind blowing over the surface of the ocean, and wave power devices extract energy directly from surface waves or from pressure fluctuations below the surface.

The Minerals Management Service within the U.S. Department of the Interior issued a technology white paper in May 2006 titled *Wave Energy Potential on the U.S. Outer Continental Shelf* [6], which included the following points:

- The total annual average wave energy off the U.S. coastlines (including Alaska and Hawaii), calculated at a water depth of 60 m has been estimated at 2100 TWh.
- Estimates of the worldwide economically recoverable wave energy resource are in the range of 140–750 TWh/year for existing wave-capturing technologies that have become fully mature.
- With projected long-term technical improvements, this could be increased by a factor of 2 to 3.
- The fraction of the wave power that is economically recoverable in U.S. offshore regions has not been estimated, but is significant even if only a small fraction of the 2100 TWh/year available is captured.
- Currently, approximately 11,200 TWh/year of primary energy is required to meet total U.S. electrical demand.

13.9.2 Technologies That Can Be Used to Harness Ocean Energy

Figures 13.11, 13.12, and 13.13 show some of the technologies that can be used to produce electricity from the ocean including:

- Tidal barrages or dams in which gates are opened during high-tide periods allowing water to flow into a bay or estuary, and then gates are closed before the tide goes back out. Then at low-tide periods, water impounded by the barrage or dam can be released through a generator to produce electricity.
- Tidal turbines, which operate similarly to wind turbines, except that they are located underwater where coastal currents can be used to generate electricity.

Wave power terminator devices are placed just offshore perpendicular to the direction of wave travel and capture or reflect the power of the wave.

- Wave power attenuator devices are placed parallel to the direction of the waves where the height of the waves causes the device to flex, with the flexing motion then driving hydraulic pumps or generators.

FIGURE 13.11 The SeaFlow tidal stream generator prototype with rotor raised (Photo Credit: Fundy, http://en.wikipedia.org/wiki/File:Seaflow_raised_16_jun_03.jpg).

FIGURE 13.12 Pelamis Wave Energy Converter on site at the European Marine Energy Test Centre (EMEC). (Photo Credit: P123, http://en.wikipedia.org/wiki/File:Pelamis_at_EMEC.jpg).

FIGURE 13.13 Ocean Power Technology's 150-kW PowerBuoy (Photo Credit: Ocean Power Technologies, http://en.wikipedia.org/wiki/File:Optbuoy.jpg.

- Wave power point absorbers are floating structures usually anchored to the ocean floor that use the relative motion of the system's components to drive electromechanical or hydraulic energy converters.
- Ocean thermal energy conversion (OTEC) systems, which use the ocean's natural thermal gradient (or temperature differential) between deep ocean water and surface ocean waters to produce electric power from various types of heat engines. Given the vast amount of ocean resources, National Renewable Energy Laboratory (NREL) indicates on their website [4] that OTEC could be used to produce 100,000 GW of baseload power generation.

13.9.3 Positive Attributes of Ocean Power

- Renewable resource that could help conserve fossil fuels.
- Huge, untapped resource, particularly wave power and OTEC.
- OTEC systems may also produce desalinated water.
- Predictable, base-load resource, but not necessarily dispatchable.
- No emissions.

13.9.4 Potentially Negative Attributes of Ocean Power

- Limited availability of economic resources, particularly for tidal power.
- Tidal power plants that dam estuaries can impede sea life migration.
- Silt buildup behind tidal power plants could impact local ecosystems.
- Tidal fences may also disturb sea life migration.
- Construction costs are high, so the cost of energy is not competitive with conventional fossil fuel power.
- Low capacity factor of tidal power (La Rance site capacity factor is about 28%).
- Wave power facilities may disrupt recreational boating or commercial shipping.
- May impact the aesthetics of ocean landscapes.

13.10 HYDROGEN

13.10.1 Overview

Hydrogen is the simplest element and most plentiful gas in the universe. Currently, most hydrogen is produced from fossil fuels, meaning that it would not be considered a renewable resource. However, there are ways to produce hydrogen from water with the aid of renewable energy sources such as wind energy or solar energy, which would allow hydrogen produced in such a manner to be considered as a renewable fuel. Figure 13.14 shows the estimated potential for hydrogen production from renewable resources throughout the United States. Methods to do this include the following:

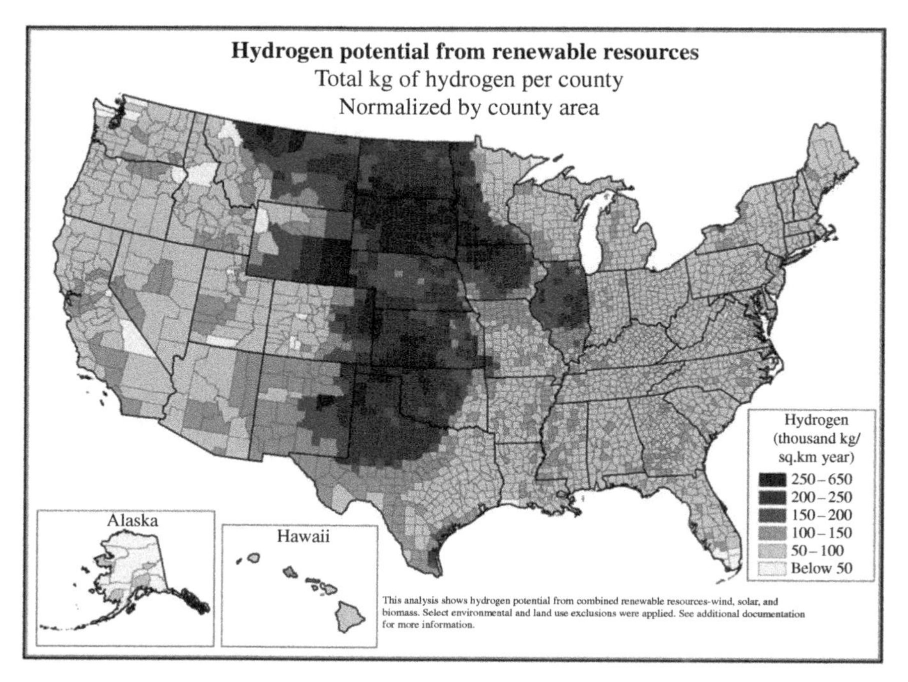

FIGURE 13.14 Hydrogen potential from renewable resources in the United States (Source: NREL [4]).

- Using wind and solar resources for electrolysis of water, a process in which water is split into hydrogen and oxygen using an electrolyzer
- Gasification of biomass such as agricultural crops, agricultural wastes, or wood wastes using partial oxidation at high temperature, producing a gaseous fuel that is then reformed to produce hydrogen
- Steam methane reforming methods for converting landfill gas, animal manure, and wastewater sludge into hydrogen

The NREL's report titled *Potential for Hydrogen Production from Key Renewable Resources in the United States* [7] contains the following conclusions:

- "About 1 billion metric tons of hydrogen could be produced from wind, solar, and biomass resources annually in the U.S.
- The Great Plains emerge as the area in the U.S. with the highest potential for producing hydrogen from these key renewable resources.
- Renewable hydrogen has the potential to displace gasoline consumption in most states.
- The infrastructure needed to enable the widespread use of hydrogen as a transportation fuel is not currently available since areas with the greatest resource are located remotely from the areas of highest demand."

13.10.2 Positive Attributes of Hydrogen

- Renewable resource that could help conserve fossil fuels and reduce imports of foreign oil.
- Hydrogen produced from biomass can provide a new revenue source for the agricultural industry.
- Hydrogen could be produced using wind power in periods of low electric load and high winds and stored until hours of high electric load, when it could be used to fuel electric generators or fuel cells.
- Could provide a dispatchable, base-load resource for electric generation.
- No emissions—hydrogen only releases water vapor and heat as its byproducts.

13.10.3 Potentially Negative Attributes of Hydrogen

- Hydrogen is highly flammable and explosive.
- Hydrogen is expensive to transport, store and produce, and takes substantial amounts of energy to produce.
- Hydrogen produced from biomass may drive up the cost of food and/or livestock feed.
- Hydrogen combustion produces oxides of nitrogen gases.
- Low conversion efficiencies when using renewable electric sources to produce hydrogen and then converting hydrogen back to electricity.

REFERENCES

[1] U.S. Energy Information Administration. *Annual Energy Review 2010*. Washington (DC): United States Govt Printing Office; 2011.

[2] U.S. Energy Information Administration. *Renewable Energy Trends in Consumption and Electricity, 2007*. Washington (DC): U.S. Energy Information Administration; 2009.

[3] Swift A, Reid R, McGraw K. Operational strategy and results for 90 degree C operation of a solar pond for electric power and desalination. In: Murphy L, Mancini T, editors. *Solar Engineering-1988*. New York (NY): ASME Press; 1988. p 107–112.

[4] National Renewable Energy Laboratory website. Golden (CO). Available at http://www.nrel.gov. Accessed August 6, 2012.

[5] U.S. Environmental Protection Agency website. Washington (DC). Available at http://www.epa.gov. Accessed August 6, 2012.

[6] U.S. Department of the Interior, Minerals Management Service. *Technology White Paper on Wave Energy Potential on the U.S. Outer Continental Shelf*. Washington (DC): U.S. Department of the Interior, Minerals Management Service; 2006.

[7] Milbrandt A, Mann M. Potential for hydrogen production from key renewable resources in the United States. National Renewable Energy Laboratory technical report NREL/TP-640-41134, National Renewable Energy Laboratory, Golden (CO); 2007.

14

NONRENEWABLE ENERGY SOURCES USED TO MEET ELECTRICITY NEEDS

14.1 COMPARING SOURCES OF ELECTRIC GENERATION

As stated at the start of the previous chapter, all sources of electric generation have positive attributes and negative attributes, and they all seem to have both supporters and detractors. Two of the sources widely used for electricity generation that seem to attract the most vocal opposition are nuclear energy and coal-fired generation, both of which are discussed in this chapter. Factors relevant to environmental and economic impacts will be considered in hopes of presenting an "apples-to-apples" comparison. The various facts, factors, impacts, and/or issues that will be evaluated include the following:

- Installed capacity and net generation in the United States/average capacity factors
- Basics of the technology/efficiency
- Primary operational mode within the electric grid
- Capital cost/variable costs including fuel/levelized cost of energy
- Availability of fuel source and price variability
- Emissions/effect on human health
- Production of solid and/or hazardous wastes and disposal options
- Cooling water requirements/other water requirements

Wind Energy Essentials: Societal, Economic, and Environmental Impacts, First Edition.
Richard P. Walker and Andrew Swift.

- Impacts to surface water and groundwater quality
- Land use requirements/compatibility with other uses
- Workforce health and safety

14.2 PROCESSES OR TECHNOLOGIES USED TO PRODUCE ELECTRICITY

14.2.1 Turbine

A turbine is a machine that uses the kinetic energy of a moving fluid to produce rotating mechanical energy by causing a bladed rotor to rotate. The moving fluid can be (i) air, such as used by a wind turbine, (ii) water, such as used by a hydroelectric generator, (iii) steam, such as used by many fossil-fueled power plants, or (iv) hot gases, such as used in gas turbines or combustion turbines (CTs).

14.2.2 Steam Generator

Thermal power plants have been used for many decades to produce electricity using a process called the Rankine cycle, in which steam is produced and injected through a steam turbine, creating rotational energy that can then be used to turn a generator to produce electricity. The steps in this process are as follows:

- Water is heated using a fuel source, which can include the following:
  - Coal
  - Natural gas
  - Petroleum
  - Nuclear energy
  - Wood or other forms of biomass
  - Municipal solid waste, which is also sometimes included as a form of biomass
- The water is heated in a boiler producing steam that is then directed through a steam turbine.
- The steam turbine is connected by the output shaft to an electric generator. High-pressure steam striking the turbine blades results in rotation of the turbine, the shaft, and the generator rotor.
- Rotation of the generator rotor within a magnetic field then produces electricity.
- Electricity produced by the generator goes through an electric transformer that increases the voltage of the electricity to that of the transmission grid (69 kV or above).

Additional information about the Rankine cycle is presented in Chapter 15, particularly with regard to efficiency and use of water.

14.2.3 Gas Combustion Turbine or Simple Cycle Combustion Turbine

Most new natural gas–fired power plants now utilize gas turbines instead of steam turbines. Gas turbines used for electricity generation use the Brayton cycle, named after an engineer from Boston named George Brayton, who first proposed the process in the nineteenth century. Gas turbines can also be used for aircraft engines. The low capital cost of gas turbines, the quick-start capabilities, and the lower emissions than coal generation have led to their widespread use for peaking and cycling (or load following) purposes.

The process by which gas turbines or combustion turbines (CTs) produce electricity is illustrated in Fig. 14.1 and described as follows:

- A compressor is used to increase the pressure of intake air.
- Fuel (most often natural gas) is mixed with the high-pressure air from the compressor in a combustion chamber or combustor and ignited.
- Ignition of the fuel/air mix increases temperature, creating higher pressures and forcing the resulting gases at a high velocity through a turbine.
- The turbine is connected by the output shaft to an electric generator. High-pressure gases striking the turbine blades result in rotation of the turbine, the shaft, and the generator rotor.
- Rotation of the generator rotor within a magnetic field then produces electricity.
- Electricity produced by the generator goes through an electric transformer that increases the voltage of the electricity to that of the transmission grid (69 kV or above).

14.2.4 Combined Cycle Power Plant or Combined Cycle Combustion Turbine

Combined cycle power plants combine steam turbines with gas turbines to improve the overall efficiency with which fuel is used. A gas turbine on its own produces significant quantities of waste heat as exhaust gasses as shown in Figure 14.1.

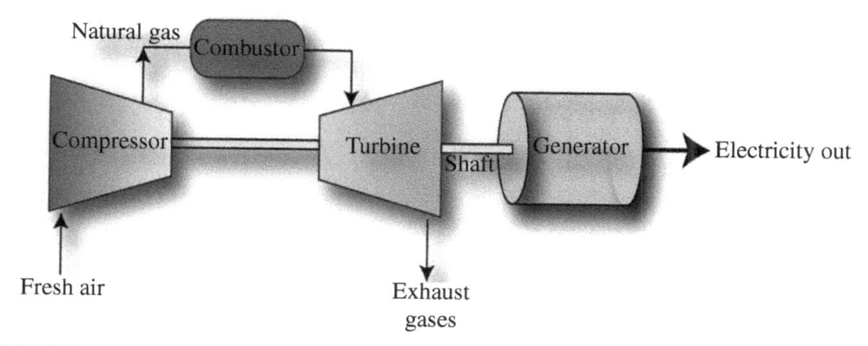

FIGURE 14.1 Example of simple Brayton cycle combustion turbine for electric generation (Source: Reproduced by permission of Amy Marcotte).

When used in a combined cycle process, instead of wasting this heat, residual heat from the CT is used to heat water to produce steam for a steam turbine, thus increasing the efficiency with which the natural gas is used. When natural gas–fired generation is needed for baseload purposes instead of only peaking and cycling purposes, a combined cycle plant is the preferred configuration due to efficiency improvement over simple cycle combustion turbine (SCCTs).

14.2.5 Integrated Gasification Combined Cycle

Integrated Gasification Combined Cycle (IGCC) systems use a coal gasification process to turn coal into synthetic gas and then remove most impurities before the gas is combusted to produce energy. The synthetic gas is then used to fuel a combined cycle power plant. Coal is gasified by heating coal or a coal/water slurry at high temperatures and pressure, producing a combustible gas composed of carbon monoxide (CO) and hydrogen. When companies talk about "clean coal plants," they are often referring to the use of IGCC technology. While the IGCC process reduces some of the emissions associated with traditional coal-fired plants, carbon dioxide (CO_2) would still be produced, so some of the plants being proposed will have processes to capture and sequester the CO_2, such as injecting it underground or into the ocean. In addition, mercury may be emitted from the stack of an IGCC plant.

14.2.6 Natural Gas–Fueled Internal Combustion Engines

Although not a significant fraction of power generation technologies, internal combustion (IC) engines are sometimes used for distributed generation purposes, including backup power supplies, while some power plants use a series of large IC engines to generate electric power.

14.3 COMPARING EFFICIENCIES OF GENERATION TECHNOLOGIES

14.3.1 Efficiency

The efficiency of a power station is indicated with the Greek character η. Mathematically, thermal efficiency is calculated as

$$\eta = \frac{\text{Energy out}}{\text{Energy in}}$$

The efficiency of a typical IC gasoline engine in an automobile is around 20–25%, while diesel engines have efficiencies of about 40%. Power plant technologies also differ in efficiency. For instance, a coal-fired steam generator may have an efficiency of around 40% compared with SCCTs with efficiencies of 35–40% or combined cycle plants, which can achieve efficiencies nearing 60%. These concepts are discussed in more detail in Chapter 15.

TABLE 14.1 Average Operating Heat Rate of US Electric Generators (Btu/kWh)

Year	Coal	Natural gas	Nuclear
2001	10,378	10,051	10,443
2002	10,314	9,533	10,442
2003	10,297	9,207	10,421
2004	10,331	8,647	10,427
2005	10,373	8,551	10,436
2006	10,351	8,471	10,436
2007	10,375	8,403	10,485
2008	10,378	8,305	10,453
2009	10,414	8,160	10,460
2010	10,415	8,185	10,452
2011	10,444	8,152	10,464
2012	10,498	8,039	10,479

Source: US-EIA [1].

14.3.2 Heat Rate

The efficiency of power plants is often referred to in terms of their heat rate. Heat rate is the ratio of energy input, measured in Btu, to electricity produced, measured in kWh. Thus, the more efficient a power plant is, the lower the heat rate will be. So remember, a low heat rate is good while a high heat rate is bad. Table 14.1 compares the average heat rates of all coal, natural gas, and nuclear generators in the United States based on data for the years 2001 through 2012 from the U.S. Energy Information Administration (US EIA). Note that the heat rates of coal-fired generation and nuclear-powered generation have remained fairly constant over this 10-year period, while the heat rate of natural gas–fired generation has improved substantially. The improved heat rate from natural gas plants can probably be attributed to retirements of old steam-generating units, technology improvements, and the increased use of combined cycle configurations.

Note that 3412 Btu = 1 kWh (thermal); thus, a heat rate of 3412 Btu/kWh represents a 100% efficient process, while a heat rate of 6824 Btu/kWh would represent a cycle efficiency of 50%. Based on US EIA data for average heat rates by prime mover and energy source found in Table 5.4 of Electric Power Annual 2010 [2], following are the average efficiencies of coal, natural gas, and nuclear power plants in the United States during 2010.

Coal-fired steam turbine	3,412 Btu/kWh ÷ 10,415 Btu/kWh = 32.8% efficiency
Gas-fired steam turbine	3,412 Btu/kWh ÷ 10,416 Btu/kWh = 32.8% efficiency
Gas-fired gas turbine	3,412 Btu/kWh ÷ 11,590 Btu/kWh = 29.4% efficiency
Gas-fired internal combustion	3,412 Btu/kWh ÷ 9,917 Btu/kWh = 34.4% efficiency
Gas-fired combined cycle	3,412 Btu/kWh ÷ 7,619 Btu/kWh = 44.8% efficiency
Nuclear power plants	3,412 Btu/kWh ÷ 10,452 Btu/kWh = 32.6% efficiency

Remember that these are averages, so certainly some newer technologies and newer plants will have better efficiencies than the numbers shown here.

14.4 GENERATION RESOURCES IN THE UNITED STATES

Table 14.2 shows electric generation capacity and annual production data from all sources for calendar year 2012 (the latest full-year data available from the US EIA as of the writing of this book). Coal-fired generation was the source used to produce the most electricity in the United States during 2012, despite this type of electricity production having the greatest environmental consequences in terms of air emissions, including greenhouse gases. The table also shows the average net capacity factor for each type of generation resource. In general, high capacity factors are beneficial since the fixed costs of the generator may be spread over more of electric energy production, helping to reduce the overall cost of energy to the consumer.

Note that the average 2012 capacity in Figure 14.2 is determined by averaging generating capacity at the beginning of the year (or end of the previous year) and capacity at the end of the year.

14.5 NUCLEAR ENERGY

14.5.1 Overview

The United States has 104 commercial nuclear power reactors in operation totaling 107 GW of generation capacity, but no nuclear plants have been completed in the United States since June 1996 when Watts Bar 1 was completed in Tennessee (Fig. 14.2). One-half of them are 30 years or older. Electricity was first produced from a nuclear reactor in 1951 from an experimental breeder reactor operated by the US government in Idaho. The first civilian nuclear power plant began generating electricity in California and Pennsylvania in 1957.

While the United States is the nation with the largest amount of nuclear generation capacity and produces the most electricity from nuclear power, it gets only about 19% of its electricity from this source, whereas several other nations have a much greater reliance on nuclear energy. According to the website of the World Nuclear Association [4], some countries getting much larger percentages of their electricity from nuclear reactors in 2012 include France (75%), Slovakia (54%), Belgium (51%), Ukraine (46%), and Hungary (46%).

Prior to the March 2011 earthquake and resulting tsunami leading to the nuclear accident at Fukushima, Japan, at least 20 proposed US nuclear reactors were undergoing permitting review by the Nuclear Regulatory Commission (NRC). There is much speculation that the incident in Japan may slow the permitting process or cause some of the owners of the proposed projects in the United States to withdraw their permit requests, causing the nuclear industry to stall in much the same way that occurred following the 1979 accident at the Three Mile Island plant near Middletown, Pennsylvania.

TABLE 14.2 Average Electric Generation Capacity, Annual Production, and Average Capacity Factor During Calendar Year 2012

	Coal	Natural gas	Nuclear	Wind	Solar thermal and PV	Wood and wood-derived fuels	Other biomass	Geothermal	Hydro-electric	Total
2011 year-end capacity (MW)	317,640	415,191	101,419	45,676	1,524	7,077	4,536	2,409	78,652	1,051,251
2012 year-end capacity (MW)	309,680	422,364	101,885	59,075	3,170	7,508	4,811	2,592	78,738	1,063,033
Average 2012 capacity (MW)	313,660	418,778	101,652	52,376	2,347	7,293	4,674	2,501	78,695	1,057,142
2012 generation (GWh)	1,514,043	1,225,894	769,331	140,822	4,327	37,799	19,823	15,562	276,240	4,047,765
Average annual capacity factor (%)	55.0	33.3	86.2	30.6	21.0	59.0	48.3	70.9	40.0	43.6

R. Walker based on data from US-EIA's *Electric Power Annual 2012* [1].

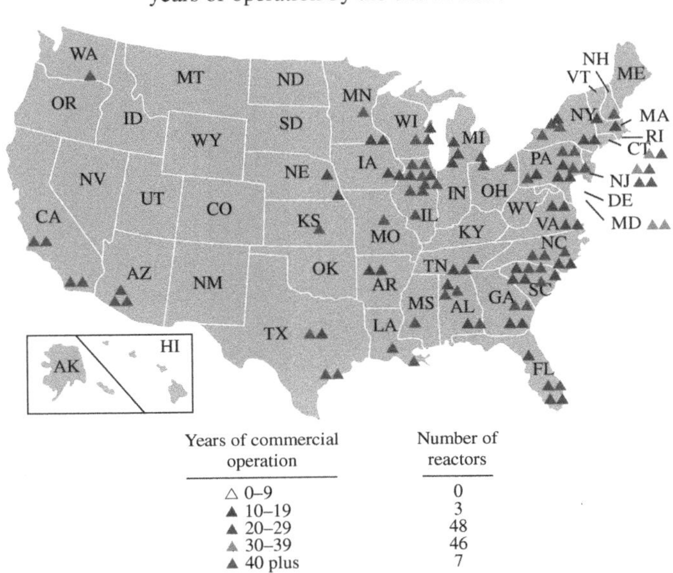

FIGURE 14.2 US nuclear power reactor locations and years of operation (Source: U.S. Nuclear Regulatory Commission [3]).

However, in February 2012, the US NRC approved construction of two new nuclear power units at Southern Company's Vogtle Plant in eastern Georgia where two nuclear units are already in operation, and the following month the NRC approved construction of two more new units at SCANA Corporation's Virgil Summer nuclear station in South Carolina.

Another significant issue facing the nuclear energy industry is the lack of an approved disposal site for spent nuclear fuel. In 2009, shortly after the election of President Barack Obama, the newly appointed Secretary of the U.S. Department of Energy, Dr. Steven Chu, essentially ruled out the use of Yucca Mountain facility for disposal of nuclear waste. This means that spent nuclear fuel will continue to be stored on-site at each of the nuclear reactors in the United States.

All 104 of the commercial reactors in the United States are fission reactors, but scientists continue to seek ways to harness nuclear fusion (the nuclear reaction that powers the sun) for safe generation of electric energy. The main difference between fission and fusion is that fission is the splitting of an atom into two or more smaller ones while fusion is the fusing of two or more hydrogen atoms into a larger helium atom using powerful magnets. A small amount of mass is lost when the hydrogen atoms combine, in the process releasing vast quantities of energy. And unlike nuclear fission, only low-level radioactive material would be left. However, despite decades of research and billions of dollars invested, nuclear fusion has not been developed into a practical energy source.

14.5.2 Statistics from US EIA

US EIA data [1] shows the following statistics for nuclear energy in the United States for calendar year 2012:

- Installed capacity: 101.8 GW.
- Electric generation: 769,331 GWh.
- Capacity factor: 86%.
- Percent of US electric production: 19%.
- Land requirements: Cooling can require about 5 acres/MW of generating capacity if a cooling lake is constructed for the project; significant additional land is needed related to the mining and processing of fuel for nuclear reactors (Figure 14.3).

FIGURE 14.3 South Texas nuclear project: 2500 MW and about 20 square miles (Map Data: Google Earth/2014 Google).

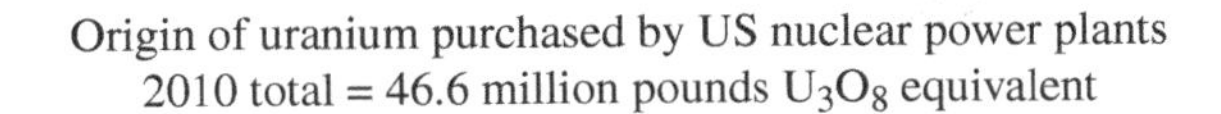

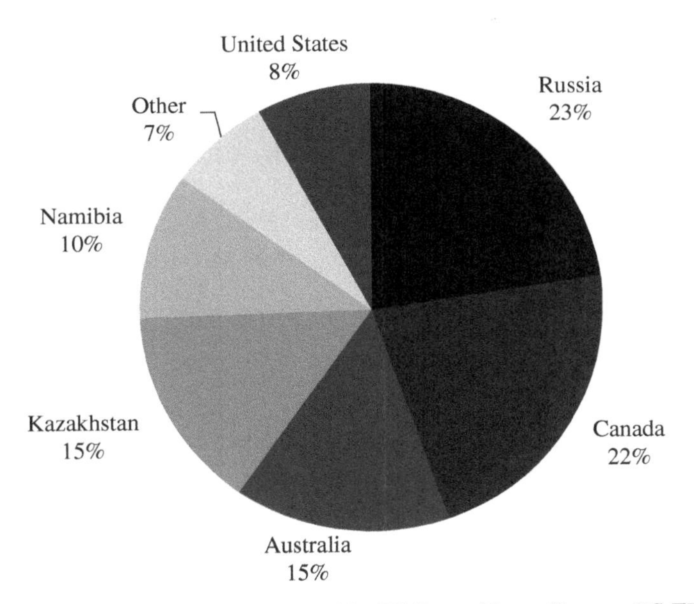

FIGURE 14.4 Origin of uranium purchased by US Power Plants (Source: US-EIA [5]).

- Water use: 400–800 gallons per MWh consumed; much more water is withdrawn and returned to a reservoir.
- Fuel reserves: The US EIA's last update of estimated uranium reserves in the United States was for year-end 2008, and its website [5] states that US reserves totaling 1227 million pounds were obtainable at a price of $100/lb, or about 23 years of usage at current rates.
- Fuel sources: US EIA statistics [5] also show that domestic production of uranium only accounts for about 8% of uranium usage in the United States in 2010, with the remainder imported from other nations as shown in Figure 14.4.

14.5.3 Positive Attributes of Nuclear Power

- No air or greenhouse gas emissions (except during plant construction, fuel production, and fuel processing).
- Power density of fuel source: one refined uranium fuel pellet about equal to the size of one's fingertip provides as much energy as one ton of coal, 150 gallons of oil, or 17,000 ft^3 of natural gas.
- Dispatchable baseload generation resource with high capacity factor.
- Very low marginal cost of operation, including fuel.

14.5.4 Potentially Negative Attributes of Nuclear Power

- Safety concerns/potential for large-scale disaster/public's "fear factor."
  - Potential target for terrorists
  - Hurricanes or tornadoes
  - Earthquakes
  - Tsunamis
  - Prior accidents at Fukushima (Japan), Chernobyl (Ukraine), and Three Mile Island (Pennsylvania, United States)
- Limited reserves of fuel source (uranium): 30- to 60-year known supply.
- Nuclear waste disposal (Yucca Mountain stalled for years).
- Life of nuclear waste (~10,000 years).
- Potential for radiation exposure.
- Cooling water requirements.
- Opposition from environmentalists.
- A very long and expensive permitting process can be expected, with a high potential for cost overruns.
- Land use and environmental issues related to mining and processing of fuel.
- Increased potential for proliferation of nuclear materials used for nonpeaceful purposes.
- High cost of decommissioning plants at the end of their useful life due to need for disassembly and disposal of highly radioactive components.
- Significant expense for government oversight and regulation, as well as catastrophic accident liability protection of plant owners.

14.6 COAL-FIRED GENERATION

14.6.1 Overview

Coal-fired generation produces more electricity in the United States than any other source, but the construction of new coal-fired plants slowed dramatically after the mid-1980s. As shown in Table 14.2, installed coal generation capacity in the United States declined by about 8000 MW during 2012, and the increasingly stringent emissions regulations will make it difficult to construct new coal-fired power plants without including carbon capture and sequestration technologies as part of the plant. In addition, strong opposition from many environmentalists and low natural gas prices has led to the cancellation of many projects and the rejection of permit applications by some state regulatory agencies in recent years. Other significant issues that the coal power industry routinely deals with are transportation costs of coal, health and safety of coal miners, and disposal of solid wastes.

14.6.2 Statistics from US EIA

- Installed capacity: 309.7 GW nameplate (end of 2012) [2].
- Electric generation: 1,514,043 GWh (calendar year 2012).
- Capacity factor: 55%.
- Percent of US electric production: 37%.
- Land requirements: Coal plants, including cooling ponds, require about 1–2 acres per MW of generation capacity. If the area of a surface mine required to fuel the coal plant is included, land use requirements are many times this amount, although much of the area used for surface mining may be reclaimed for its traditional use after the coal deposits have been removed.
- Water use: 300–500 gallons of water is consumed per MWh generated, with much more water withdrawn and returned to a reservoir.
- Fuel reserves: According to the National Energy Technology Laboratory, at current rates of consumption, coal could meet US needs for more than 250 years.
- Fuel sources: The United States is a net exporter of coal. Table 14.3 shows leading coal-producing states.

14.6.3 Positive Attributes of Coal-Fired Generation

- Dispatchable baseload generation resource with high capacity factor
- Low marginal cost of operation, including fuel
- Centuries of domestic fuel reserves, thus not contributing to trade deficits
- Adds significant amounts of jobs and tax base in areas where plants are located

TABLE 14.3 Top Coal-producing States in 2012

Rank	State	2012 Production (thousand short tons)
1	Wyoming	401,442
2	West Virginia	120,425
3	Kentucky	90,862
4	Pennsylvania	54,719
5	Illinois	48,486
6	Texas	44,178
7	Indiana	36,720
8	Montana	36,694
9	Colorado	28,566
10	North Dakota	27,529
11	Ohio	26,328
12	New Mexico	22,452
13	Alabama	19,321
14	Virginia	18,965
15	Utah	17,016

Source: R. Walker based on data from US-EIA's *Annual Energy Review 2012* [6].

14.6.4 Potentially Negative Attributes of Coal-Fired Generation

- Emissions: Coal generation accounts for about 40% of all CO_2 emissions in the United States plus large amounts of NO_2, SO_2, mercury, particulate matter, and radioactive trace elements such as uranium and thorium.
- Some emissions can be removed at a reasonable price, but disposal can be a problem; adding carbon capture and sequestration to coal plants will significantly increase the cost of energy.
- Large amount of water required for cooling and large volumes of high-temperature water may be returned to rivers or lakes (thermal pollution).
- Transportation of coal by rail can mean coal plant operations are vulnerable to rail strikes or work stoppages.
- Impacts of coal mining on the environment are as follows:
  - Eliminates existing vegetation.
  - Displaces or destroys wildlife and habitat.
  - Degrades air quality.
  - Alters current land uses.
  - May result in a scarred landscape with no scenic value if site remediation is not done properly.
  - Mine tailings (waste leftover after certain processes, such as from an ore-crushing plant or in milling grain) produce acidic mine drainage, which can seep into waterways and aquifers.
  - Mining operations can cause releases of methane.
  - Health and safety of coal miners (pneumoconiosis or black lung disease).
  - Collapse of underground mine tunnels can cause subsidence of land surface.

14.6.5 Emissions Associated with Coal-Fired Generation

Atmospheric emissions and hazardous solid wastes from incineration of coal are the largest environmental issues faced by the coal power industry. Emissions from coal-fired generation can include the following:

- Carbon dioxide (CO_2), a greenhouse gas.
- Nitrogen dioxide (NO_2) emissions, a contributor to smog in urban areas and to acid rain.
- Sulfur dioxide (SO_2) emissions, which can cause acid rain.
- Mercury emissions, which can harm the brain, heart, kidneys, lungs, and immune system of people of all ages. Research shows that high levels of methylmercury in the bloodstream of unborn babies and young children may harm the developing nervous system, making a child less able to think and learn.

TABLE 14.4 Emission Rates From Fossil-fueled Plants and Price Impacts of Adding Carbon Capture and Sequestration

	Emissions (kg/MWh)				
Technology	Greenhouse gases (CO_{2-e})	NO_X	SO_x	Particulate matter	Levelized cost of energy ($/kWh)
Natural gas combined cycle (NGCC)	467	0.37	0.01	0.01	0.09
NGCC with carbon capture and sequestration	137	0.43	0.02	0.01	0.13
Integrated coal gasification combined cycle (IGCC)	931	0.30	0.03	0.08	0.12
IGCC with carbon capture and sequestration	217	0.27	0.04	0.06	0.16
Super critical pulverized coal plant (SCPC)	943	0.32	0.38	0.07	0.09
SCPC with carbon capture and sequestration	241	0.44	0.04	0.10	0.16
Existing pulverized coal plant (EXPC)	1109	2.06	2.38	0.70	0.03
EXPC w/carbon capture and sequestration	444	0.46	1.31	0.04	0.13

Source: NETL [7].

Notes: Greenhouse Gas (GHG) emissions include: CO_2, CH_4, N_2O, and SF_6 in CO_2 equivalences.

Natural gas price assumption = $6.76/MMBtu in year 1 escalating at 1.87% annually.

Coal price assumptions = $1.51/MMBtu ion year 1 escalating at 1.87% annually.

- Particulate matter, which contributes to asthma, increased respiratory problems, and cardiac mortality.
- Radioactive trace elements in coal such as uranium and thorium, making many coal plants much greater emitters of radioactive materials than comparable-sized nuclear power plants.

Table 14.4 shows average emission rates of various fossil-fueled power plants, including the effect of adding carbon capture and sequestration. The emission rates are in kilograms per MWh. Note that an average home consumes about 1 MWh of electricity per month, so the information in this table can provide individuals an idea of how much their electricity use may affect the atmosphere. The information in this figure also is indicative of the much greater emissions from existing coal plants as compared with combined cycle natural gas–fired plants.

Note the significant impact that adding carbon capture and sequestration systems to fossil-fueled plants has on the levelized cost of energy. This can increase the cost of electricity from natural gas–fired generation by about 40% and the cost of electricity from new coal-fired generation by almost 80%.

14.7 NATURAL GAS–FIRED GENERATION

14.7.1 Overview

Natural gas is an abundant energy source that is created by the decomposition of organic matter. Natural gas can be imported, either by pipeline or by shipping in liquefied natural gas. Only about 12–15% of the natural gas consumed in the United States is imported, with Canada providing the most imports to the United States. Texas is the nation's largest producer and consumer of natural gas, providing one-fourth of US supplies and consuming one-sixth, primarily in the industrial and electricity-generation sectors.

Most natural gas generators are designed to change output either up or down very quickly (called "cycling") and can be started up more quickly than coal-fired power plants or nuclear power plants. Therefore, they are often the resource used by utilities to meet peak load requirements, as depicted in Figure 14.5. Natural gas–fired plants also produce significantly less emissions per unit of energy produced when compared with coal-fired plants.

As discussed earlier in the chapter, electricity can be produced from natural gas generators in several different ways or configurations that may have a large impact on how efficiently natural gas is burned, how much water is required, and the amount of air emissions produced. This can include the use of steam generators; gas turbines or CTs; combined cycle generators; small distributed generation devices such as stand-alone engine-generator sets; microturbines; or fuel cells that allow homes, businesses, or industries to install their own on-site generation. Reasons for doing so may include cost control, backup power supply, premium power quality, or environmental reasons.

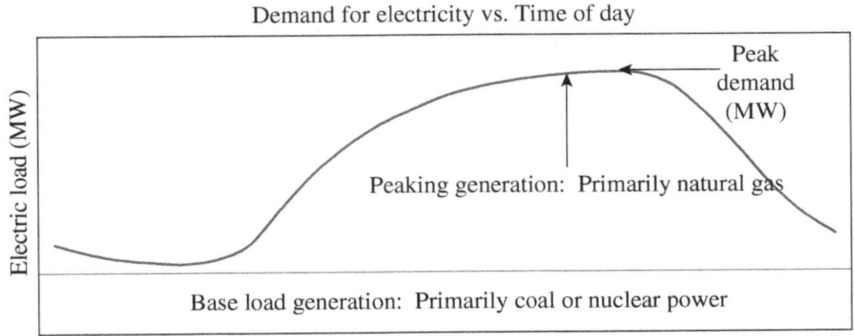

FIGURE 14.5 Example of natural gas generation being used to meet peak load (Source: R. Walker).

In 1978, Congress passed the Powerplant and Industrial Fuel Use Act restricting construction of power plants using oil or natural gas as a primary fuel and encouraged the use of coal, nuclear energy, and other alternative fuels, thinking that supplies of oil and gas were constrained. But in the following decade, it became clear that there were ample supplies of natural gas, allowing for the repeal of those sections of the Fuel Use Act that restricted the use of natural gas by electric utilities.

14.7.2 Statistics from US EIA

- Installed capacity: 422 GW nameplate (end of 2012) [2].
- Electric generation: 1,225,894 GWH (calendar year 2012).
- Capacity factor: 33.3%.
- Percent of US electric production: 30.3%.
- Land requirements: Natural gas power plants such as SCCTs may require less than 1 acre per MW of generation capacity; steam units that may have cooling ponds or lakes may require about 1–5 acres per MW of generation capacity. In addition, land is required for gas wells, pipelines, and compressor stations.
- Water use: 100–200 gallons per MWh consumed, although some plants may withdraw large amounts of water returning most of it to a reservoir.
- Fuel reserves: As shown in Figure 14.6, proven reserves of natural gas in the United States have been increasing steadily in recent years due in part to shale gas development in Louisiana, Arkansas, Texas, Oklahoma, and Pennsylvania. US proved reserves were estimated to be 350 trillion cubic feet at the end of 2011, or over 13 times the annual consumption of natural gas.

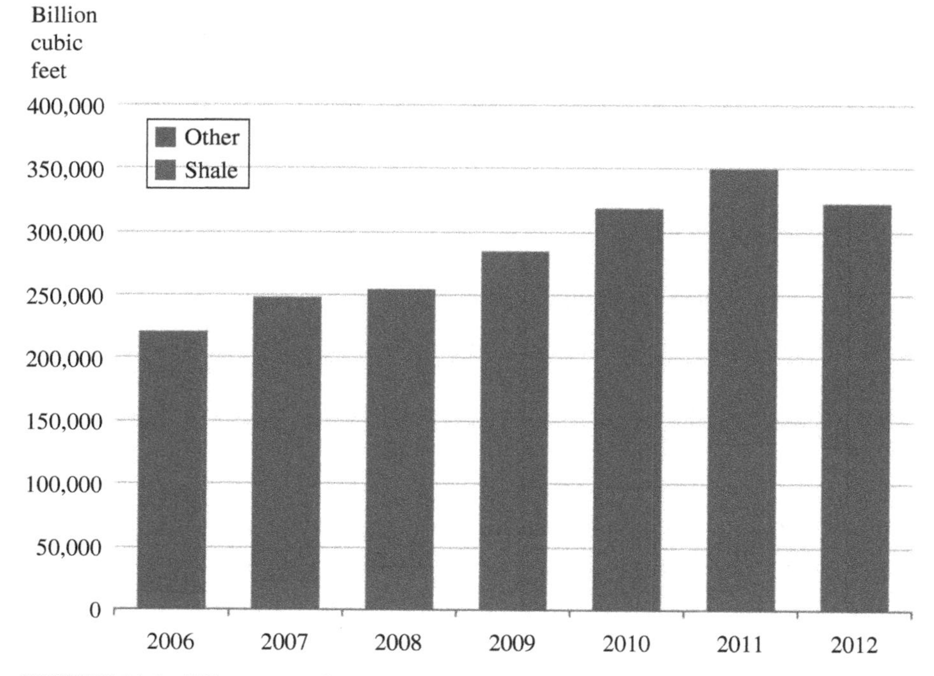

FIGURE 14.6 US wet natural gas proved reserves (Source: US-EIA [8]).

- Fuel sources: The United States typically produces about 85–88% of the natural gas consumed in the country, with Canada providing most of the imported gas. Only a small amount of natural gas is exported by the United States, which amounted to less than 1% of the amount produced during 2012.

14.7.3 Positive Attributes of Natural Gas–Fired Generation

- Fast-starting dispatchable generation resource capable of being cycled to react to changes in electric load or changes in the output of other generators, such as when power plants unexpectedly go out of service or when winds driving wind turbines rapidly increase or decrease.
- Primarily relies on domestic fuel sources, although reliance on imports can contribute to trade deficits.
- Significantly lower emissions than coal generation.
- Scalable—microturbines to large multiunit plants.
- Low capital cost:
  - $650/kW for CT (heat rate ~ 10,000 Btu/kWh)
  - $950/kW for combined cycle (heat rate ~ 7000 Btu/kWh)

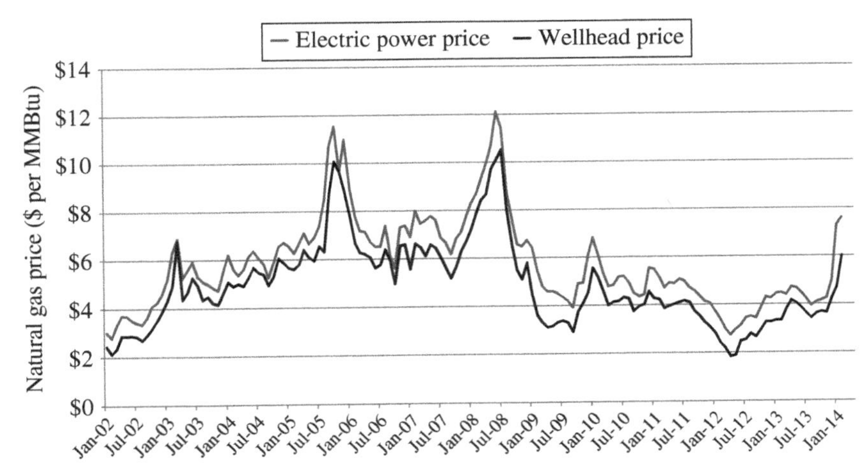

FIGURE 14.7 Monthly average natural gas price variability (wellhead price and price for electric generation in $ per MMBtu) (Source: R. Walker based on data from U.S. Energy Information Administration, monthly average prices of natural gas for electric power [9]).

14.7.4 Potentially Negative Attributes of Natural Gas–Fired Generation

- Variability of natural gas prices causes problems for consumers and utility companies (see Fig. 14.7).
- Rapid growth of natural gas use in the United States as well as developing nations such as China could reduce proven reserves, putting upward pressure on price.
- Poor heat rate of older units, many of which are used only in emergency situations.
- Emissions including CO_2 and NO_X (although significantly less than coal-fired generation).
- Cooling water requirements.

14.8 OIL-FIRED GENERATION

14.8.1 Overview

Reliance on petroleum to generate electricity in the United States has declined steadily since 1979, as shown in Figure 14.8. In 1979, almost 17% of electricity in the United States was produced using petroleum compared with less than 1% in 2011. The Powerplant and Industrial Fuel Use Act of 1978 discussed in Section 14.7 was largely responsible for this decline. Since very little of the nation's electric energy is generated with oil, one has to be careful when discussing the ability of renewable resources such as biomass, wind energy, or solar energy to reduce energy imports. Petroleum imports are primarily a function of our transportation needs.

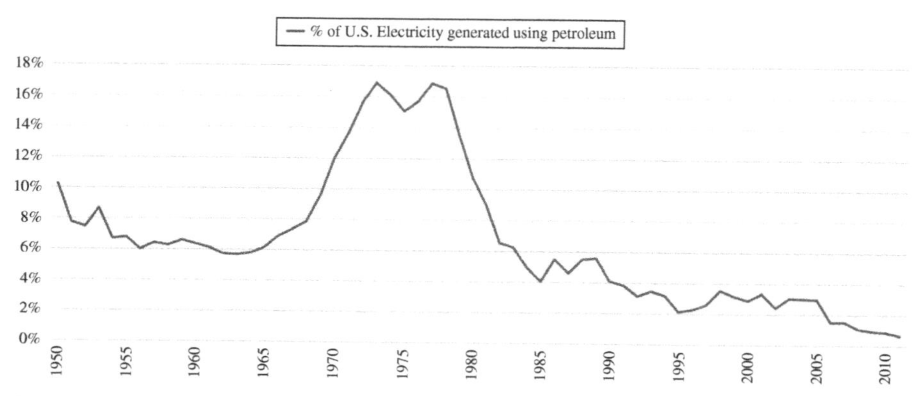

FIGURE 14.8 Percentage of US electricity produced using petroleum (Source: R. Walker based on data from US-EIA's Electricity Annual (2012) page on US-EIA website [1]).

In order to make an argument that renewable energy can reduce petroleum imports, one has to bring up other changes that would need to occur such as the following:

- Increased use of wind or solar energy for electricity generation combined with less use of natural gas for electricity generation, allowing more use of natural gas as a transportation fuel
- Improved electric vehicle and battery technologies that enable a transition away from gasoline vehicles to electric and hybrid-electric vehicles
- Increased use of biofuels for transportation purposes

14.9 METHANE HYDRATES

14.9.1 Overview

A huge, virtually untapped source of fossil fuel is methane hydrates, gas hydrates, or methane clathrate. These are crystalline solids consisting of gas molecules, usually methane, each surrounded by a cage of water molecules. They look very much like water ice and are found in deep-sea sediments several hundred meters thick directly below the sea floor and in association with permafrost in the Arctic. They are not stable at normal sea-level pressures and temperatures, which makes them a challenge to study.

The US EIA's website [10] includes the following statement regarding the potential amount of energy available from methane hydrates: "According to the United States Geological Survey, the world's gas hydrates may contain more organic carbon than the world's coal, oil, and other forms of natural gas combined. Estimates of the naturally occurring gas hydrate resource vary from 10,000 trillion cubic feet to more than 100,000 trillion cubic feet of natural gas. Tapping such resources would require significant additional research and technological improvements."

The U.S. Geological Survey (USGS) believes that study of methane hydrates is important for the following three reasons:

- They may contain a major energy resource.
- They may pose a significant hazard because they can alter seafloor sediment stability, influencing collapse and landslides.
- The hydrate reservoir, if released, may have a strong influence on the environment and climate because methane is a significant greenhouse gas, as discussed in Chapter 16.

Mapping conducted by the USGS off the coasts of North Carolina and South Carolina showed large accumulations of methane hydrates [11]. Two of these areas, each about the size of the State of Rhode Island, show intense concentrations of gas hydrates, which USGS scientists estimate may contain more than 1300 trillion cubic feet of methane gas, an amount representing more than 50 times the natural gas consumption of the United States in 2011.

While such an abundant supply of fossil fuel may sound appealing, many people are convinced that the release of greenhouse gases into our atmosphere is causing global climate change, so these huge deposits of methane might be viewed by many as an immense danger to our planet. The U.S. Department of Energy's Methane Hydrate R&D Program includes research on the potential implications of the global carbon cycle, long-term climate change, seafloor stability, and future energy policy.

Japan, being a nation with few onshore domestic natural resources, is highly interested in being able to tap into this huge potential energy resource, and in March 2013, a 6-day offshore research project resulted in production of about 120,000 m^3 of gas from methane hydrate [12].

14.9.2 Positive Attributes of Using Methane Hydrates for Electricity Generation

- Apparent huge supply
- Cleaner-burning than oil or coal, so if used to displace coal-fired generation and petroleum as a transportation fuel, could help reduce the rate of carbon emissions

14.9.3 Potentially Negative Attributes of Using Methane Hydrates for Electricity Generation

- Emissions including CO_2 and NO_X.
- May inhibit transition to renewable energy.
- Given vast supply, could allow for a significant increase in methane emissions adding to the total greenhouse gases in the atmosphere.
- No proven method for safe, cost-effective extraction.

- Significant safety concerns associated with drilling through hydrate zones, which could destabilize supporting foundations for platforms and production wells.
- Drilling for methane hydrates could cause disruption of the ocean floor, surface slumping or faulting, which could endanger work crews and the environment.

REFERENCES

[1] U.S. Energy Information Administration website. Electricity annual. Available at http://www.eia.gov/electricity/annual/. Accessed December 18, 2013.

[2] U.S. Energy Information Administration. *Electric Power Annual 2010*. Washington (DC): U.S. Energy Information Administration; 2011.

[3] Nuclear Regulatory Commission website. Available at http://www.nrc.gov. Accessed August 12, 2012.

[4] World Nuclear Association website. Available at http://www.world-nuclear.org/info/Facts-and-Figures/. Accessed December 18, 2013.

[5] U.S. Energy Information Administration website. Nuclear page. Available at http://www.eia.gov. Accessed August 12, 2012.

[6] U.S. Energy Information Administration website. Annual Coal Report, 2013. Available at http://www.eia.gov/coal/. Accessed December 18, 2013.

[7] National Energy Technology Laboratory. Life cycle analysis: power studies compilation report. U.S. Department of Energy/National Energy Technologies Laboratory report DOE/NETL-2010/1419, Pittsburg (PA); 2010.

[8] U.S. Energy Information Administration. Table 1 from U.S. crude oil and natural gas proved reserves 2011, Washington (DC): U.S. Energy Information Administration; 2013, April 10, 2014.

[9] U.S. Energy Information Administration website. Monthly average prices of natural gas for electric power. Available at http://www.eia.gov/dnav/ng/ng_pri_sum_a_epg0_peu_dmcf_m.htm. Accessed November 24, 2013.

[10] U.S. Energy Information Administration website. Potential of gas hydrates is great, but practical development is far off. Available at http://www.eia.gov/todayinenergy/detail.cfm?id=8690. Accessed December 19, 2013.

[11] Dillon, W. (1992) *Gas (Methane) Hydrates—A New Frontier.* U.S. Geological Survey, Marine and Coastal Geology Program factsheet, Reston (VA).

[12] Tumagai, T. Japan urges US to move forward methane hydrate cooperation agreement, Platts; 2013.

15

ENERGY EFFICIENCY AND CONSERVATION: "LESS IS MORE" AND "DOING WITHOUT"

15.1 OVERVIEW

Previously in this text we have referred to energy consumption, the cost of electricity, energy efficiency, conservation, and related concepts. For example, in Chapter 1, which the reader may want to refer back to, energy efficiency was discussed on a societal scale to include energy use, quality of life, and Jevon's Paradox related to energy efficiency and public policy. Also in Chapter 2, billing for residential electricity consumption was discussed with how to read an electric bill, the cost to operate various home appliances, and the concept of electricity generation and peak demand from the utility. Additionally, in Chapter 14, generation efficiencies were discussed and the concept of calculating efficiency as the ratio of the energy out of a process to the energy in and the concept of heat rate from thermal generation resources were discussed. This chapter will discuss in more detail these topics as well as introduce other topics related to energy efficiency and conservation. This chapter will include the following:

- Discussion of concepts
- Consumer energy efficiency and demand-side management (DSM)
- Generation and transmission efficiency
- Conserving resources and reducing emissions
- Positive and limiting attributes of efficiency and conservation

Wind Energy Essentials: Societal, Economic, and Environmental Impacts, First Edition.
Richard P. Walker and Andrew Swift.

15.2 CONCEPTS

How can we reduce consumption of energy resources while providing an adequate level of services from our energy resources? Saving energy is generally considered a plus for both the individual and the society. Saving energy saves money, reduces the use of natural resources, and reduces demand for energy services (such as electricity) making more available for others on the system. One can save energy by being more efficient with energy use or by conservation—both save energy but in different ways. *Energy efficiency* means to use less energy to provide equal or more of the desired energy service. *Energy conservation* (for purposes of the discussion in this chapter) means to use less energy by doing without the desired energy service. Everyone is familiar with saving gas for driving; so let us begin with that as an example to illustrate these two concepts and then apply them to the electricity sector, wind energy generation, and society in general.

15.2.1 An Example: Driving and Saving Gasoline

We all want to save money by using less gas in our vehicles, but we also need mobility for getting to work, to the store, or other places. The easiest approach to reduce the amount of gas consumed in our vehicles is to do without by not driving, such as cutting out a trip to the store—or perhaps walking or riding a bicycle instead. This is the *conservation* approach. It conserves gasoline (energy) and saves money by doing without the services provided by your car. Being a well-informed driver, you know that if you keep your tires properly inflated, avoid quick starts and stops, and drive at slower speeds, you can save gas by operating your car more efficiently. You do not have to *do without*, but you are still able to reduce the consumption of fuel and thus save both money and natural resources. You do not save as much as doing without (i.e., not driving), but you still do the things you want and need to do with the services provided by your car. This is *operational energy efficiency*.

Next, you realize you can really save a lot of gas money while still using your car if you trade in your gas guzzler for a new hybrid gas–electric vehicle. Instead of getting 15 miles/gallon with your old car, you could be getting 40 or 50 miles/gallon. The hybrid can provide excellent fuel efficiency even while doing all of the driving you want and saving lots of money on gas. If you purchased an electric plug-in vehicle, the gas savings could be even higher. (Although your electric bill would increase as you consume more electrical energy to power your car.) This is *replacement efficiency*—replacing an energy-consuming device with one that is more efficient. In this example, however, you may find that you cannot afford the hybrid because it is just too expensive. You consider maybe a high mile per gallon compact car that is cheaper to purchase, good for the environment since there are reduced emissions, saves a lot on gas, but it has little room to carry your things and it is not very "cool." What to do? These are energy decisions made by consumers every day and can be applied to all aspects of electrical energy production, transmission, and consumption.

Figure 15.1 shows wind-powered vehicles that use no fuel and although some travel at high speeds in a good wind, they are not very practical for general transportation. Figure 15.2 shows students in Texas Tech University's "Run on the Wind" summer camp land-sailing and working on wind turbine models.

(a)

(b)

FIGURE 15.1 (a) Land-sailing is a popular sport on dry lake beds in the West with enthusiasts pointing to high-speed racing without fuel costs or emissions. The Ecotricity "Greenbird" holds the land speed record for a wind-powered land-sailer of 126 mph in a 40+ mph wind; (b) Graduate students at Texas Tech University use land-sailing to demonstrate wind power principles to younger students in the annual "Run on the Wind" summer camp on the runways at Reese Technology Center in Lubbock, Texas (Photo Credit: (a) Peter Lyons, peterlyonsphoto.com; (b) A. Swift).

(a)

(b)

FIGURE 15.2 Wind energy students at Texas Tech's "Run on the Wind" summer camp land-sailing (a) and working on their wind turbine models (b) (Photo Credit: (a) Texas Tech University IDEAL office; (b) Liz Inskip-Paulk, TTU/NWI).

15.2.2 Principles of Energy Efficiency

The technical definition of energy efficiency (typically represented as "η," the Greek letter "eta") for a process is the ratio of useful energy output to the total energy input, as shown in Figure 15.3.

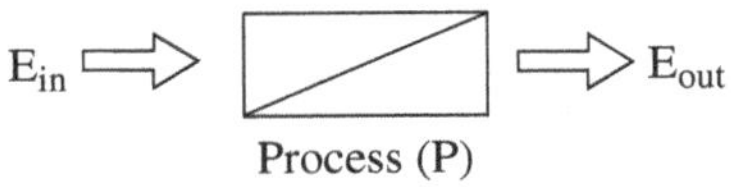

E_{in} ⇨ ⇨ E_{out}

Process (P)

FIGURE 15.3 Definition of energy efficiency process diagram.

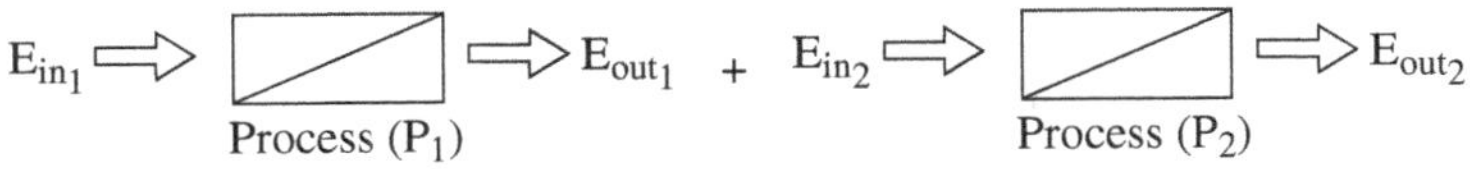

E_{in_1} ⇨ ⇨ E_{out_1} + E_{in_2} ⇨ ⇨ E_{out_2}

Process (P_1) Process (P_2)

FIGURE 15.4 Series of energy processes.

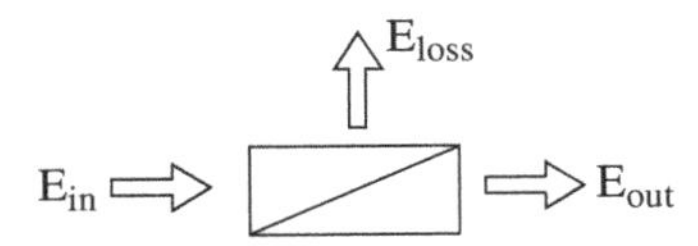

FIGURE 15.5 Process diagram for the total energy flow in a process.

In equation form, the definition of energy efficiency is written as $\eta = E_{out}/E_{in}$

A *process* can be a process of energy generation, transmission (transport), or consumption. Efficiency is a dimensionless quantity and is often given as a fraction, decimal, or percent ($\eta = ½ = 0.5 = 50\%$). Its value is always less than 1 (100%) since one cannot get out of a process all the energy that one puts into the process. There are always losses. We will return to this topic later.

Sometimes, one is interested in the overall efficiency of a series of process elements as shown in Figure 15.4.

The overall efficiency for the sequential process aforementioned is

$$\eta = \frac{E_{out_2}}{E_{in_1}} = \eta(P1) \times \eta(P2)$$

The overall efficiency of a sequence of processes is calculated by multiplying the sequential efficiency values. Return to the car example. If fuel in your car is burned with 40% efficiency and your car's mechanical efficiency (engine output to driving wheels efficiency) is 80%, the overall efficiency (fuel energy to driving energy at the wheels) is then:

$$\eta = \left(\frac{40\%}{100}\right) \times \left(\frac{80\%}{100}\right) = 0.4 \times 0.8 = 0.32 \text{ or } 32\%$$

15.2.3 The First Law of Thermodynamics (The Study of Thermal Processes and Heat Engines)

Losses are related to efficiency. The First Law of Thermodynamics, often called the Law of Conservation of Energy, states that the energy in a process is neither created nor destroyed and the total energy remains constant (Fig. 15.5).

Note: Terminology can be confusing. The scientific "Law of Energy Conservation" is different from the social policy of "Energy Conservation."

By the Law of Energy Conservation, one can write

$$E_{in} = E_{out} + E_{loss} = E_{constant}$$

Dividing by E_{in}, one can write

$$1 = \frac{E_{out}}{E_{in}} + \frac{E_{loss}}{E_{in}}$$

We now define a loss fraction, L, as E_{loss}/E_{in}. Then we have

$$\eta = 1 - L$$

or

$$L = 1 - \eta$$

For example, if your car loses 68% of the energy in the fuel burned as heat through the radiator, tire friction, and loss factors, what is the overall efficiency of the car? The answer is:

$$\eta = 1 - L = 1 - 0.68 = 0.32 = 32\%$$

Loss factors are calculated differently than efficiency values in a sequential process. In a sequential process, such as shown in Figure 15.3, loss factors must first be converted to efficiency values for each element in the process sequence and then an overall efficiency value is calculated by multiplying sequential process values, and finally an overall loss factor can be calculated. **Note that loss factors cannot be calculated by addition or multiplication of the series process elements.** This is demonstrated in the following example provided.

1. Assume that your car's engine burns fuel with losses of 60%.
2. Assume that the gears and drive train have 20% losses.

One can calculate the overall loss as shown in the table below.

Process	Loss factor	Efficiency
Engine	0.60	$=1-L=0.40$ (step 1)
Drive train	0.20	$=1-L=0.80$ (step 2)
Overall	$=1-\varepsilon=1-0.32=0.68=68\%$ Answer (step 4)	$=0.4\times0.8=0.32$ (step 3)

Where does the lost energy go? It is converted to heat energy and lost to the environment through friction or in cooling. If the lost heat energy was captured and added to the useful energy produced it would total to the original input energy.

15.3 CONSUMER EFFICIENCY AND DSM

15.3.1 Electric Energy Efficiency and the Consumer

Saving electricity at home or work is similar to the gasoline-car example, but there are some differences due to the way that electricity is generated, transported, and used. Electricity is charged to the consumer by the kilowatt-hour (kWh)—1 kW (1000 W) of electric power used for 1 h. The cost of electricity is currently about 10–15 cents/kWh used, and bills are typically issued on a monthly basis. Consumer use of electric energy is registered in an electric meter that is read monthly, and the customer typically pays at the end of the month, although some utility companies are now offering prepaid electric services. Smart meters, now becoming prevalent throughout the United States, can facilitate prepaid electric service as well as allowing the consumer to see how much he or she is using at a given point in time compared with the cost of electricity at that time, thus allowing them to choose whether to use or not use electrical devices such as air conditioners, water heaters, dishwashers, clothes washers, or dryers. Smart meters are discussed more fully in later chapters.

Gasoline for your car, however, is purchased ahead of time and stored in a tank, then used. Gasoline consumption per purchase is limited to one full tank. Utility electricity cannot be effectively stored—it must be used as generated—but there are no limits to consumption other than the electrical capacity of the grid connection. One can turn on every household appliance and run them continuously, and the owner will pay the bill at the end of the month.

As shown in Chapter 2, monthly electricity use typically is metered, billed, and then paid for the next month. If the bill is too high, one can conserve by turning things off and "doing without" the next month. While it has historically been very difficult to know one's consumption levels and electricity cost as it is being used, smart meters are now making this possible. To save electricity, one can turn off lights or increase the temperature setting of the thermostat and be a little warmer on a hot day. All of these options are "doing without" the service provided by electrical energy and meet our definition of conservation of electricity.

Increasing the efficiency of electric energy use is another option. The consumer is interested in the services provided by electricity, not in the electricity itself. Reducing the energy used, and energy costs, for the same service is to increase energy efficiency. There are two approaches. First one can operate appliances more efficiently. For example, closing the windows and doors when the air conditioner is running so conditioned air is maintained in the house and does not escape or purchasing a more energy-efficient air conditioner both increase energy efficiency. The first we have called *operational efficiency*, the second *replacement efficiency*.

FIGURE 15.6 Light bulbs: LED in foreground, standard incandescent in middle, and compact fluorescent in back (Source: eartheasy.com [1]).

Consider light bulbs as an example. Almost every home in the United States has them. Invented in 1879 by Thomas Edison in the United States, they were the first consumer electric device. The first light bulbs were installed and powered with electricity from the Pearl Street Station in New York City through the electric utility company called The Edison Illuminating Company, the first electric utility in the nation. The incandescent light bulb has modestly improved over the years and has now been a provider of lighting for a century. Office buildings have used more efficient florescent lighting for years. In the 1980s, florescent light technology was applied to a compact florescent bulb that could replace a regular incandescent light bulb, providing similar light levels with significantly less heat and less energy use. Compact florescent lights (CFLs), as shown in Figure 15.6, are very energy efficient but will be a transition technology to very high-efficiency lighting now available using light-emitting diode (LED) technology. LED bulbs use approximately one-sixth of the electricity of an incandescent bulb, deliver the same amount of light, and last about 40 times longer. However, although prices are expected to decrease over time, the LED bulbs presently cost about 30 times more than the incandescent to buy.

Table 15.1 [1] provides a comparison of the bulbs in first cost, bulb life in hours, and energy costs over the life of the LED bulb. The data indicate that the LED bulb is the most expensive to purchase but may be the best buy when all costs are considered—first cost, lifetime replacement costs, and energy costs.

These types of life cycle analyses, as they are called, are of value but also have their limits. For example, the analysis aforementioned does not take into account the time value of money. Spending about $900 today for 25 LED bulbs to replace 25 incandescent bulbs will provide projected savings of over $6600 during the expected lifetime of the LED bulbs.

TABLE 15.1 Cost, Lifetime, and Energy Use Comparison Between LED Bulbs, CFLs, and incandescent Light Bulbs[a]

	LED	CFL	Incandescent
Light bulb projected life span (h)	50,000	10,000	1,200
Watts per bulb (equivalent 60 W)	10	14	60
Cost per bulb ($)	35.95	3.95	1.25
kWh used over 50,000 h	500	700	3,000
Cost of electricity (@ $0.10/kWh) ($)	50	70	300
Bulbs needed for 50,000 h of use	1	5	42
Equivalent 50,000 h bulb expense ($)	35.95	19.75	52.50
Total cost for 50,000 h ($)	85.75	89.75	352.50
Energy savings over 50,000 h, assuming 25 bulbs per household			
Total cost for 25 bulbs ($)	2,143.75	2,243.75	8,812.50
Savings by switching from incandescent ($)	6,668.75	6,568.75	0

[a] Source: eartheasy.com [1].

Notes

- Cost of electricity will vary. The figures used in the table are for comparison only and are not exact. Residential energy costs among the various states range from 28.53 cents (Hawaii) to 6.34 cents (Idaho) per kWh.
- The cost per bulb for LED bulbs may vary; $35.95 was used in the example (for a 60-watt equivalent LED bulb) as an average among lighting retailers.
- Estimates of bulb life span are projected, since it would take about 6 years of continuous lighting to test. Some manufacturers claim the new LED bulbs will last up to 25 years under normal household use, but this is not proven.
- Bulb breakage and bulb replacement costs have not been factored into this comparison chart. Incandescent bulbs and CFL bulbs are more easily broken than LED bulbs, which increases their cost of use.
- Most LED bulbs come with a minimum 2-year guarantee. Any defective LED bulb will usually fail within this time.

This analysis assumes a reasonable use factor for the light bulbs to reach the 50,000-h lifetime of the LED bulbs. If the 25 lights are on for 6 h a day, 365 days/year, it will take over 22 years for the LED bulb projected savings of $6668 to be realized, although it would only take about 2.5 years for the energy savings to have offset the higher cost of the LED bulbs. Another consideration in switching from incandescent bulbs to CFL or LED bulbs is the reduction of heat generated by lighting. This reduction of heat in the home or building can also reduce the electricity used by air conditioners.

However, CFL or LED bulbs may not be cost-effective in all situations. Consider if one of these bulbs were to be used in an attic space, where one wants adequate lighting to safely move about. The bulb is typically used about 1 h per month to store or retrieve items from the attic space. It would take over 4000 years to reach the lifetime of the LED bulb and to recoup the initial investment savings. Obviously, for this application, one would use the cheapest possible bulb, the incandescent, since the savings would, for all practical purposes, never be realized.

A number of governments around the world, including the U.S. Federal Government, have imposed lighting efficiency standards that will phase out the incandescent bulb since it will not be able to meet the new standards [2]. In the United States, a 2007 lighting efficiency rule was to take effect in 2012 that would essentially eliminate the use of the standard 100 W incandescent bulb since these bulbs could not meet the efficiency standard imposed. The rule was not funded for enforcement in 2012 by the U.S. Congress, due to concerns of many that the government should not be determining the types of light bulbs that consumers can purchase; rather these decisions should be made by the consumer based on free market principles.

Given the complexity of energy economics and the perceived value of saving energy and resources while protecting the environment, the US government, through the Environmental Protection Agency and the Department of Energy, has implemented programs to assist consumers in making energy decisions through an Energy Guide Tag system and an *Energy Star* [3] rating for appliances, computers, air conditioners, washing machines, light bulbs, and so forth. Upon completion of specified government tests, each appliance will have an *Energy Star* rating to help the consumer buy products that are more energy efficient and cost-effective based on its initial cost, energy use, impact on the environment, and cost of comparable products. A sample tag is shown with consumer advice in Figure 15.7.

The light bulb example and Table 15.1 illustrate these complexities very clearly. Based on first cost only, and without additional information, the incandescent bulb is the obvious first choice of the consumer. However, both the economics and convenience of the product are quite different when the entire life cycle replacement costs and cost of operation are taken into account. For a lighting application requiring long hours of use and difficult bulb replacement, the LED bulb has the lowest life-cycle cost, with the highest convenience since it will not burn out for many years. Most CFL and LED bulbs are *Energy Star* rated.

Finally, and as pointed out at the beginning of this section, consumers are interested in the service and the cost of that service as provided by a particular electrical appliance. Consumers are usually not concerned about the generation, transmission, and distribution of the electrical energy to power the device. Thus, saving electrical energy is generally a matter of personal economics, although more and more people are becoming concerned about the environmental consequences of how the power they consume is produced. The primary value in saving kilowatt-hours or electrons is the money that is saved in obtaining the desired energy service. What we really want to do is get a service, such as a certain amount of light, from a device for the minimum cost while also doing what we can to conserve natural resources and protect the environment.

15.3.2 Demand-Side Management

Our discussion of electric energy efficiency has thus far been about saving energy. However, as was shown in Chapter 2, the utility must not only deliver electric energy, it must also be able to meet the electric capacity demand at all times—including what is called the peak demand—which usually occurs in summer for a utility with high air-conditioning loads. Reducing customer electric energy use and peak reduction are

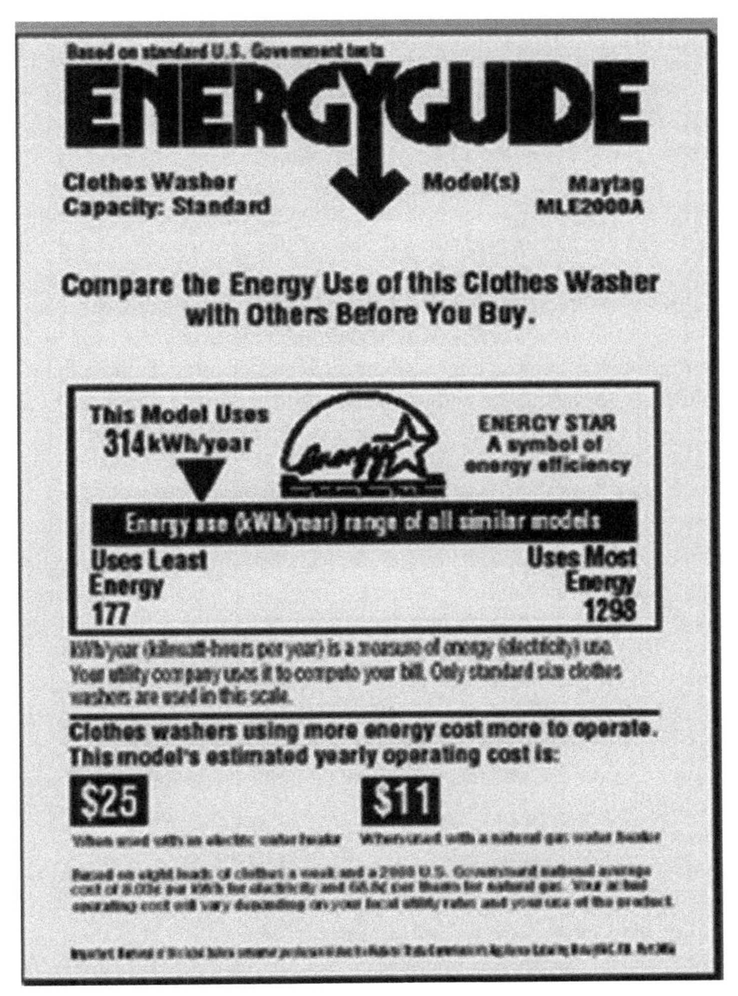

FIGURE 15.7 Energy guide/energy star label that can be found on appliances that are subject to minimum efficiency standards set by the federal government (Source: State of Delaware).

considered "demand-side management" (DSM) since conservation and efficiency are strategies that reduce both electric energy consumption and the peak demand on a utility system.

Wind Energy's Role in Meeting Peak Electricity Demand in Texas

Winds on the Texas Gulf Coast are often quite strong on summer afternoons, providing substantial amounts of wind-generated electricity during times of consumers' peak demand for electricity in the state. West Texas, however, as with most of the Great Plains, typically has its strongest winds in the nighttime hours, having less of an impact in meeting peak electric demand in the region.

Large commercial and industrial customers often have two electric meters, one for which they pay for the peak electric power demand that they use in a month (averaged over a 15-min interval) and a second meter where, like residential customers, they also pay for the energy consumed during the billing cycle. Residential rates combine these together into a single energy rate for ease of billing and simplicity.

There are several strategies for reducing the peak demand, also called *peak shaving*, which includes rotating power blackouts to large customers who in return pay a reduced rate, knowing that their power may be interrupted during peak demand times. As will be pointed out in the last chapter of the book there are new strategies for peak shaving and energy savings by individual customers with the installation of what are called *smart meters* and a *smart grid* system that could potentially have significant effects for both consumers and the utility.

15.4 GENERATION AND TRANSMISSION EFFICIENCY

In order to meet the peak demand load of electric customers, there has to be sufficient power plant capacity available to generate enough power to meet that demand. Additionally, there must be sufficient fuel resources to keep the plant operating to supply the energy needed.

Most major sources of energy for electric power generation involve heat and are called *thermal power plants*. They are powered by coal, natural gas, and nuclear energy as are described elsewhere in the book. Some sources of electrical energy conversion do not require heat. These include the renewable sources such as hydropower, wind power, and solar power. The laws and principles of energy conservation, efficiency, and losses described before also apply to the generation and transmission of electric energy. However, there are some differences. First, consider conservation: "doing without" is not really an option for a utility since, without generating power, they have nothing to sell and will go out of business. Utilities can cut back, however, to conserve resources by instituting rolling blackouts or interruptible power, but this is unusual, and actually it is the customer that is curtailed, not the generator.

Next, consider generation efficiency. Here the mix of thermal sources and renewable sources can lead to some interesting differences when considering the social impacts of electric power generation. Compare coal and wind power generation—increased efficiency at a coal power plant reduces coal consumption and emissions per unit of electrical energy produced. This conserves coal resources for later use and reduces air pollution. It also saves money for the operator since less coal is purchased per unit of electricity delivered. The same is not true for wind. Increased efficiency at a wind power plant conserves wind energy that is free to the user and cannot be stored for later use. Increased efficiency or reduction of losses at a wind plant would reduce the cost to deliver electricity since there would be more output for the same capital investment. However, no fuel resources would be conserved and no emissions would be avoided. From a technical and economic view point, the efficiency of electrical power generation from all sources is important in minimizing cost and optimizing production.

The efficiency of thermal power plants is limited by what is called the *Carnot efficiency*. This principle was first explained by Nicholas Carnot, a French physicist and engineer who, in 1824, gave the first successful theoretical account of heat engines and the concept of a maximum efficiency of conversion of thermal (heat) energy to mechanical (and electrical) energy. All thermal power plants need a source of heat energy at high temperatures and a thermal sink (source of cooling) at a low temperature. Heat is a process that moves from high to low temperature. Extracting energy from that process—that is converting it to rotating mechanical energy and driving a generator—is the process of thermal electric generation. Figure 15.8 illustrates the process.

Carnot showed that the theoretical maximum efficiency of this process is mathematically related to the temperatures involved, when measured on an absolute scale, Kelvin or Rankine. Carnot's thermal power efficiency equation is as follows:

$$\eta = 1 - \frac{T(\text{cold})}{T(\text{hot})}$$

where T is the temperature of the cold and hot reservoirs measured in degrees Rankine or Kelvin.

$$R = 460 + {}^{\circ}\text{F}$$

$$K = 273 + {}^{\circ}\text{C}$$

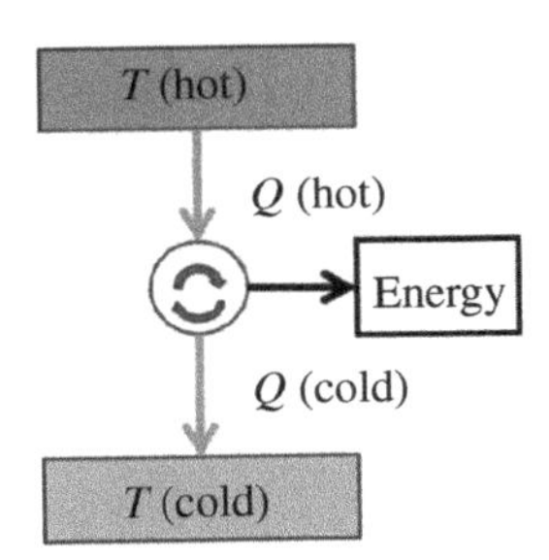

FIGURE 15.8 Shown is a schematic diagram of a thermal engine. T represents temperature, and Q represents heat energy. Heat moves from the high-temperature source T (Hot), in a controlled fashion, extracting useful energy from the process with the remaining heat energy going to the cold sink, T (Cold). Since energy is conserved by the First Law of Thermodynamics Q(Hot) = E(Energy extracted) + Q(Cold). This process can be reversed: if mechanical energy is put into the system, one can move heat from the cold reservoir to the hot reservoir. This is how a refrigerator works (Source: A. Swift).

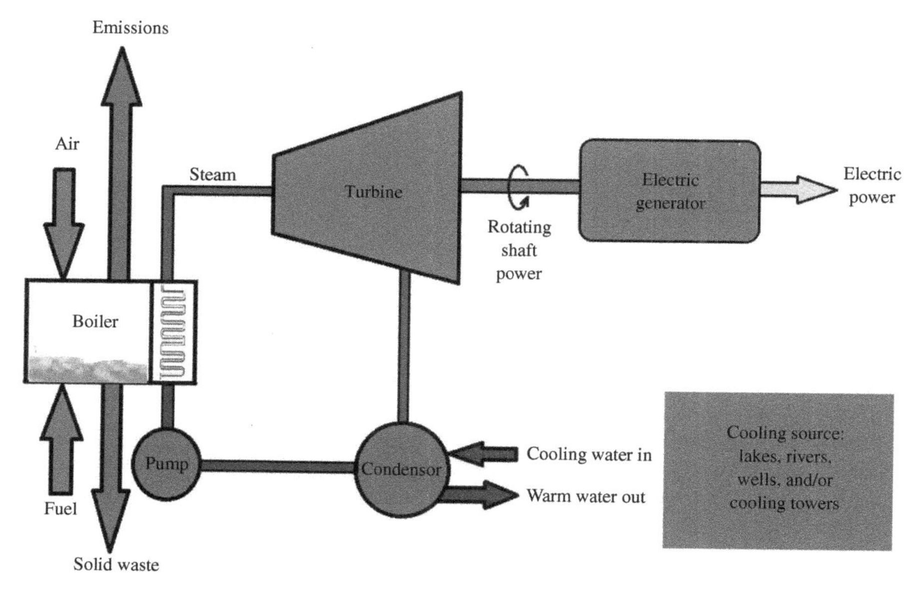

FIGURE 15.9 Schematic of a thermal power plant Rankine cycle (Source: R. Walker).

What the formula says is that the maximum possible efficiency of a thermal–mechanical process is related to the temperatures involved and that there are practical limits to this efficiency. The efficiency can only be 100% when either the absolute temperature of the cold sink is absolute zero (−273 °C or −460 °F) or the hot source is infinitely hot—neither of which is possible. Note also that if *T*(cold) = *T*(hot), the efficiency will be zero. There must be a temperature difference for the system to work.

Figure 15.9 illustrates in practical manner how a thermal power plant operates. The cycle shown is called a Rankine cycle, named after the inventor, and forms the basis of most steam power plants operating today.

Water is heated in the boiler making steam, which is expanded through the turbine. The low-energy steam is converted to liquid water by condensation in a cooling unit, which typically uses river, lake, or underground water to provide the cooling. Emissions from the fuel combustion go into the atmosphere, and solid waste, such as coal ash, is removed for disposal.

A sample problem may help clarify these concepts. A coal power plant furnace and boiler provide steam at 600 °F (1060 degrees R) to the turbine and generator and cooling water at 50 °F (510 degrees R) to the cooling unit. Calculate the maximum theoretical plant efficiency.

$$\text{Carnot efficiency} = 1 - \left(\frac{510}{1060}\right) = 0.48 = 48\%$$

TABLE 15.2 List of Various Sources of thermal Power Generation With Efficiencies[a]

Coal-fired steam turbine	32.8%
Gas-fired steam turbine	32.8%
Gas-fired gas turbine	29.4%
Gas-fired internal combustion	34.4%
Gas-fired combined cycle	44.8%
Nuclear power plants	32.6%

[a] Source: US EIA [4].

For comparison, actual coal power plants operate at 30–40% efficiency. **This means that 60–70% of the heat energy in the coal (when burned) is lost in the process.**

In summary, all thermal power plants are limited by the Carnot efficiency—as well as other factors such as electrical generator efficiency, mechanical efficiency, and various other loss factors. The electric utility power plant owner, although limited by the principles of energy conversion, has energy efficiency decisions similar to the consumer, as outlined in the previous section. The plant can be operated more efficiently making various changes in operating strategies, or there can be replacement efficiencies by installing new and more efficient equipment. Even wind energy project developers have to consider energy efficiency when evaluating turbines with different drive trains (use of turbines with gearboxes vs. direct-drive turbines) and when designing or selecting transformers and collection system conductors. Table 15.2 lists the efficiencies of various types of thermal electrical generating plants as derived in Chapter 14, Section 14.3.2.

To complete this section, we will discuss the efficiency of delivering electricity from the electric generator to the consumer. This is usually called the *transmission and distribution system*. Because all electric lines and electric equipment have electrical resistance, there are losses. These losses are related to the current and the electrical resistance using Ohm's Law, discussed previously, where the loss is related to the current squared times the line resistance as shown in the following equation:

$$\text{Loss}(\text{Watts}) = \left[\text{Current}(\text{Amps})^2\right] \times \text{Resistance}(\text{Ohms})$$

These losses represent one reason that the transmission system uses alternating current (AC power). AC transformers allow voltages to be increased at the generator transformer, resulting in a directly proportional decrease in current for the long-distance transmission line—thereby reducing the losses in the lines according to the formula aforementioned. It is significant that these losses go with the current squared so doubling the current results in a quadrupling of transmission losses. Typically, transmission losses are estimated at 10% of the energy that flows from the generator to the load but are very dependent on the wiring size, configuration, and distance from the generator to the customer load.

15.5 CONSERVING RESOURCES AND REDUCING EMISSIONS

When energy is conserved through more efficient operation, new equipment, or direct conservation—on the part of either the consumer or the electric power utility—energy resources are conserved, since most power plants are thermally powered by consuming a fuel, such as coal, natural gas, petroleum, or nuclear fuels. Fossil fuel resources of coal, natural gas, and petroleum are a result of a very long decomposition process of organic materials taking place eons in the past. These are limited in quantity; thus, conserving these resources makes them available for future use as well as limiting emissions of various gases and combustion products when these fuels are burned.

Reducing emissions from fossil fuel–fired power plants is considered an important strategy for improving human health by reducing air pollution and reducing the emission of greenhouse gases into the atmosphere (namely carbon dioxide and other combustion products). These topics will be discussed more completely in Chapter 16. Table 16.2 quantifies the major emissions from fossil fuel electric generators.

15.5.1 The Energy–Water Nexus

As diagrammed in Figure 15.9, thermal power plants use water for cooling by withdrawing it from a nearby source (such as a river, stream, lake, underground aquifer, cooling pond, or ocean) and returning it slightly hotter, to the source. While some units are air-cooled, this is the exception. This can cause thermal pollution in lakes, streams, and coastal areas, which in turn can affect wildlife and natural systems influenced by these artificially heated water sources. Power plants can also withdraw water and use *cooling towers* to evaporate water to provide cooling. Large clouds of water vapor can often be seen above a power plant cooling tower on a cool day, as shown in Figure 15.10.

The cooling process uses water in large quantities. The U.S. Geological Survey (USGS) estimates water withdrawals in the United States for various uses. Water withdrawal for thermal electric generation is the single largest use for water in the nation, even greater than agriculture irrigation, as shown in Figure 15.11. The ratio of total water withdrawals to electric energy produced, in gallons per kilowatt hour, can be calculated using the USGS estimate of thermoelectric water use with the US Energy Information Administration records of historic net power production. This ratio decreased from an average of 63 gallons per kWh during 1950 to 23 gallons/kWh during 2005.

Not all of the water withdrawn is consumed. Some water is used and returned to the source (so called *once-through cooling*), while some of the water evaporates. Water consumption for power generation in arid areas of the nation, such as the Great Plains and Southwest, is a major issue with power plant siting and operation. Thermal power plants typically consume between 0.5 and 1 gallon of water per kWh of electricity produced—much less than the withdrawal rate per kWh cited earlier.

FIGURE 15.10 Cooling tower water vapor emissions from a thermal nuclear power plant in France (Photo Credit: Stefan Kühn, http://en.wikipedia.org/wiki/File:Nuclear_Power_Plant_Cattenom.jpg).

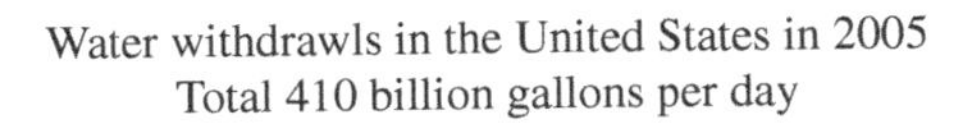

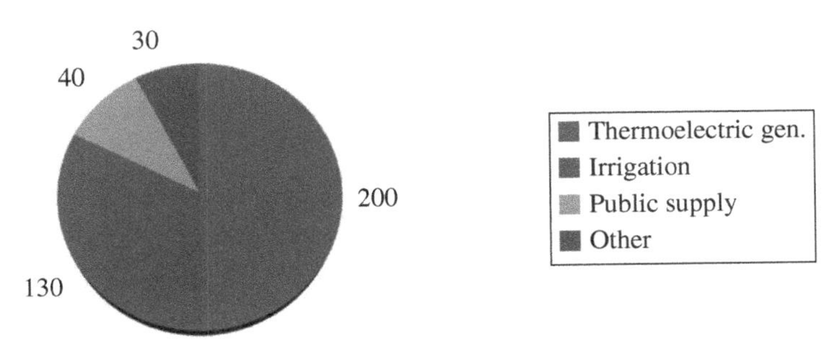

FIGURE 15.11 Water withdrawals in the United States by end use (Source: USGS [5]).

Let us look at an example. A typical US household uses 1000 kWh of electricity per month. Using an average water consumption value of ¾ gallon of water per kWh and withdrawal value of 63 gallons/kWh, calculate the amount of water consumed and the amount withdrawn by a thermal power plant to generate the electricity required by a typical US household over a month.

Example
Water consumption:

$$1000\,\text{kWh/month} \times 0.75\,\text{gallons/kWh} = 750\,\text{gallons/month} - \text{about } 1\text{ gallon/h.}$$

Water withdrawal:

$$1000\,\text{kWh/month} \times 63\,\text{gallons/kWh} = 63{,}000\,\text{gallons/month} - \text{about } 88\,\text{gallons/h.}$$

15.6 POSITIVE ATTRIBUTES OF ENERGY EFFICIENCY AND CONSERVATION

There are both positive aspects of electric energy efficiency and conservation as well as limits.

Positive attributes include the following:

- Saving money for the consumers of electrical power and increasing the efficiency of the entire electric system by maximizing services to the customer while minimizing cost of service.
- Electric load reduction through conservation and efficiency as well as improvements in generation and transmission efficiency result in the following:
  - Conservation of fossil fuel resources
  - Reduction in emissions and cooling water use from thermal power plants
  - Reduction of nuclear wastes for storage and disposal from nuclear power plants
- Peak shaving reduces the peak for electric load demand, which in turn can postpone the construction of new power plants, which can be costly and difficult to site.
- Additional job creation through the development of new higher-efficiency products and services to include installation, replacement of older equipment, and operation.

15.7 LIMITS TO ELECTRIC ENERGY EFFICIENCY AND CONSERVATION

There are also limits associated with electric energy efficiency and conservation, although they are not usually discussed because most people typically consider energy efficiency and conservation as common goods. However, a list of limits would include the following:

- Increased costs for new high-efficiency equipment, which may or may not deliver a return on the investment depending on the particular situation and consideration of life cycle effects.

- Severe efficiency and conservation measures can result in a reduction in service to a point where the service provided drops below an adequate level causing more harm than good.
- Unintended consequences of large-scale energy efficiency and conservation measures may occur if these measures are not instituted in a careful and planned manner. For example, Jevon's paradox (as discussed in Chapter 1) indicates that the wide-scale application of energy efficiency and conservation measures to reduce fuel use may actually increase the use of fuel if public policies are not instituted properly.

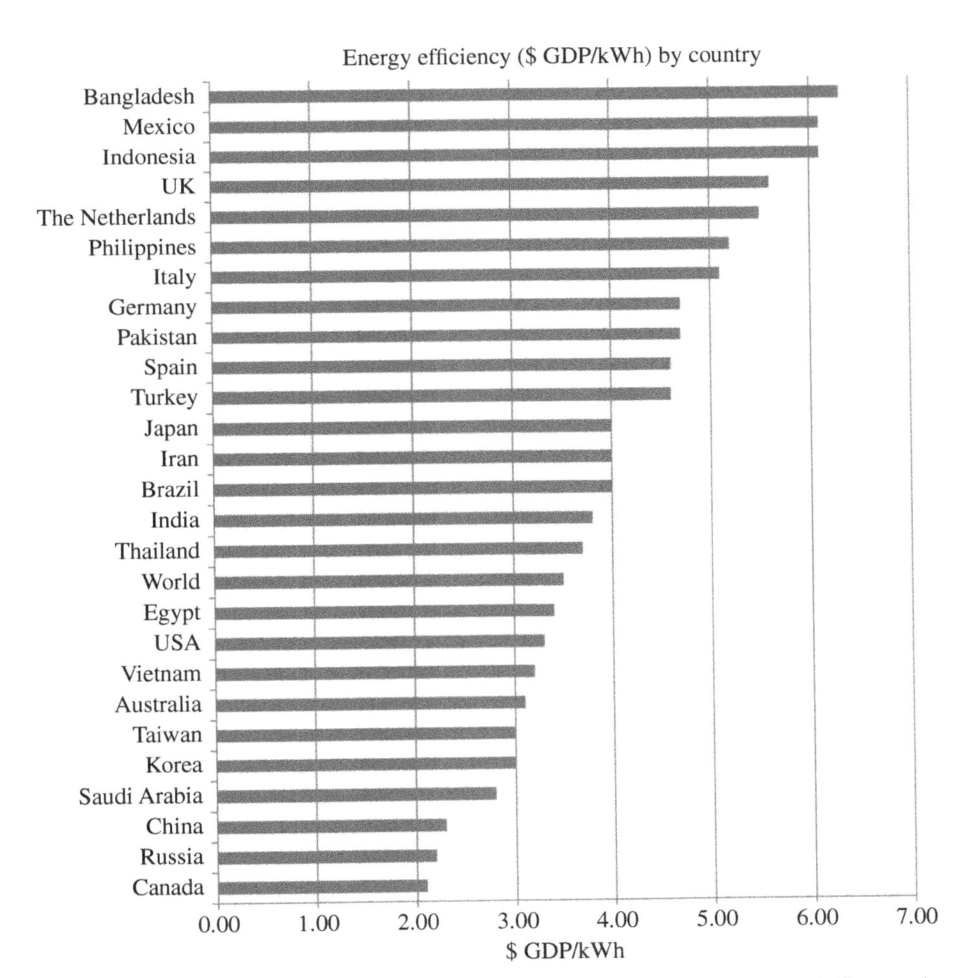

FIGURE 15.12 National energy efficiency data for countries listed in Table 1.4 (Source: A. Swift based on data from Table 1.4).

- Public policy to reduce power plant emissions, such as a *Cap and Trade* emissions program as proposed in the United States in 2009, did not obtain congressional approval due to its potential negative consequences on the economy and job creation in the United States during a time of recession.
- Energy efficiency and conservation have inherent limits. For example, a thermal power plant cannot exceed the Carnot efficiency—although there is obviously significant room for improvement from current thermal efficiency levels to the ideal level of the Carnot cycle.
- There are limits to both energy efficiency and conservation measures at which point new sources of generation have to be identified and put into use to provide the necessary electric services required by customers.

Before completing this chapter on energy efficiency, it is useful to consider the overall effects of energy efficiency on a country-by-country basis. Figure 15.12 shows GDP per kWh, a measure of national energy efficiency, for countries listed in Chapter 1, Table 1.4. Notice that the United States is in the lower third of this list indicating that there is still room on a national basis for energy efficiency improvements as compared with other industrialized countries.

Next, quality-of-life index data from Chapter 1, Table 1.3 for each country can be compared with the calculated national efficiency data and is shown in Figure 15.13. The scatter of the data shows that there is no direct correlation between the quality-of-life index and energy efficiency. Additionally, one can calculate GDP per person for each country as a measure of personal wealth and compare that data with national energy efficiency. These data are shown in Figure 15.14 and also indicate that there is no direct correlation between GDP per person and energy efficiency. [Note that PPP refers to "purchasing power parity."]

These results demonstrate that energy efficiency in and of itself should be a limited and carefully considered national objective, and national goals of energy

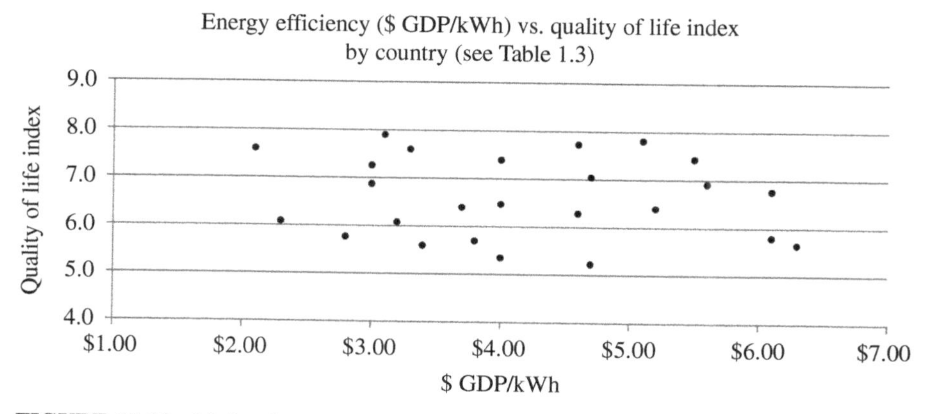

FIGURE 15.13 National energy efficiency and quality-of-life index data from Table 1.3 (Source: A. Swift).

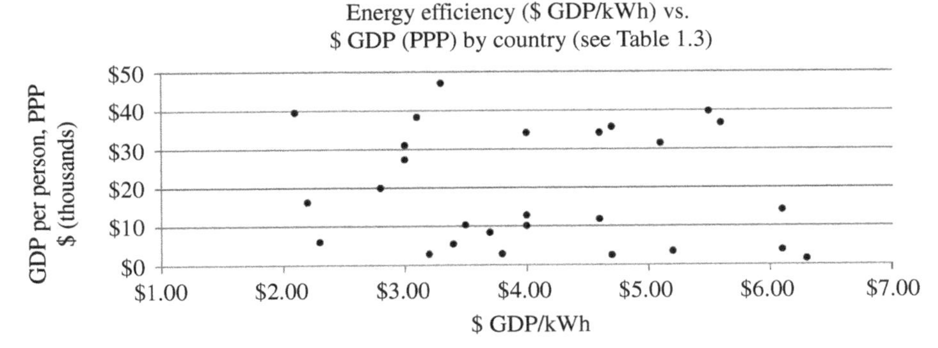

FIGURE 15.14 National energy efficiency data versus personal wealth indicated by GDP per person from Table 1.3 (Source: A. Swift).

efficiency have to be carefully balanced with other public policies to obtain the best overall electric utility system. Developing countries will obviously want to achieve a better quality of life through increased use of energy, but at some point, the efficiency with which energy is used must be considered in order to improve economic competitiveness and to reduce adverse effects on the environment.

REFERENCES

[1] Eartheasy: Solutions for Sustainable Living website. Energy efficient lighting page. Available at http://eartheasy.com/energy-efficiency/energy-efficient-led-lighting. Accessed August 8, 2012.

[2] J Snyder. Congress Spares Incandescent Bulbs in Victory for U.S. Tea Party, Bloomberg L.P., Bloomberg; 2011.

[3] Energy Star website. Available at www.energystar.gov/index.cfm. Accessed August 8, 2012.

[4] U.S. Energy Information Administration. *Electric Power Annual 2010*. Washington (DC): U.S. Government Printing Office; 2011.

[5] U.S. Geological Survey. *Estimated Use of Water in the United States in 2005*. Reston (VA): U.S. Geological Survey; 2009. Circular 1344.

16

GREENHOUSE GAS EMISSIONS

16.1 OVERVIEW

As has been mentioned in previous sections, the use and combustion of fossil fuels to generate electrical power produces emissions, some of which are called *greenhouse gases* (GHGs) because they contribute to what is called the *greenhouse effect*. This chapter will examine the following:

- The science of the greenhouse effect and related global warming
- The relation to fossil fuel emissions
- Global climate change and global warming, including public policy and public opinion
- Mitigation strategies related to carbon sequestration, a process by which carbon is captured when burning fossil fuels rather than being emitted and then stored deep underground or deep in the ocean

16.2 THE GREENHOUSE EFFECT

The fundamental science of the greenhouse effect has to do with the selective transmission and absorption of energy depending on the wavelength of the source. All bodies emit electromagnetic radiation proportional to the fourth power of their

Wind Energy Essentials: Societal, Economic, and Environmental Impacts, First Edition.
Richard P. Walker and Andrew Swift.

temperature and in a spectrum of wavelengths also proportional to their temperature. High-temperature bodies, like the Sun, emit significant amounts of what is called short-wavelength radiation, while cooler bodies such as the Earth emit large quantities of what is called longer-wavelength radiation.

Because longer-wavelength radiation is closer to the red end of the spectrum, it is often called infrared radiation. It turns out that regular window glass is a material that selectively absorbs and transmits energy of different wavelengths, as depicted in Figure 16.1. For a selective absorber such as glass, short-wavelength energy from a high-temperature body, like the Sun, is easily transmitted whereas longer-wavelength energy closer to the infrared part of the spectrum is absorbed. This effect is used quite effectively in the greenhouse industry to let solar radiation in and block longer-wavelength radiation, keeping the temperature inside the greenhouse warmer than the outside air temperature, which can protect plants from freezing and maintain a warmer climate within the greenhouse. You can observe this phenomenon on a hot summer day in your car when you park in the Sun. The temperature inside the car rises substantially from the greenhouse effect of the Sun's short-wavelength radiation passing through the windows while the long-wavelength radiation emitted inside the car is blocked by the glass. As you know, the temperature can get very high inside a car sitting in the Sun.

Certain gases in the Earth's atmosphere exhibit this same property of selective absorption and are called the GHGs, as shown in Figure 16.2. The presence of these gases and the greenhouse effect itself are vital to life on Earth. Without them, the average temperature of the Earth would be about 30°C (54°F) cooler and the Earth would be covered in ice.

Four major GHGs—namely water vapor, carbon dioxide, methane, and nitrous oxide—are the result of the use and combustion of fossil fuels. Although water vapor is the most prevalent of these gases in the atmosphere and has the most pronounced greenhouse effect as compared with the other gases, it is usually not considered in this discussion, due to the large quantities naturally in the atmosphere.

In Figure 16.3, from the Barrett Bellamy Climate Web site [2], the red area in the figure represents the Sun's spectrum absorbed by the atmosphere and the Earth's

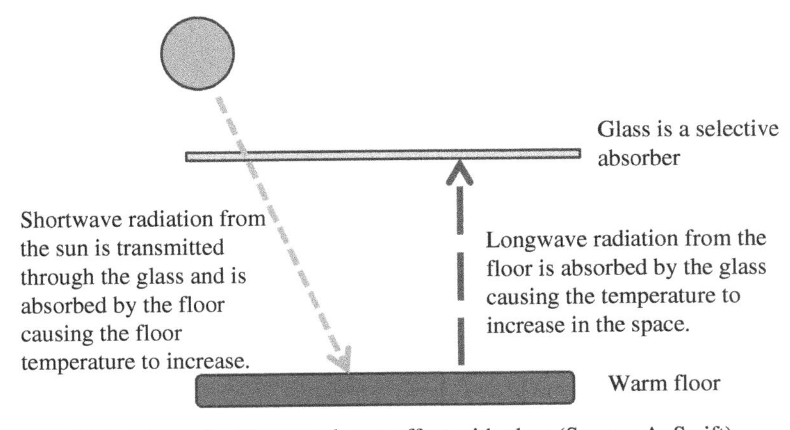

FIGURE 16.1 The greenhouse effect with glass (Source: A. Swift).

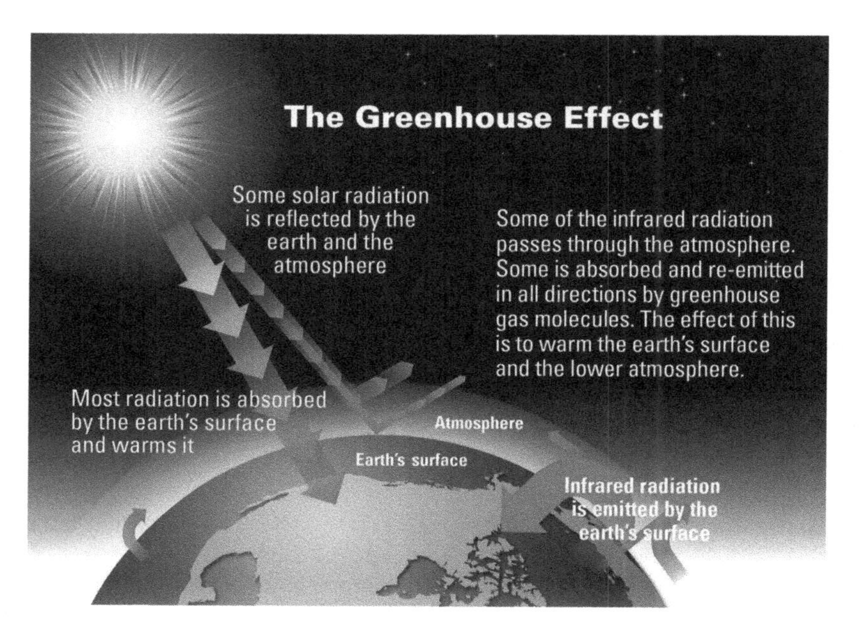

FIGURE 16.2 The atmospheric greenhouse effect (Source: US EPA [1]).

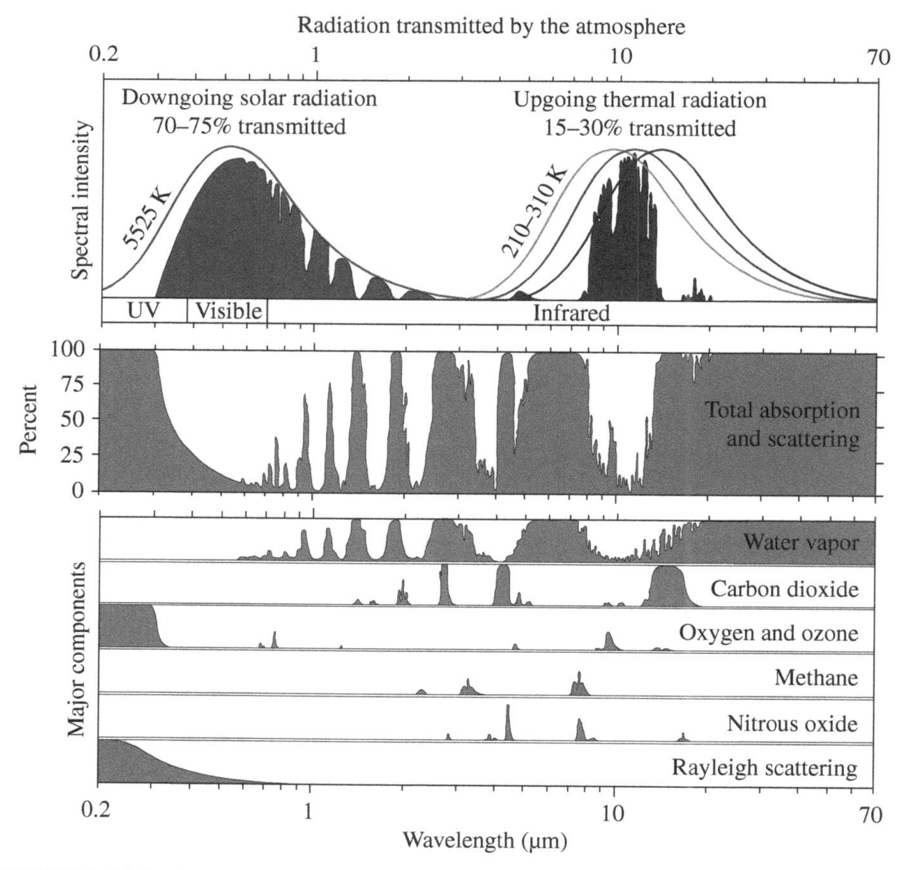

FIGURE 16.3 Radiation spectra of the Sun and Earth with GHG absorption bands (Source: www.barrettbellamyclimate.com [2]).

surface, while the grey areas are the parts of the spectra that are absorbed by the atmosphere. It shows the actual radiation absorption bands for these gases and the dominance of water vapor. Note also that the most significant carbon dioxide absorption band falls near the peak wavelength for blocking long-wavelength thermal radiation from the Earth.

16.3 FOSSIL FUELS AND THEIR CONTRIBUTION TO GHG LEVELS

The GHG that drives the discussion about the effect on the potential increase in warming of the atmosphere is carbon dioxide. This is due to the large quantities of carbon dioxide emitted in fossil fuel combustion and its relatively long residence time in the atmosphere—which makes the process reversible only on a long time scale. The global warming potential (GWP) of the other GHGs, natural gas or methane and the oxides of nitrogen, sometimes called NO_X, can be related to the same mass of carbon dioxide in what is called GWP. Table 16.1 gives the GWP relationship of methane and NO_X to carbon dioxide, based on a 100-year time scale for residence time in the atmosphere, which is the standard used by regulators.

For example, the 100-year GWP of methane is 25, which means that if the same mass of methane and carbon dioxide were introduced into the atmosphere, methane will trap 25 times more heat than the carbon dioxide over the next 100 years. Thus, methane has a significantly higher warming potential than carbon dioxide on a unit mass basis, but it exists in the atmosphere in much smaller quantities and has a much shorter residence time than carbon dioxide and thus has overall a smaller effect.

In the 1950s, Charles Keeling began to measure carbon dioxide concentrations in the atmosphere in Hawaii at the Mauna Loa Observatory. The results of his measurements show increasing carbon dioxide concentrations in the atmosphere, as shown in Figure 16.4, with an annual cycle superimposed on the increasing annual average. The scale to the right shows the concentration in parts per million by volume. Notice that the scale does not go to zero and therefore somewhat exaggerates the 21% increase in carbon dioxide concentration over the period shown; however, these measurements are considered accurate and are not disputed in the scientific community.

Although sources of carbon dioxide in the atmosphere are part of the carbon cycle (see Fig. 16.5), which includes natural sources (such as volcanic eruptions and animal and human respiration) and natural carbon storage (in the oceans, plants, and organic materials), it is generally considered that these increases are at least partially due to

TABLE 16.1 Global Warming Potential (GWP) For 100-year Time Horizon[a]

Common name	Chemical formula	100-year GWP
Carbon dioxide	CO_2	1
Methane	CH_4	25
Nitrous oxide	N_2O	298

[a]Source: Intergovernmental Panel on Climate Science [3].

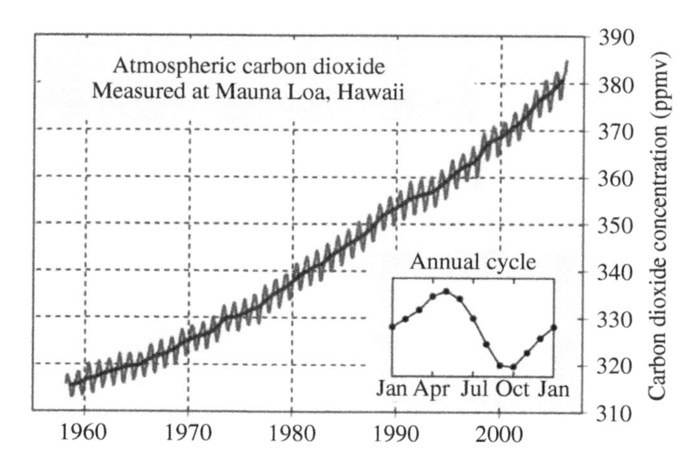

FIGURE 16.4 The keeling curve measuring atmospheric carbon dioxide in parts per million at the Mauna Loa Observatory, Hawaii (Photo Credit: Narayanese, Sémhur and the NOAA, http://en.wikipedia.org/wiki/File:Mauna_Loa_Carbon_Dioxide_Apr2013.svg).

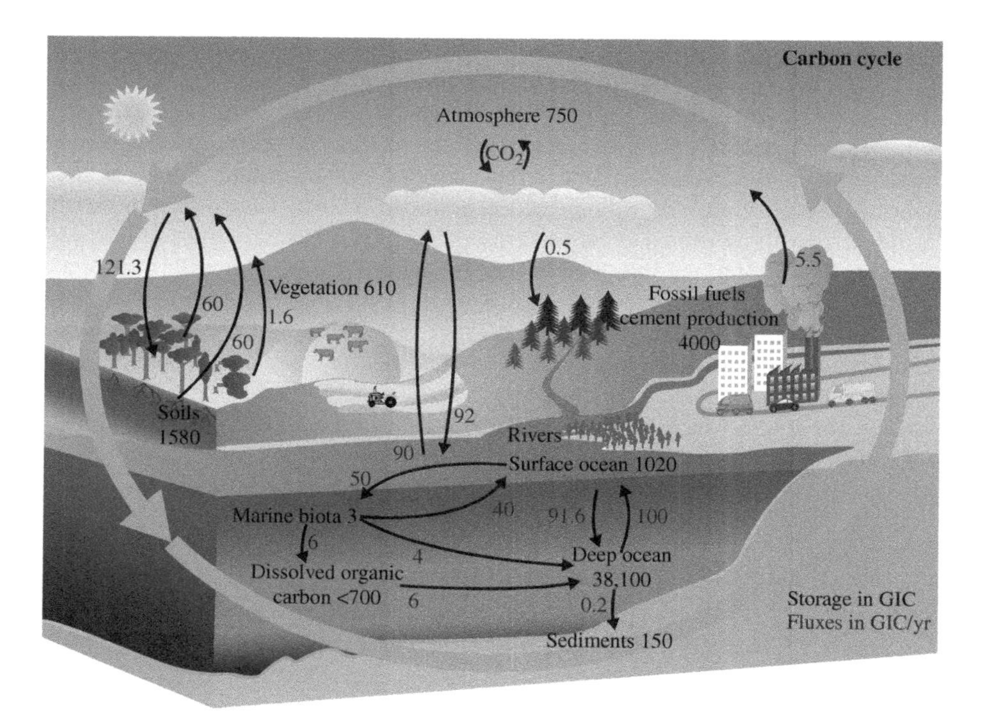

FIGURE 16.5 The carbon cycle (Source: Illustration courtesy NASA Earth Science Enterprise [4]).

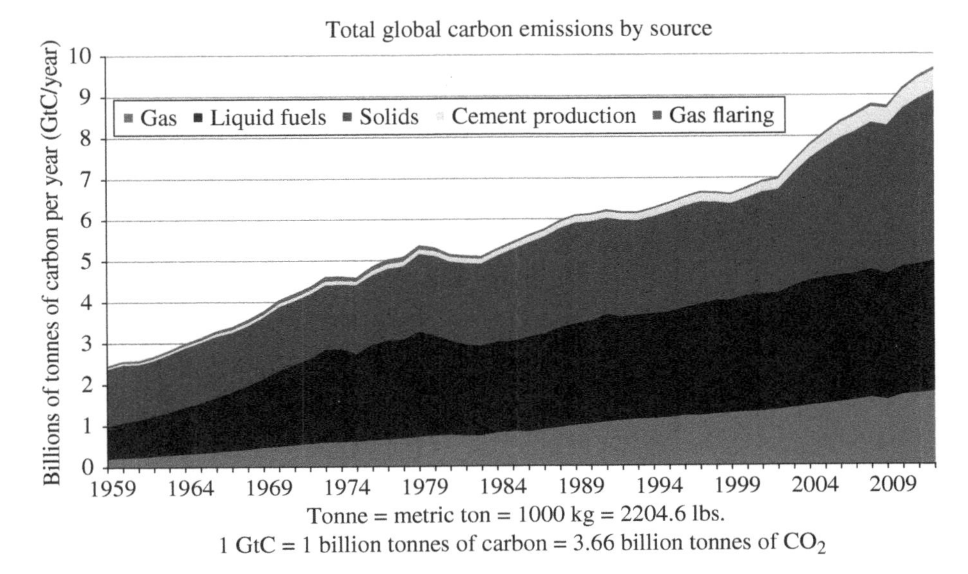

FIGURE 16.6 Global carbon emissions by source in billion tonnes of carbon (GtC) (Source: K. Jay based on data from the Global Carbon Project [5]).

the growing consumption of fossil fuels and the related carbon dioxide emissions. Persons exhale approximately 1 kg of carbon dioxide per day or about one-third of a metric ton per year, less than 2% of the US per capita emissions.

Figure 16.6 illustrates the growth in fossil fuel consumption and related carbon emissions from 1959 through 2010. World carbon dioxide emissions are expected to increase by 1.4% annually between 2006 and 2030. In the United States and other industrialized countries, the production of electricity is a major source of carbon emissions. Note the significant growth in carbon emissions from "solids," primarily coal used for electricity production. Much of this growth is occurring in developing nations trying to meet the rapidly growing demand for electricity.

Figure 16.7 shows the significant fraction of carbon emissions from electricity production in the industrial, commercial, and residential sectors of the United States. Note that the transportation and agricultural sectors use almost no electricity and therefore contribute to carbon emissions by direct combustion of fuels.

Coal power plants, such as that shown in Figure 16.8, produce about 40% of the electricity generated in the United States and are a major source of carbon emissions. Table 16.2 quantifies the levelized cost of energy and the carbon dioxide and other emissions from various types of coal power plants. Notice the significant reductions in carbon dioxide emissions possible with carbon capture and sequestration. However, these technologies add significantly to the cost of energy from these power plants. In addition, the energy necessary to power the additional pollution-control equipment can reduce the conversion efficiency (or increase the heat rate) of the plants.

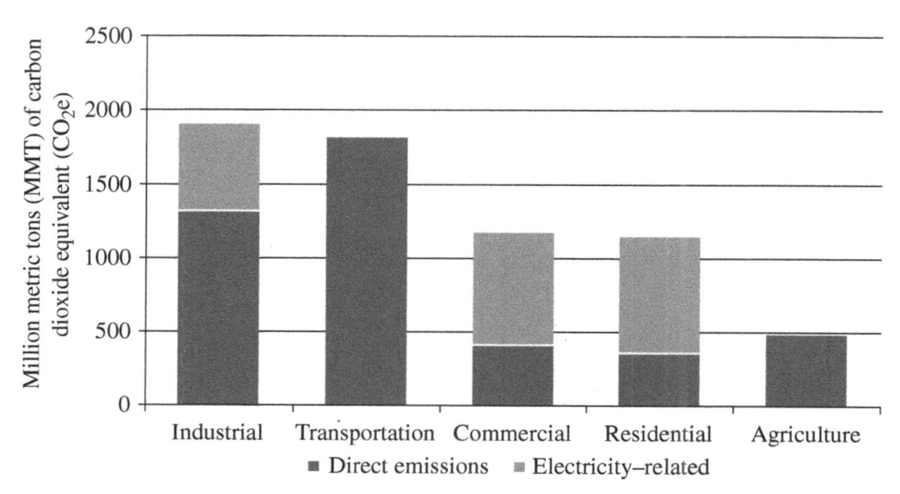

FIGURE 16.7 U.S. GHG emissions from direct combustion and electricity by end-use sector (2009) (Source: US EPA [6]).

FIGURE 16.8 Rail cars deliver coal to an electric power generating plant near Wheatland owned by Tri-State Generation and Transmission (Source: Allen Best/WyoFile (an independent, nonprofit news service) [7]).

TABLE 16.2 Emissions and Cost of Energy From Coal and Natural Gas-Fired Generation[a]

	Emissions (kg/MWh)				
Technology	Greenhouse gases (CO_{2-e})	NO_x	SO_x	Particulate matter	Levelized cost of energy ($/kWh)
Natural gas combined cycle (NGCC)	467	0.37	0.01	0.01	0.09
NGCC with carbon capture and sequestration	137	0.43	0.02	0.01	0.13
Integrated coal gasification Combined cycle (IGCC)	931	0.30	0.03	0.08	0.12
IGCC with carbon capture and sequestration	217	0.27	0.04	0.06	0.16
Super critical pulverized coal plant (SCPC)	943	0.32	0.38	0.07	0.09
SCPC with carbon capture and sequestration	241	0.44	0.04	0.10	0.16
Existing pulverized coal plant (EXPC)	1109	2.06	2.38	0.70	0.03
EXPC w/carbon capture and sequestration	444	0.46	1.31	0.04	0.13

[a]Source: NETL [8].

Notes: Greenhouse gas (GHG) emissions include: CO_2, CH_4, N_2O, and SF_6 in CO_2 equivalences.

Natural gas price assumption=$6.76/MMBtu in year 1 escalating at 1.87% annually.

Coal price assumptions=$1.51/MMBtu ion year 1 escalating at 1.87% annually.

16.4 INTERNATIONAL POLITICAL AND ECONOMIC ISSUES OF CONTROLLING GHG EMISSIONS

Table 16.3 shows that the global emissions of carbon dioxide exceeded 31 billion tons/year in 2010 with China and the United States the leading countries, emitting approximately 26% and 18% of the total, respectively. If the European Union is considered as an entire block, it becomes a significant emitter also. Also shown in this table are the per capita emissions of carbon dioxide in 2010. On this basis, some very small nations such as Gibraltar, Qatar, and Bahrain are among the top emitters per person, while China drops to 70th position in the world and the United States drops to 19th. US carbon dioxide emissions per capita have remained relatively constant at around 17–21 metric tons/person/year for the last 30 years, although total emissions in the United States have generally increased due to the increase in population.

Given the large disparities between the world's nations in the amount of carbon they emit in total or on a per capita basis, the negotiation of treaties to address how to collaboratively reduce emissions of carbon dioxide and other GHGs is incredibly complex and difficult. Since much of the emission of GHGs is related to industrial processes that are key drivers of the economy in highly industrialized nations, politicians fear that agreeing to significant reductions in GHG emissions will be detrimental to their nation's economy or that businesses in their country may be placed at a disadvantage to competing businesses in other nations.

16.5 PUBLIC OPINION AND POLICY

As the Keeling data became available to the scientific community along with the realization that carbon dioxide had a significant GWP, scientific research and debate has increased ever since. In the late 1980s, this debate was extremely vigorous and resulted in the United Nations forming the Intergovernmental Panel on Climate Change, also known as the IPCC. Reports from the IPCC, which was awarded the Nobel Prize for its scientific work several years ago, and conclusions published by the National Academies of Science in the United States and other nations led to strong opinions that worldwide public policy initiatives were needed to address concerns related to human-induced global warming. The text box provides a joint statement by the world academies of science issued in 2005.

The growing consensus and public opinion on the topic led to the Kyoto Protocols for intergovernmental agreement on reducing GHG emissions. The Protocol was initially adopted in December 1997 in Kyoto, Japan, and went into effect in February 2005. Under the Protocol, Annex I countries (including the United States) collectively agreed to reduce their GHG emissions (carbon dioxide, methane, nitrous oxide, and sulfur hexafluoride) by 5.2% on average for the period 2008–2012, relative to their base year emissions, usually 1990. As of 2011, 191 states have signed and ratified the protocol with the United States being the only remaining signatory not to have ratified the Kyoto treaty [10].

TABLE 16.3 World Carbon Dioxide Emissions in 2010: Top 20 By Total Emissions On Top/top 10 By Per Capita Emissions On Bottom[a]

Nation	Population (millions)	2010 CO_2 emissions (million metric tons)	CO_2 emissions per capita (tons/person)	Rank by total emissions	Rank by per capita emissions
World	6,853.0	31,780.4	4.6		
China	1,330.1	8,321.0	6.3	1	70
United States	310.2	5,610.1	18.1	2	19
India	1,173.1	1,695.6	1.4	3	142
Russia	139.4	1,633.8	11.7	4	36
Japan	126.8	1,164.5	9.2	5	46
Germany	82.3	793.7	9.6	6	40
Korea, South	48.6	579.0	11.9	7	33
Iran	76.9	560.3	7.3	8	62
Canada	33.8	548.8	16.3	9	21
United Kingdom	62.6	532.4	8.5	10	52
Saudi Arabia	25.7	478.4	18.6	11	18
South Africa	49.1	465.1	9.5	12	43
Brazil	201.1	453.9	2.3	13	125
Mexico	112.5	445.3	4.0	14	101
Italy	58.1	416.4	7.2	15	63
Australia	21.5	405.3	18.8	16	16
France	63.3	395.2	6.2	17	71
Indonesia	243.0	389.4	1.6	18	140
Spain	46.5	316.4	6.8	19	66
Taiwan	23.0	305.4	13.3	20	27
Gibraltar	0.0	3.9	135.3	130	1
Virgin Islands, U.S.	0.1	11.9	108.9	96	2
Qatar	0.8	64.7	76.9	51	3
Bahrain	0.7	30.7	41.6	76	4

(Continued)

TABLE 16.3 (*Contiued*)

Nation	Population (millions)	2010 CO_2 emissions (million metric tons)	CO_2 emissions per capita (tons/person)	Rank by total emissions	Rank by per capita emissions
Trinidad and Tobago	1.2	49.9	40.6	62	5
United Arab Emirates	5.0	199.4	40.1	26	6
Netherlands Antilles	0.2	8.8	38.6	104	7
Singapore	4.7	172.2	36.6	30	8
Kuwait	2.8	81.3	29.2	43	9
Montserrat	0.0	0.1	28.7	210	10

[a]Source: R. Walker, based on data from US-EIA's International Energy Statistics [9].

Joint Science Academies' statement: *Global response to climate change, 2005* [11]
There will always be uncertainty in understanding a system as complex as the world's climate. However, there is now strong evidence that significant global warming is occurring. The evidence comes from direct measurements of rising surface air temperatures and subsurface ocean temperatures and from phenomena such as increases in average global sea levels, retreating glaciers, and changes to many physical and biological systems. It is likely that most of the warming in recent decades can be attributed to human activities. This warming has already led to changes in the Earth's climate.

Following the Kyoto Protocols, climate change meetings were held in Copenhagen, Denmark, in 2010 to address the next steps in global policies to reduce GHG emissions. Just before the Copenhagen meetings, however, e-mail communications indicated that a number of researchers had either modified data or worked to exclude data that was not in line with their own views on planetary temperature measurements and analysis.

This has led to recent controversy over the issue at both ends of the spectrum—with the human-induced global warming community insisting this was an urgent problem based on valid data—to the other extreme, charging the whole issue was a hoax and of little scientific merit.

Results of public surveys in 2009 showed that while 84% of scientists concurred that global temperatures were increasing because of human activity (as opposed to natural cycles or other natural phenomena), only 49% of the public shared those same views. A Rasmussen poll [12] in that same year showed that 69% of the population felt that it was either "somewhat" or "very likely" that some scientists had falsified global warming research data.

Also in 2009, the United States House of Representatives passed an emissions cap-and-trade bill that would put caps on GHG emissions for energy producers. The bill was not passed in the senate, and the law has not been enacted. The Copenhagen meeting of the next year also did not provide the progress toward global solutions that many of those in the scientific and environmental community had anticipated prior to the release of the communications showing data manipulation and exclusion.

Provided here are bulleted lists summarizing arguments of those both for regulation to control GHG emissions and those opposing such regulations.

- Factors expressed by those in favor of public policy to regulate GHGs are as follows:
  - The consensus of the scientific community to include the National Academy of Sciences issuing public statements as late as May 2010 affirming that human activity had contributed to the increasing global temperatures measured to date.
  - No one questions the rise in CO_2 levels as indicated on the Keeling curve.
  - There is general agreement and understanding that the scientific principles behind the greenhouse effect are real, namely that atmospheric gases such as carbon dioxide selectively transmit and absorb thermal radiation.

  - No one disputes the fact that our planet would be uninhabitable with life in its current form without the greenhouse effect driven in a large part by water vapor and other gases. Estimates are that the temperature would be 30 °C colder and we would essentially be a planet covered in ice.
- Factors expressed by those opposing regulation of GHGs are as follows:
  - The increased costs associated with regulating fossil fuel consumption are quite large and will negatively affect the global economy.
  - Since this is a global problem, any public policy would have to be an "all in" policy. Large emitters such as China or the United States could not opt out.
  - The public release of the e-mails showing data manipulation by climate scientists made many suspect of the entire scientific process concerning GHG emission effects on Earth temperatures.
  - CO_2 is a trace gas in the atmosphere, currently at about 380 parts/million.
  - Carbon dioxide would be hard to regulate. There is a question as to how one can distinguish between human-induced global warming and natural climate cycles, such as those that brought the ice age and then the retreating glaciers as the planet warmed.
  - There are a number of natural sources of carbon dioxide, such as volcanoes, and it is a natural product of respiration by humans and animals.
  - Methane, a very strong GHG, is a ubiquitous gas of decomposition.

16.6 POTENTIAL EFFECTS OF GLOBAL CLIMATE CHANGE

If, in fact, the atmosphere is heating significantly, the consequences will be global and could lead to local climate extremes, changing the planet in very significant ways. It is fair to say that no one knows for sure the effect of these changes, whether human-induced or natural, but they could be profound. The world is extremely different now than during the ice age several millennia ago.

Consequences of a warming planet could be both positive and negative on a regional level. Warming could increase the growing season in some areas, for example. However, some of the more catastrophic effects could be a rising of sea levels causing islands with low elevations to literally disappear—to severe droughts—to more severe and frequent storms.

Although scientists and economists predict that there will be both positive and negative impacts from global climate change, if warming exceeds 2–3 °C (3.6–5.4 °F) over the next century, the consequences of the negative impacts are likely to be much greater than the consequences of the positive impacts. A summary of these impacts follows [1, 13]:

- Melting of ice sheets and glaciers, combined with the thermal expansion of seawater as the oceans warm, is causing sea level to rise. Seawater is beginning to move onto low-lying land and to contaminate coastal freshwater sources and beginning to submerge coastal facilities and barrier islands.

- Climate plays an important role in the global distribution of freshwater resources. Changing precipitation patterns and temperature conditions will alter the distribution and availability of freshwater resources, reducing reliable access to water for many people and their crops. Winter snowpack and mountain glaciers that provide water for human use are declining as a result of global warming.
- Incidents of extreme weather—heat waves, severe cold, intense precipitation, and prolonged drought—are projected to increase as a result of climate change.
- Increasing carbon dioxide levels in the atmosphere is causing ocean water, which absorbs carbon dioxide, to become more acidic, threatening the survival of shell-building marine species.
- Ecosystems on land and in the ocean have been and will continue to be disturbed by climate change. Animals, plants, bacteria, and viruses will migrate to new areas with favorable climate conditions.
- Human health and mortality rates will be affected to different degrees in specific regions of the world as a result of climate change. Although cold-related deaths are predicted to decrease, other risks are predicted to rise.
- Agriculture and forestry: The supply and cost of food may change as farmers and the food industry adapt to new climate patterns. A small amount of warming coupled with increasing CO_2 may benefit certain crops, plants, and forests, although the impacts on vegetation depend also on the availability of water and nutrients.
- Energy: Warmer temperatures may result in higher energy bills for air-conditioning in summer and lower bills for heating in winter.
- Wildlife: Warmer temperatures and precipitation changes will be likely to affect the habitats and migratory patterns of many types of wildlife. The range and distribution of many species will change, and some species that cannot move or adapt may face extinction.

More extreme scenarios include positive feedback mechanisms—events due to warming that lead to other events that amplify the warming. For example, one possible scenario predicted by permafrost scientists is that global warming will result in melting of the Arctic permafrost, which in turn releases large quantities of methane gas and carbon dioxide into the atmosphere, adding to the atmospheric GHG levels and creating additional warming [14].

The insurance industry is very interested in these effects, whether man-made, natural, or a combination, since their business of insuring people and property is based on actuarial tables derived from historical data. If the future is to be significantly different than the past, these companies are extremely interested in trying to understand these changes. In any event, given the potential severity and global nature of these effects, it is little wonder that there is vigorous debate about global policies and potential mitigating strategies.

16.7 METHODS AND COSTS OF MITIGATION

The most comprehensive mitigation strategy is to move the energy economy from one based on fossil fuels to one based on energy efficiency and renewables. The incentives to make this transition are many, as pointed out in the previous section. The global warming debate intensifies the arguments to energy policies with a focus on energy efficiency and renewables, to include wind power. However, a worldwide energy transition of this magnitude will take decades, during which time population growth and emerging economies will demand increased energy use.

Some energy planners include nuclear power as a major part of the global warming solution since nuclear plants emit negligible carbon dioxide in the production of electricity. New nuclear power plant designs are now available, and some energy planners talk of a nuclear power renaissance. There has not been a new nuclear power plant constructed in this country in several decades. However, the recent earthquake and tsunami in Japan that damaged the Fukushima large nuclear power facility—with a near-meltdown scenario—has put on hold many potential nuclear power projects. After the Fukushima incident, many people speculated that it would be decades before a new nuclear plant would be built in the United States; however, the U.S. Nuclear Regulatory Commission has approved permits for at least four new nuclear units since that time.

Another mitigation strategy is to increase the use of natural gas. Natural gas emits about one-half the amount of carbon dioxide as compared with coal, per unit of energy delivered. Natural gas—combined with energy efficiency and renewables—can provide the backup energy for the intermittent renewable sources, and thus that energy mix can deliver reliable energy with substantial reductions in carbon emissions. Recent improvements in natural gas drilling technologies that include directional and horizontal drilling combined with hydraulic fracturing (also called “fracking”) of extensive oil and gas shale formations have provided large increases in natural gas supplies worldwide.

Since the consumption of carbon-based fuels is a major contributor to the increasing carbon dioxide levels in the atmosphere, mitigation strategies include investigating ways to capture and sequester carbon dioxide. These methods include capturing carbon dioxide at the source and storing it in caverns under the ocean, injecting it into oil and gas wells to enhance extraction, and other schemes of capture and store or conversion to other materials. Most of these schemes are quite expensive. The US Department of Energy (DOE) planned a project called Future Gen to establish several pilot-scale coal-fired power plants that included carbon sequestration technology. These projects were eventually cancelled due to technological concerns and high costs. As pointed out previously in Table 16.2, pulverized coal electric power goes from cheapest to highest cost when carbon capture and sequestration is included.

16.8 CONCLUSIONS

As we have presented in this book, energy production, distribution, and consumption are foundational elements in society—and have been for millennia. Presently, the energy economies of the world are dependent on carbon-based fuels. Additionally, the growth in world population will require growth in energy demand to provide a reasonable quality of life. Reducing the use of carbon fuels in the face of these challenges and doing so without detriment to the economies of the world is a significant challenge.

In summary, there are a number of technical mitigation strategies as outlined in this section. However, the sheer magnitude of the carbon-based worldwide energy economy will require more than technological improvements to address this issue. Ultimately, successful approaches will depend on a global consensus of the problem and the threat, as well as viable options that will address that threat. Next, there will need to be agreement as to the best mitigation strategies and finally multilateral global policies with public support and enforcement mechanisms to implement the agreed-upon strategies.

REFERENCES

[1] U.S. Environmental Protection Agency. *Frequently Asked Questions About Global Warming and Climate Change: Back to Basics*. Washington (DC): U.S. Government Printing Office; 2009.

[2] Barrett J, Bellamy DJ. 2012. Greenhouse gas spectra year. Available at http://www.barrettbellamyclimate.com. Accessed August 13, 2012.

[3] Intergovernmental Panel on Climate Science. *Climate Change 2007: Working Group I: The Physical Science Basis*. New York: U.S. Government Printing Office; 2007.

[4] Lorentz K. 2007. Humans and the global carbon cycle: A Faustian bargain? U.S. Government Printing Office. Available at http://www.nasa.gov/centers/langley/news/researchernews/rn_carboncycle.html. Accessed 29 November, 2014..

[5] Le Quéré C, Peters GP, Andres RJ, Andrew RM, Boden TA, Ciais P, Friedlingstein P, Houghton RA, Marland G, Moriarty R, Sitch S, Tans P, Arneth A, Arvanitis A, Bakker DCE, Bopp L, Canadell JG, Chini LP, Doney SC, Harper A, Harris I, House JI, Jain AK, Jones SD, Kato E, Keeling RF, Klein Goldewijk K, Körtzinger A, Koven C, Lefèvre N, Maignan F, Omar A, Ono T, Park G-H, Pfeil B, Poulter B, Raupach MR, Regnier P, Rödenbeck C, Saito S, Schwinger J, Segschneider J, Stocker BD, Takahashi T, Tilbrook B, van Heuven S, Viovy N, Wanninkhof R, Wiltshire A, and Zaehle S. 2014. Global Carbon Project, 2013. Copernicus Publications. Available at http://www.globalcarbonproject.org/carbonbudget. Accessed November 1, 2014.

[6] U.S. Environmental Protection Agency. *Inventory of U.S. Greenhouse Gas Emissions and Sinks: 1990–2009*. Washington (DC): U.S. Environmental Protection Agency; 2011.

[7] Best A. 2011. A reluctant move away from coal. WyoFile News Service.

[8] National Energy Technology Laboratory. NETL life cycle analysis: power studies compilation report. Report nr DOE/NETL-2010/1419. Pittsburg (PA); 2010.

[9] U.S. EIA International Energy Statistics. Available at http://www.eia.gov. Accessed August 7, 2012.

[10] United Nations Framework Convention on Climate Change. Report of the Conference of the Parties on its third session, Kyoto, Japan; March 25, 1998.

[11] Joint Science Academies' statement: global response to climate change (2005). Available at http://nationalacademies.org/onpi/06072005.pdf. Accessed August 13, 2012.

[12] Rasmussen Reports, July 29–30, 2011.

[13] U.S. Climate Change Science Program. *Climate Literacy: The Essential Principles of Climate Science*. Washington (DC): U.S. Global Change Research Program; 2009.

[14] Seth Borenstein (2011). *Study: Thawing permafrost will Worsen Global Warming*; Associated Press, USA Today.

17

SITING AND PERMITTING OF ELECTRIC GENERATION PROJECTS

17.1 FEDERAL LAW, REGULATIONS, AND PERMITTING REQUIREMENTS

The development and construction of a power plant, whether it is a wind project, nuclear power plant, coal-fired plant, or natural gas-fired plant, can involve interaction with several federal agencies. Permits can be required from several of them, and there are existing laws in place that, while not necessarily requiring acquisition of a permit, can impact siting, construction, and operations decisions. The most significant of the many regulations or laws that can come into play when developing or constructing power plants include the one mentioned as follows.

17.1.1 US Code Section 44718: Structures Interfering with Air Commerce

Probably the most important step for construction of a wind energy project is obtaining a "Determination of No Hazard to Air Navigation" from the Federal Aviation Administration (FAA) for each wind turbine. These are frequently described as "FAA permits." As discussed in Chapter 10, the FAA oversees and manages navigable airspace in the United States, develops plans and policies for its use, and is charged with ensuring the safety of aircraft and the efficient use of the airspace. The FAA requires that persons or entities wishing to build structures of certain heights and/or proximity to military or civilian airports or heliports provide public notice of the construction by

Wind Energy Essentials: Societal, Economic, and Environmental Impacts, First Edition.
Richard P. Walker and Andrew Swift.

filing FAA Form 7460-1, Notice of Proposed Construction or Alteration. Generally, public notice is required if the structure is more than 200 ft in height above ground level or near or on an airport or heliport, either military or civilian. Therefore, construction of almost all utility-scale wind turbines, some small wind turbines, and even some meteorological towers will require review and approval from the FAA. Knowingly and willingly violating FAA notice criteria for a structure requiring notice can result in civil penalties of $1000 per day until the notice is received. Traditional power plants may also need to obtain Determinations of No Hazard to Air Navigation for tall structures, such as the smokestacks or even for very tall transmission towers.

17.1.2 National Environmental Policy Act

According to The Citizen's Guide to the National Environmental Policy Act (NEPA), this law was enacted in 1970 as one of the first major environmental laws in the United States [1]. As such, it is sometimes referred to as the "Magna Carta" of environmental laws. It established national environmental policies in the United States, requiring agencies to prepare environmental assessments (EAs) and environmental impact statements (EISs) prior to making major decisions or undertaking significant projects. Under NEPA, Federal agencies must assess the environmental effects of their proposed actions prior to making decisions, such as granting Federal permits or allowing the use of Federal lands for projects. The Bureau of Land Management (BLM) oversees large amounts of land in the western United States, some of which has been leased to wind project developers. Such projects will likely require an EIS.

17.1.3 Endangered Species Act

According to the US Environmental Protection Agency's (EPA's) website, the Endangered Species Act (ESA) was passed in 1973 and provides a program for the conservation of threatened and endangered plants and animals and the habitats in which they are found [2]. The lead federal agencies for implementing ESA are the U.S. Fish and Wildlife Service (USFWS) and the U.S. National Oceanic and Atmospheric Administration (NOAA) Fisheries Service. The FWS maintains a worldwide list of endangered species that includes birds, insects, fish, reptiles, mammals, crustaceans, flowers, grasses, and trees. The law requires federal agencies, in consultation with the USFWS and/or the NOAA Fisheries Service, to ensure that actions they authorize, fund, or carry out are not likely to jeopardize the continued existence of any listed species or result in the destruction or adverse modification of designated critical habitat of such species. The law also prohibits any action that causes a "taking" of any listed species of endangered fish or wildlife.

17.1.4 Bald and Golden Eagle Protection Act

Enacted in 1940 and amended several times since then, this Act prohibits anyone, without a permit issued by the Secretary of the Interior, from "taking" bald eagles, including their parts, nests, or eggs [3]. The Act provides criminal penalties for persons who "take, possess, sell, purchase, barter, offer to sell, purchase or barter, transport, export or import, at any time or any manner, any bald eagle or any golden eagle, alive or dead, or any part, nest, or egg thereof." The Bald and Golden Eagle Protection Act (BGEPA) defines "take" as "pursue, shoot, shoot at, poison, wound, kill, capture, trap, collect, molest or disturb." A violation of the Act can result in a fine of $100,000 ($200,000 for organizations), imprisonment for one year, or both, for a first offense. Penalties increase substantially for additional offenses, and a second violation of this Act is a felony.

17.1.5 Migratory Bird Treaty Act

According to the website of the USFWS, which enforces it, this 1918 act was implemented for the protection of birds migrating between the United States and Canada [4]. Similar US agreements with Mexico, Japan, and the USSR further expanded the scope of international protection of migratory birds. These treaties established Federal responsibilities for the protection of nearly all species of birds, their eggs and nests, and made it illegal for people to "take" migratory birds, their eggs, feathers, or nests. "Take" is defined in the Migratory Bird Treaty Act (MBTA) to include "by any means or in any manner, any attempt at hunting, pursuing, wounding, killing, possessing or transporting any migratory bird, nest, egg, or part thereof."

In total, 836 bird species are protected by the MBTA, although many of them are not listed as either threatened or endangered, and over 50 of the species can be legally hunted as game birds. A migratory bird is any species or family of birds that live, reproduce, or migrate within or across international borders at some point during their annual life cycle. An MBTA review may be undertaken concurrently with an ESA review. When migratory birds are an issue, consultation is required. There is no permit issued for accidental impacts, and violations can even be construed as criminal.

17.1.6 National Historic Preservation Act

The National Historic Preservation Act (NHPA) was enacted in 1966 and governs the preservation of cultural and historic resources in the United States, establishing a national preservation program and a system of procedural protections that encourage the identification and protection of cultural and historic resources of national, state, tribal, and local significance [5]. Primary components of the act include a national policy governing the protection of historic and cultural resources, establishment of a comprehensive program for identifying historic and cultural resources for listing in the National Register of Historic Places, creation of a federal–state/tribal–local partnership for implementing programs established by the Act, requiring federal agencies to take into consideration actions that could adversely affect historic properties listed or eligible for listing on the National Register of Historic Places, and establishment of the Advisory Council on Historic Preservation. Other laws concerning treatment of archaeological sites include the Archaeological Resources Protection Act, the Native American Graves Protection and Repatriation Act, and the Federal Land Policy Management Act.

17.1.7 Clean Air Act

Enacted in 1970, the Clean Air Act defines the EPA's responsibilities for protecting and improving the nation's air quality and the stratospheric ozone layer [6]. Major changes in the law occurred in 1990 with enactment of the Clean Air Act Amendments, and several minor changes have been made since that time. The Clean Air Act is probably the most significant barrier to installation of new fossil-fueled power plants, and coal plants in particular. Wind power plants may require permits under the Clean Air Act related to emissions produced during the construction phase of the project.

17.1.8 Clean Water Act

Passed in 1972, the Clean Water Act (CWA) regulates discharges of pollutants into the waters of the United States and quality standards for surface waters [7]. Under the CWA, EPA has implemented pollution-control programs such as setting wastewater standards for industry and setting water-quality standards for all contaminants in surface waters. It makes it unlawful to discharge any pollutant from a point source, such as pipes or man-made ditches, into navigable waters, unless a permit is obtained. EPA's National Pollutant Discharge Elimination System (NPDES) permit program controls discharges. Since many fossil-fueled and nuclear power plants take in and discharge large volumes of water for cooling purposes, permits are required under the CWA for most such projects. Offshore wind farms may require permits from the U.S. Army Corps of Engineers (COE) under this act.

17.1.9 Resource Conservation and Recovery Act

Resource Conservation and Recovery Act (RCRA) gives EPA the authority to control hazardous waste from "cradle-to-grave" including the generation, transportation, treatment, storage, and disposal of hazardous waste [8]. Coal combustion wastes that are disposed of in landfills or surface impoundments are regulated under RCRA.

17.1.10 Coastal Zone Management Act

The U.S. Congress passed the Coastal Zone Management Act in 1972 to recognize the importance of continued growth in the nation's coastal zone [9]. The NOAA administers the Act, which provides for management of the nation's coastal resources, including the Great Lakes, and balances economic development with environmental conservation. Development of offshore wind farms are likely to be regulated under this act.

17.1.11 Energy Reorganization Act of 1974

The Energy Reorganization Act of 1974 created the Nuclear Regulatory Commission (NRC), which began operations in January 1975. The NRC's mission is to ensure the safe use of radioactive materials for beneficial civilian purposes while protecting people and the environment. It regulates commercial nuclear power plants and other uses of nuclear materials (such as in nuclear medicine) through licensing, inspection, and enforcement of its requirements. Its oversight of nuclear power plants includes nuclear reactor safety, security, licensing, radioactive material safety, and management of spent nuclear fuel. Before the NRC was created, nuclear regulation was the responsibility of the Atomic Energy Commission, which Congress first established in the Atomic Energy Act of 1946, which was later replaced with the Atomic Energy Act of 1954. The latter act made the development of commercial nuclear power possible for the first time.

17.1.12 Executive Order 13212

President George W. Bush issued this order titled "Actions to Expedite Energy-Related Projects" in 2001, which directed federal agencies involved in reviewing energy-related projects to streamline their internal approval processes and established an interagency task force to coordinate federal efforts at expediting approval mechanisms.

17.1.13 Energy Policy Act of 2005 and the National Energy Policy Act of 2001

These acts provided additional guidance to federal agencies and project developers to promote the development of domestic renewable energy supplies.

17.1.14 Federal Governmental Agency Interaction

Federal governmental agencies that wind project developers are more likely to interact with and the laws or circumstances most relevant to these agencies include the following:

- EPA
  - NEPA
  - Clean Air Act
  - CWA
  - RCRA
  - Oil Pollution Act
- Council on Environmental Quality
  - NEPA
- U.S. Department of the Interior's USFWS
  - ESA
  - BGEPA
  - MBTA
  - USFWS' *Land-Based Wind Energy Guidelines* [10] (released in March 2012, this is a new set of guidelines that wind developers need to be aware of)
  - USFWS' *Draft Eagle Conservation Plan Guidance* [11] (awaiting final approval by the U.S. Secretary of the Interior as of the writing of this book)

- U.S. Department of the Interior's Bureau of Ocean Energy Management, Regulation and Enforcement
  - Responsible for overseeing the safe and environmentally responsible development of energy and mineral resources on the Outer Continental Shelf
- Advisory Council on Historic Preservation
  - NHPA
- U.S. Army COE
  - CWA, Rivers and Harbors Act of 1899
- FAA
  - US Code Section 44718: Structures Interfering with Air Commerce
- U.S. Department of Defense (DOD) / Department of Homeland Security (DHS)
  - As discussed in Chapter 10 of this book, part of the FAA process for reviewing requests for determinations regarding wind turbines or other tall structures involves coordination with DOD and DHS to identify potential conflicts with airspace vital to national security.
- Federal Interagency Wind Siting Collaboration
  - An interagency working group that has evolved out of national initiatives to facilitate the coordination among federal agencies regarding wind energy specifically and to develop a federal agency wind energy information center.
- National Telecommunications and Information Administration (NTIA)
  - NTIA is responsible for managing the federal spectrum and is involved in resolving technical telecommunications issues for the federal government and private sector; it can provide information to aid in siting wind turbines so they do not cause interference in radio, microwave, radar, and other frequencies, disrupting critical lines of communication.

17.2 FEDERAL PERMITS, APPROVALS, OR CONSULTATIONS

Regulatory approvals such as permits, certificates, or consultations that may be required for generation projects will vary by the type of project (i.e., wind, solar, coal, nuclear, gas, etc.) and by location (offshore, remote, rural, suburban, and urban). Such regulatory approvals may include the following:

- Determination of No Hazard to Air Navigation from the FAA: Generally required for objects over 200 ft in height, which would include almost all utility-scale wind turbines and some meteorological towers, as well as smokestacks of fossil-fueled power plants. Shorter objects within a few miles of an airport may also require such determinations.
- NPDES Permit under CWA Section 402: To discharge wastewater to surface waters, this permit is required prior to operation and should be obtained prior to commencing construction.

- NPDES Storm Water General Permit for Construction Sites: This permit addressing storm water runoff from construction areas is required before construction.
- NPDES Storm Water General Permit for Operational Sites: This permit addressing industrial storm water runoff after the project goes into operation must be obtained prior to beginning operation.
- Water Quality Certificate under Section 401 of the CWA: This requirement may be triggered by application for a U.S. Army COE Construction Permit.
- RCRA Small-Quantity Hazardous Waste Generator Identification Number: The presence of hazardous waste in quantities greater than a specified threshold amount requires the granting of this identification number. Some of the lubricating oils or hydraulics fluids used in wind turbines may be considered as hazardous waste, so procedures need to be in place for handling, storage, disposal, and spill cleanup of these materials.
- EA/EIS: May be required before a U.S. Army COE Construction Permit can be issued for projects with significant environmental impacts.
- Federal Endangered Species Consultation: This may be required before a U.S. Army COE Construction Permit can be issued for projects that have potential effects to federally listed species or critical habitat.
- Federal Operating Permit pursuant to Clean Air Act Title V Permit: May be required for continuing operation following the period of initial operation allowed by a Construction Permit.
- EISs: For activities involving federal action, NEPA specifies when an EIS must be prepared, including major federal actions with the potential for a significant impact on the quality of the environment. The EIS describes the effects of a proposed activity on the environment, which includes potential impacts on the land, water, air, structures, wildlife, endangered species, wetlands, historical and cultural sites, antiques, fragile ecosystems, and regional economic factors.
- USFWS Consultation: Consultation with the USFWS for any large generation project is recommended, specifically in regard to the ESA, the BGEPA, and the MBTA. In some instances, this may result in a requirement for an "Incidental Take Permit" or if a project will require the relocation of nests, a permit for this will be required. The USFWS *Land-Based Wind Energy Guidelines* [10] and the USFWS *Draft Eagle Conservation Plan Guidance* [11] are good places to start in understanding the expectations of the USFWS. The U.S. Army COE may require such consultations for issuing a Construction Permit for projects that have potential effects to federally listed species or critical habitat.
- Advisory Council on Historic Preservation Consultation: Projects seeking other Federal permits or approvals may be required to consult with the Advisory Council on Historic Preservation under terms of the NHPA. An initial consultation with the State Historic Preservation Office (SHPO), which is strongly recommended for such projects, may help to determine if consultation with the Advisory Council on Historic Preservation is needed.

- U.S. Army COE Section 404 Permit: Section 404 of the CWA enables the Army COE to grant permits for certain activities within waterways and wetlands. Construction projects affecting wetlands in any state cannot proceed until a Section 404 permit has been issued. The COE reviews permit applications to determine if practical alternatives to the project exist, and it can also impose mitigation requirements on the developer and perform a public interest review. The COE also determines if other environmental laws must be addressed, including the NEPA, ESA, and the NHPA. If the COE's review reveals that the project should not proceed, they have the authority to either deny the project or place conditions upon it.

17.3 STATE AND LOCAL PERMITTING

State permitting processes applicable for wind energy projects will have much variation between states. In some states, such as California or New York, the permitting process for most types of electric generation projects, including wind projects, can be very time-consuming and expensive. Issues in which state officials are likely to get involved include endangered species, wetlands areas, historic preservation of cultural resources, storm water management, agricultural production, use of state-owned lands, or use of state-controlled coastal waters. Some states may even require State Environmental Impact Reviews, commonly referred to as "Little NEPAs," particularly if the project involves use of state-controlled lands.

States such as Wisconsin have very well-defined permitting processes. Wisconsin's process specific to wind projects is outlined in the document *Application Filing Requirements for Wind Energy Projects in Wisconsin* [12] jointly produced by the Public Service Commission of Wisconsin; the Wisconsin Department of Natural Resources; the Wisconsin Department of Agriculture, Trade, and Consumer Protection; and the Wisconsin Department of Transportation. In November 2005, the Idaho Department of Water Resources' Energy Division produced guidelines for wind developers titled *Permitting of Small and Medium Sized Wind Turbine Projects in Idaho: A Handbook Guide with Specific Examples for Counties of Bonneville, Cassia, Elmore, Jerome and Twin Falls* [13].

Other states, such as Texas or Oklahoma, may have significantly less permitting requirements or processes, particularly for projects that will not be generating air emissions or using large volumes of water, such as wind plants or solar projects. One aspect of generation project development that may require permits or approvals from state regulators in many states is the construction of electric transmission lines associated with such projects.

Good resources for understanding the regulatory and permitting requirements in states where you may be considering a project or on Federally controlled properties include the following:

- *Wind Power Siting Regulations and Wildlife Guidelines in the United States* [14], by the Association of Fish & Wildlife Agencies and the USFWSs in April 2007

- BLM's *Wind Development Policy* [15], which provides guidance on implementing the Record of Decision for the Programmatic Environmental Impact Statement on Wind Energy Development and guidance on processing right-of-way applications for wind energy projects on public lands administered by the BLM

The location of one's project and the type of project will impact how many and what types of permits, approvals, or consultations from state and local governmental officials will be needed for the development and construction of electric generation projects. Projects releasing air emissions or using large volumes of water are likely to encounter significantly more approvals, as are projects being constructed within the city limits of a municipality. Such permits, approvals, or consultations can include the following:

- Air quality permits or operating permits: Fossil fuel plants (and even some renewable resources such as biomass plants, landfill gas electric generation projects, and geothermal plants that release emissions into the atmosphere) are likely to require air quality permits. These are generally issued by state and/or local authorities, although EPA may issue permits for projects on Native American lands and some other situations.
- Certificates of convenience and necessity (CCN) or certificates of need: These may be required for a power plant, particularly if the owner of the project is a regulated utility company. State utility commissions have traditionally controlled who could provide utility services, build power plants, or construct transmission lines by the issuance of CCNs because utilities have operated as monopolies. Such state commissions may approve a CCN application if they find that the certificate is necessary for the public's convenience or safety.
- CCNs for associated transmission lines: These may be required in some states, particularly if the owner of the transmission line is a regulated utility company.
- Preapproval of cost recovery: Even if a CCN is not required under state law, regulated utilities may still desire to seek preapproval of a planned generation project in order to increase the likelihood of being able to include the project in their cost-of-service that can be passed along to their ratepayers.
- SHPO consultation
- State Department of Transportation
  - Oversize and/or overweight permits from the State Department of Transportation will be required for transporting wind turbine components to turbine construction sites.
  - High-structure permits may also be required.
  - Permits required for placement of utility lines within state rights-of-way, that is, "highway-crossing permits."
  - Permits required for construction of access roads onto state roads.
- County Road Use Permit: This may be required for oversize or overweight vehicles; developers are often required to give assurances that any damage to roads or bridges caused by construction of the project will be repaired.

- County road-crossing permits or right-of-way use permits: Will likely be required to cross over or under local roads with electric and communications lines
- Water use permits
- Wastewater permits
- Waste disposal permits
- Storm water discharge permits: The NPDES is often delegated to states.
- Septic permits are often required for projects not able to interconnect to a local sewage system.
- Noise permits may be required, particularly as more counties begin to implement ordinances for wind turbines or wind plant development.
- State NEPA equivalent ("Little NEPA").
- State wildlife authorities may issue permits for impacts to protected wildlife or habitat.
- County, city, or township may have zoning ordinances requiring that a Conditional Use or Special Use Permit be obtained.
- Building permits may be required by a county, city, or township.

17.4 PUBLIC OPPOSITION TO POWER PLANTS AND TRANSMISSION LINES

Electric generation projects of all types may experience opposition or difficulty when trying to obtain permits or regulatory approvals needed for their project. Electric transmission line projects being proposed to facilitate the construction of new power plants, to increase the reliability of the electric grid, or to enable more cost-effective interchange of electric energy are also frequently opposed.

Permitting of coal plants, in particular, is becoming increasingly difficult. A February 2008 brief by the Council of State Governments [16] noted the following:

> In 2007, nearly 60 proposed coal-fired plants in over 20 states were either put on hold, denied permits, or cancelled by developers, who cited rising construction costs and uncertainty about future climate regulations as central complicating factors. Although not all of these projects were directly quashed because of climate concerns, many were located in the more than half of all U.S. states that have adopted policies to address greenhouse-gas emissions.

The same brief later stated that

> 2007 also saw the U.S. Supreme Court rule that carbon dioxide (CO_2) meets the Clean Air Act's definition of a pollutant. Scientists consider CO_2 to be a key contributor to global warming. As a result of these developments, even the most advanced coal-fired plants ended up looking more like the problem than the solution in the eyes of many

state officials, and their permitting has stirred intense debate. In an effort to address the climate impacts of coal, industry and government are investing in carbon capture and storage (CCS) technologies, so that this abundant and relatively inexpensive fuel might be used without undermining efforts to lower atmospheric levels of CO_2.

17.5 EXAMPLES OF RENEWABLE ENERGY PROJECTS EXPERIENCING DIFFICULTY IN OBTAINING REQUIRED PERMITS OR APPROVALS

Following are examples of wind or solar energy projects that faced or are currently facing opposition or difficulty in obtaining permits or regulatory approvals, including some of the reasons for opposition to the projects.

17.5.1 Cape Wind Offshore Wind Energy Project

As of the writing of this book, the United States has no offshore wind energy projects in operation, although several are under development including the Cape Wind project being developed in Horseshoe Shoal, a shallow portion of Nantucket Sound located south of Cape Cod. Permitting for the project began in 2001, and construction of the project is finally anticipated to begin in 2014. It has taken this long to move the project forward due to opposition from various sources and for various reasons, which include the following:

- Concerns about visibility from the shore, including from the Kennedy family compound
- Native American groups that assert that the wind turbines will block views across the Sound, disturb ancestral burial grounds, and perhaps disturb cultural relics.
- Electric ratepayers opposed to the project due to the relatively high cost of energy from the project.
- Concerns about impacts on ocean navigation and marine life.

17.5.2 Goshen North Wind Farm in Idaho

This project, now in operation, had early requests for permits denied by the Bonneville Planning and Zoning Commission based on concerns that the project would disrupt the scenic view of foothills in the area and lower property values.

17.5.3 Permit Denied for East Haven Wind Farm in Vermont

In 2006, the Vermont Public Service Board denied the application for a "certificate of public good" by East Haven Wind Farm, finding that the proposed project would not promote the public good of the State of Vermont. The Hearing Examiner recommended that the Public Service Board deny a Certificate of Public Good for the proposed project, mainly because it would be located in the "heart of tens of thousands of acres of undeveloped, conserved lands."

17.5.4 Opposition to Wind Projects in San Miguel, New Mexico

San Miguel County first enacted a wind energy ordinance in 2003, but in 2010 it placed a moratorium on new developments so that a task force could work on an updated version of the ordinance. The San Miguel County Commission initially considered including a provision requiring a 3-mile setback between wind turbines and residential structures at the urging of concerned citizens but later reduced the setback requirement to one-half mile.

17.5.5 Ivanpah Solar Project Opposed due to Potential Impacts to Tortoises

Solar energy projects are considered by many to be even more environmentally friendly than wind energy projects, yet even these can be opposed on the basis of environmental concerns. The Ivanpah Solar Energy Generating System in California experienced many difficulties in obtaining regulatory approvals. Concerns about animal species native to the Mojave Desert (such as the desert tortoise) prompted Senator Dianne Feinstein (Democrat - California (D-CA)) to propose a ban on solar power projects in the Mojave Desert. Environmentalists feared that some tortoises would be missed during an attempt to relocate them, which could later result in their death during project construction.

17.6 EXAMPLES OF FOSSIL-FUELED OR NUCLEAR GENERATION PROJECTS EXPERIENCING DIFFICULTY IN OBTAINING REQUIRED PERMITS OR APPROVALS

While renewable energy projects can encounter opposition to their construction, coal-fired and nuclear generation projects will generally receive far more criticism, opposition, and permitting difficulties. Acquiring the necessary air quality or emission permits can be extremely hard for coal-fired plants, but even natural gas–fired plants can encounter difficulties, particularly if they are to be located near an urban area already dealing with significant air-quality issues. An increasingly important issue that both fossil-fueled and nuclear power plants are encountering, particularly in the western half of the United States, is large volumes of water required for cooling purposes.

17.6.1 Water Concerns for White Stallion Energy Center near Bay City, Texas

Water usage and availability is becoming a huge issue in many parts of the United States, particularly in the western half of the country, and electric power generation is one of the largest uses of water in the United States and many other nations. The

United States Geological Service estimates that water used for thermoelectric power generation in the United States was over 200,000 million gallons of water per day in 2005, most of which comes from surface water sources such as lakes or rivers. These withdrawals are estimated to make up almost one-half of total water use (see additional discussion in Chapter 15). The proposed White Stallion Energy Center coal-fired power plant, which would be located about 75 miles south of Houston, would require about 8 trillion gallons of water each year if it uses a wet cooling system, leading many farmers, ranchers, and other concerned citizens to oppose regulatory approvals for the project. A more expensive dry steam system would still require almost 1 billion gallons of water per year.

17.6.2 Plans for Eight Coal Plants in Texas Cancelled

In 2007, as part of an agreement with state regulators and intervening parties allowing for their purchase of the assets of the largest electric utility in Texas, Energy Future Holdings agreed to an environmental platform that included the following actions: (i) terminated plans for the construction of 8 of 11 coal-fired power plants the utility had planned to build; (ii) cancelled plans to expand coal operations in other states; (iii) endorsed the U.S. Climate Action Partnership platform, including the call for a mandatory federal cap on carbon emissions; and (iv) implemented plans to reduce the company's carbon dioxide emissions to 1990 levels by 2020.

17.6.3 Regulatory Approval of Red Rock Coal Plant in Oklahoma Denied

In 2007, the Oklahoma Corporation Commission (OCC) voted to reject a request for preapproval of plans to build a 950-MW coal-fired power plant near Red Rock, Oklahoma. The OCC ruled that American Electric Power and Oklahoma Gas & Electric did not prove that they had sufficiently explored alternative forms of energy in planning their proposed 950-MW power plant. The proposed $1.8 billion plant faced opposition from other utilities, consumer groups, environmentalists, and lawmakers.

17.6.4 Holcomb Coal Plant in Kansas

In 2007, the Kansas Department of Health and Environment rejected an air permit for two 700-MW coal-fired generators because of their potential carbon dioxide emissions and contribution to global climate change. This was one of the first instances of a regulatory agency rejecting a permit for this reason. The State Secretary of the Department of Health and Environment said in a statement, "I believe it would be irresponsible to ignore emerging information about the contribution of carbon dioxide and other greenhouse gases to climate change and the potential harm to our environment and health if we do nothing." He cited a Supreme Court ruling, "Massachusetts v. the EPA," in which the court found that carbon dioxide was a pollutant and could be regulated.

In 2010, with a new Governor and Secretary of Health and Environment in office, Sunflower Electric was able to obtain approval and permits to construct one 895-MW coal-fired unit by agreeing to significant levels of carbon mitigation. These permits were challenged in court by the Sierra Club on the premise that the permit falls short of the minimum requirements of the Clean Air Act, will not adequately protect human health and the environment, did not include enforceable limits on nitrogen oxides and sulfur dioxide pollution, and did not require "best available control technology" on new pollution sources, as required by law. The Sierra Club also contended that the power plant will emit massive amounts of air pollutants, including mercury, sulfur dioxide, nitrogen oxides, and particulate matter on downwind Kansans and will rely on water from the declining Ogallala Aquifer.

17.6.5 Permit for South Dakota Coal Plant Denied

In 2009, the US EPA's regional office in Denver denied a renewed operations permit for a coal plant in South Dakota, citing air-quality issues. The State of South Dakota had already awarded a permit to continue operations of the 34-year-old 475-MW Big Stone Power Plant in Big Stone City, South Dakota. The permit would also have allowed a portion of emissions from that plant to be integrated into a new 500- to 580-MW plant to be built alongside it in 2015. The EPA raised several objections saying that granting the permits would not fulfill the requirements of Prevention of Significant Deterioration and New Source Performance Standards as required under the Clean Air Act, did not include an adequate analysis of nitrogen oxide and sulfur dioxide emissions, and did not specify emission-monitoring measures that would guarantee compliance with regulations.

17.6.6 Permit for New Mexico Coal Plant Withdrawn by EPA

In 2009, the EPA withdrew a permit for a 1500-MW coal-fired power plant to be built on the Navajo Nation. The EPA withdrew the air permit that was issued in 2008 for the proposed 1500-MW Desert Rock power plant in northwestern New Mexico. The EPA found the permit was issued before complete analysis of its emissions and impact on endangered species.

17.6.7 Virginia Coal Plant Permits Delayed due to Concern about Mercury Pollution

Mercury emissions associated with coal-fired power plants are a growing concern. In 2009, a Virginia judge ruled that a state permit for a coal-burning power plant in Wise County did not adequately limit mercury pollution, saying that the state Air Pollution Control Board erred in issuing the permit for the $1.8 billion Dominion Virginia Power plant.

17.6.8 New York Determines Indian Point Nuclear Power Plant Violates CWA

In 2010, New York state regulators ruled that outmoded cooling technology at the Indian Point Nuclear Power Plant kills so many Hudson River fish and consumes and contaminates so much water that it violates the federal CWA. Entergy Corporation's request for a water-quality certification was denied, setting back its efforts for a 20-year renewal of its license to operate the plant. Entergy said the ruling could force it to spend $1.1 billion over 19 years to build new cooling towers. Critics of the plant have long complained that a disaster, whether resulting from plant operations or an act of terrorism, could threaten the safety of millions of people.

17.6.9 Opposition to Oyster Creek Nuclear Plant in New Jersey

Power plant cooling systems taking in huge volumes of water can be a source of harm to turtles, fish, and shellfish, along with their eggs and larvae. In addition, water returned to rivers, lakes, or bays may be significantly warmer than when it was withdrawn, which can result in ecological impacts on these resources. In 2010, the New Jersey Department of Environmental Protection issued a draft CWA permit that requires the company to install a closed-loop cooling system on the Oyster Creek Nuclear Plant at the urging of a coalition of environmental groups, fishing interests, and members of the public concerned about the health of Barnegat Bay and the economic livelihood associated with the bay. The closed-loop cooling system upgrade would greatly reduce the major effects of the plant on the Barnegat Bay including its massive intake of over 1.4 billion gallons of water a day.

17.7 TRANSMISSION LINE PROJECTS EXPERIENCING DIFFICULTY IN OBTAINING REQUIRED PERMITS OR APPROVALS

In 2005, the Texas Legislature increased the goal of the state's Renewable Portfolio Standard and authorized the state Public Utility Commission (PUC) to oversee construction of new transmission lines to Competitive Renewable Energy Zones (CREZs), to facilitate meeting the increased renewable energy goal. A CREZ is an area where wind generation facilities or other forms of renewable energy will be installed and from which transmission facilities will be built to various other areas of the state to deliver renewable power to end-user consumers in the most cost-effective manner.

In most states, transmission lines need permits, including Texas, where a CCN is required before regulated transmission service providers can build new lines. The CREZ plan calls for construction of about 2500 miles of 345-kV transmission lines between the windy areas in West Texas and the large urban areas of the state in central, east, and south Texas, where demand for electricity is high. In between West Texas and the large urban areas lies the property of many landowners who may not always desire to have transmission lines crossing their property. The epicenter of opposition to the CREZ lines was the Hill Country region of Texas, where residents feared that their scenic views would be ruined by the large transmission towers. Due to such opposition, the PUC rejected the route of some of the line segments and required alteration of several others.

The Atlantic Wind Connection is a proposed offshore electrical transmission line being planned by Trans-Elect Development Company and financially backed by Google. Construction of the line could begin in 2013 off the East Coast of the United States, where it would provide transmission service for proposed offshore wind farms. Several concerns have been raised about the proposed transmission line including potential ecological damage, high costs, and the possibility that the line could be used to transport power from coal-fired power plants if the planned offshore wind farms do not get built in the near future.

17.8 CHECKLIST FOR GOVERNMENTAL APPROVALS

Excellent organizational skills are one of the key attributes of an effective project manager. In developing a wind energy project or practically any other kind of large development or construction project, there are so many laws or regulations that can apply and so many moving parts to the process, it is easy to overlook some key permits or approvals if one does not have a "checklist" of items that need to be investigated before expending significant amounts of money. The last thing a project developer would want to do is spend several million dollars on a project and discover that operation of the project will be in violation of some Federal, state, or local law, code, or ordinance. Following are examples of things that should go into such a checklist, but by no means is this intended as an exhaustive list:

- Is the project located within the boundaries of a municipality, and if so, what permits will be required?
- Are there noise, height, or setback ordinances in the county or municipality?
- If so, are variances allowed and what is the likelihood of obtaining such variances?
- Does the project extend into multiple counties or jurisdictions? This is not uncommon for large wind projects.
- Is there a Water District in the project area? Groundwater Protection District? Hospital District? Community College District?
- Are there ordinances specific to wind turbines or wind projects in the county or municipality?
- Is there a comprehensive or master plan for land use or development in the area?
- Are there local building codes or electrical codes that may apply to the project?
- Is development plan review or site plan review required?
- Are wind projects in rural areas allowed with no zoning approval or building permits required?
- Do local zoning regulations or ordinances provide for or apply specifically to wind energy projects?
- If the local zoning regulations or building codes do not address wind projects, is there an opportunity for the developer to work with the community to enact or amend the zoning regulations to include provisions that are favorable to the development of wind projects?
- Is a special permit, special exception, or variance required for use of property as a wind energy project in the applicable zoning district?

- Are there landscaping or screening requirements?
- Are county road use permits required?
- Are there weight, length, or height limits on any of the roads and bridges to be used for the project?
- Do access roads need to comply with certain standards for construction of roads?
- Do local ordinances or zoning regulations apply to a meteorological tower?

REFERENCES

[1] Council on Environmental Quality. *Citizen's Guide to the National Environmental Policy Act: Having Your Voice Heard*. Washington (DC): Executive Office of the President; 2007.

[2] U.S. Environmental Protection Agency. Summary of the Endangered Species Act. Available at http://www.epw.senate.gov/esa73.pdf. Accessed August 7, 2012.

[3] U. S. Fish and Wildlife Service. Bald Eagle Management Guidelines and Conservation Measures: The Bald and Golden Eagle Protection Act. Available at http://www.fws.gov/southdakotafieldoffice/NationalBaldEagleManagementGuidelines.pdf. Accessed August 7, 2012.

[4] U. S. Fish and Wildlife Service. Migratory Birds & Habitat Programs. Available at https://www.fws.gov/laws/lawsdigest/migtrea.html. Accessed August 7, 2012.

[5] General Services Administration. *Office of Real Property Utilization and Disposal Fact Sheet: National Historic Preservation Act*. Washington (DC): U.S. Government Printing Office; 2009.

[6] U.S. Environmental Protection Agency. History of the Clean Air Act. Available at http://www.epw.senate.gov/envlaws/cleanair.pdf. Accessed August 7, 2012.

[7] U.S. Environmental Protection Agency. Summary of the Clean Water Act. Available at http://www.epw.senate.gov/water.pdf. Accessed August 7, 2012.

[8] U.S. Environmental Protection Agency. Summary of the Resource Conservation and Recovery Act. Available at http://www.epw.senate.gov/rcra.pdf. Accessed August 7, 2012.

[9] U.S. Department of Commerce, National Oceanic and Atmospheric Administration. Ocean & Coastal Zone Management Act. Available at http://coastalmanagement.noaa.gov/about/media/CZMA_10_11_06.pdf. Accessed August 7, 2012.

[10] U. S. Fish and Wildlife Service. *Land-Based Wind Energy Guidelines*. OMB Control No. 1018-0148. Arlington: U.S. Government Printing Office; 2012.

[11] U. S. Fish and Wildlife Service. *Draft Eagle Conservation Plan Guidance*. Arlington: U.S. Government Printing Office; 2011.

[12] Public Service Commission of Wisconsin, Wisconsin Department of Natural Resources, Department of Agriculture, Trade, and Consumer Protection, Department of Transportation. *Application Filing Requirements for Wind Energy Projects in Wisconsin*. Madison: U.S. Government Printing Office; 2008.

[13] Idaho Department of Water Resources' Energy Division. *Permitting of Small and Medium Sized Wind Turbine Projects in Idaho: A Handbook Guide with Specific Examples for Counties of Bonneville, Cassia, Elmore, Jerome and Twin Falls*. Boise: U.S. Government Printing Office; 2005.

[14] Association of Fish & Wildlife Agencies and the U.S. Fish and Wildlife Services. *Wind Power Siting Regulations and Wildlife Guidelines in the United States*. Arlington: U.S. Government Printing Office; 2007.

[15] Bureau of Land Management. *Wind Development Policy*. Washington (DC): U.S. Government Printing Office; 2008.

[16] Council of State Governments. *Climate Increasingly a Factor in Coal-Plant Permitting Decisions; CSG/ERC Energy Brief*. Lexington: U.S. Government Printing Office; 2008.

18

ECONOMICS OF ELECTRICITY GENERATION

18.1 SOURCES OF ELECTRICITY GENERATION

As discussed in Chapter 3 and shown in the Figure 18.1, petroleum is the largest source of energy used in the United States, followed by natural gas and coal. However, very little petroleum is used to produce electricity. According to the U.S. Energy Information Administration (USEIA), the generation of electric power in the United States during 2012 required the use of about 38.1 quadrillion Btu (or quads) out of total US energy consumption of 95 quads, or about 40% of the total energy use.

Only about 1% of the electricity is produced using petroleum, while coal, nuclear energy, and natural gas are currently the fuel sources for most of our nation's demand for electricity, accounting for 41%, 24%, and 21%, respectively, in 2012. The United States has abundant reserves of coal and lignite, but the use of coal for electricity generation also presents the largest environmental issues. Only about 12% of electricity came from renewable energy sources in 2012, with most of this coming from hydroelectric facilities. As discussed in Chapter 3, wind energy accounted for only about 3.5% of electric generation in the United States in 2012, but its contribution has been growing very fast since the mid-2000s.

Wind Energy Essentials: Societal, Economic, and Environmental Impacts, First Edition.
Richard P. Walker and Andrew Swift.

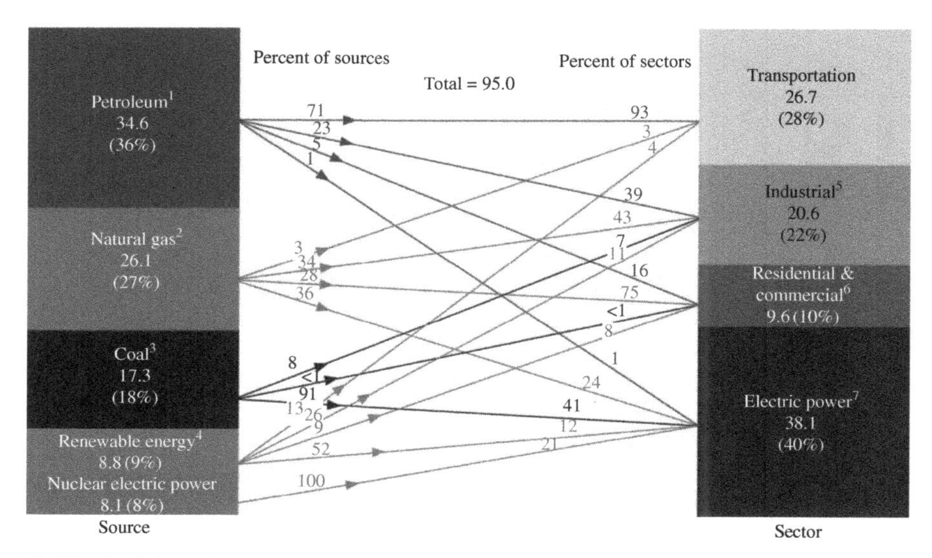

FIGURE 18.1 US primary energy flow by source and sector, 2012 (Quadrillion Btu): 1, does not include biofuels that have been blended with petroleum—biofuels are included in "Renewable Energy"; 2, excludes supplemental gaseous fuels; 3, includes less than 0.1 quadrillion Btu of coal coke net exports; 4, conventional hydroelectric power, geothermal, solar/PV, wind, and biomass; 5, includes industrial combined-heat-and-power (CHP) and industrial electricity-only plants; 6, includes commercial combined-heat-and-power (CHP) and commercial electricity-only plants; 7, electricity-only and combined-heat-and-power (CHP) plants whose primary business is to sell electricity, or electricity and heat, to the public. It includes 0.1 quadrillion Btu of electricity net imports not shown under "Source." (Source: US-EIA [1]).

18.2 ELECTRICAL DEMAND VERSUS SUPPLY

Figure 18.2 depicts how the demand for electricity changes throughout the day. In summer months, electricity usage at night will be low, particularly in hours when most people are asleep and most businesses are closed. As people get up in the morning, they begin turning on various appliances and opening up their businesses or going into work, and electric load begins to increase. As temperatures increase during the day, the electric usage from air conditioning rises, significantly contributing to the utility's peak load. Utilities in the southern part of the United States often see their highest demand for electricity in the summer months right around 4 or 5 P.M.

As customer usage grows throughout the course of the day, the electric utility will bring on more and more generating plants. Nuclear and most coal-fired plants are generally used to provide "baseload" power since they are usually not designed to be frequently ramped-up or -down. On the other hand, most modern natural gas generators are designed to quickly ramp-up or ramp-down as demand for electricity changes. Some of the more efficient natural gas plants may be operated as baseload units, particularly when natural gas prices are low, but many natural gas plants are

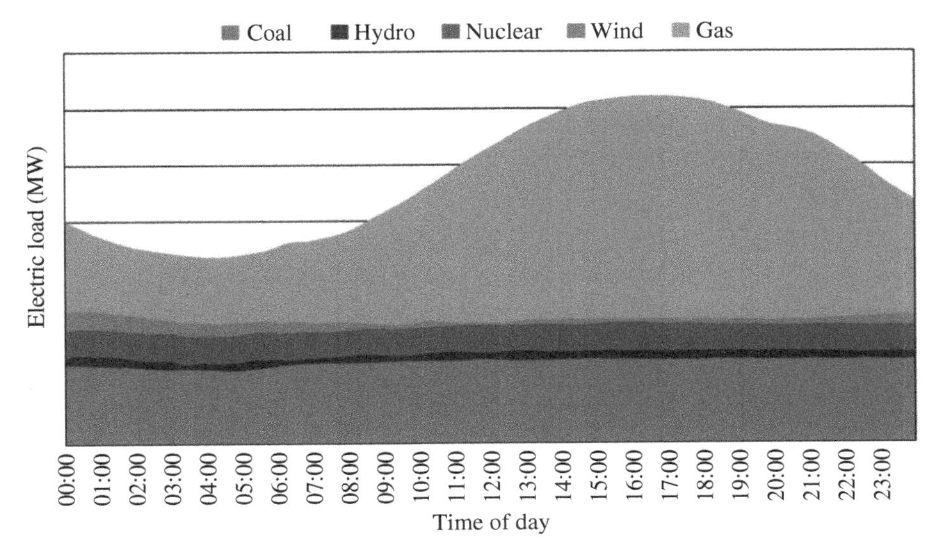

FIGURE 18.2 Example of changing demand for electricity throughout the day and resources the electric system operator may use to meet this demand (Source: R. Walker).

used primarily for peaking purposes. Some of the oldest and least efficient natural gas plants may only be used a few hours out of the year at times when demand for electricity is extraordinarily high.

The electric load profile will differ by season and by region. Electric companies in the northern part of the United States may set their annual peak load during winter months, driven primarily by heating loads.

18.3 MARGINAL COST OF ELECTRICITY/ENERGY VALUE

Marginal cost is the increase or decrease in costs as a result of one more or one less unit of output. In terms of electricity, the marginal cost is usually directly related to the cost of fuel and the efficiency of the power plant used to produce the next kilowatt-hour. The marginal cost may also include a very tiny increase in costs associated with maintenance of the power plant.

As discussed in Chapter 2, storing electricity in large quantities is not currently cost-effective due to the cost of the various technologies currently available and due to the energy losses caused by storing and subsequent use of the energy. Therefore, electric system operators must ensure that electricity is being produced at the same rate as electricity is being consumed on their system. They do this by watching parameters such as grid frequency and voltage and usually rely on changes in the output of natural gas–fired generators to match output with demand. When natural gas generators are being used for this purpose, the cost of natural gas and the efficiency of the least efficient natural gas plant being used at the time will establish the marginal cost of producing the next kilowatt-hour of electricity. In regions such as the Electric Reliability Council of Texas (ERCOT), which use natural gas to produce a large percentage of their electricity, natural gas generation may almost always set the marginal cost or be "on the margin." In other regions that may have a heavier reliance on coal or nuclear power plants, there can be periods of time when other types of generation, most likely coal-fired generation, will be on the margin.

Since natural gas–fired power plants are predominately the generation resource used to balance the fluctuating demand for electricity, they are also often the resource used to balance out the fluctuations in wind energy production. Therefore, much of the value of wind energy may be considered as "the value of fuel not burned," and the price that electric utilities will be willing to pay for wind energy is often highly correlated to the cost of natural gas. This is especially true in the ERCOT electricity region, where natural gas fuels close to one-half of electric generation. As previously shown in Chapter 14 (Fig. 14.7), the cost of natural gas is highly variable. When the price of wind energy is high for sustained periods of time, demand for new wind power purchase agreements (PPAs) by electric utilities is usually also high, but when gas prices are low, it is significantly more difficult to find a willing long-term purchaser of wind energy. While gas prices have historically been volatile, the increasing production of shale gas and the potential for importing liquefied natural gas may help to suppress the price of natural gas for some time to come and to reduce its volatility.

Since wind energy is an intermittent, nondispatchable resource, it is sometimes referred to as an "energy resource," meaning that its primary value to the system is in providing energy (as measured in kWh or MWh) to users of electricity while at the same time reducing the amount of fossil fuels that would otherwise be used to meet the need for electricity.

The relationship between natural gas prices and the incremental cost of electricity can be clearly seen in Figure 18.3 depicting the relationship between electricity prices, as reflected by the Market Clearing Price of Energy in the part of Texas surrounding Dallas-Fort Worth and the price of natural gas. A 12-month rolling average was computed for both prices from January 2007 through November 2010, when ERCOT changed from a zonal electricity pricing system to a nodal pricing system; it is clear that the two are strongly correlated.

As previously stated, the marginal cost of electricity is a function of fuel prices and the efficiency of the power plant the fuel is being used in. A power plant's efficiency can be measured using a statistic called "heat rate," which refers to the average number of Btu required to produce 1 kWh of electricity; so a lower heat rate is indicative of higher efficiency (i.e., less fuel required to produce a kWh of electricity). And as discussed in Chapter 14, the average efficiency of the fleet of natural gas–fired power plants in the United States has improved significantly over the past several years. USEIA statistics [4] show that, in 2001, it took an average of 10,051 Btu to produce 1 kWh, whereas in 2012, it only took an average of 8039 Btu, representing an efficiency improvement of about 20%.

Much of this improvement is due to retirement of older, simple-cycle gas generators with high heat rates (i.e., low efficiency) and the increased use of efficient

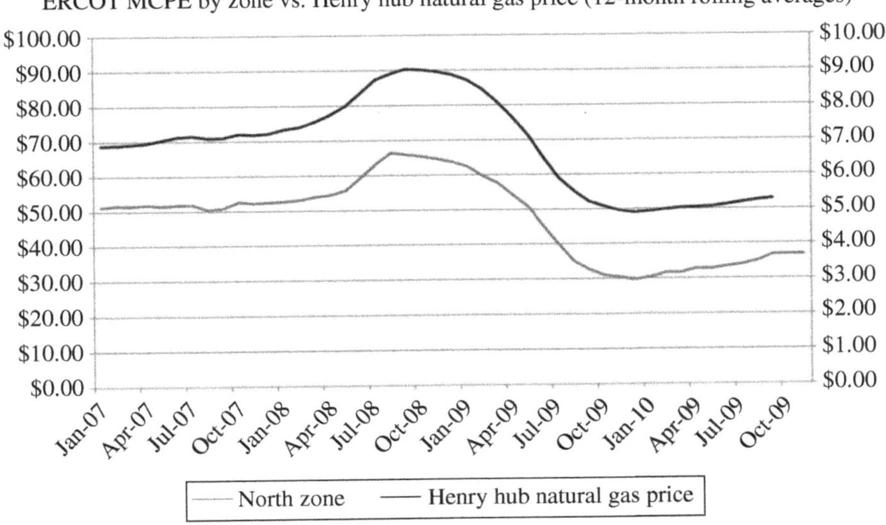

FIGURE 18.3 ERCOT North Zone marginal clearing price of electricity versus Henry Hub natural gas price (12-month rolling averages) (Source: R. Walker based on data from US-EIA [2] and ERCOT [3]).

combined cycle gas-fired units. Since the value of wind energy to the electric utility is largely based on the displacement of natural gas and any variable operations and maintenance (O&M) expenses associated with operating the gas-fired unit, the combination of low natural gas prices and significantly improved natural gas plant efficiencies makes it imperative that wind energy technology keep pace with natural gas generation in terms of improving its efficiency and providing high value to electric utilities. On the other hand, the efficiency of coal and nuclear electric generation remained fairly constant over that period of time, since very few new plants of these types have been built in the United States in recent years, and because the additional emissions controls being added to many coal-fired plants actually reduce their overall efficiency due to the energy required to power the additional equipment.

18.4 EFFECT OF GENERATION RESOURCE DIVERSITY ON ELECTRICITY MARKET PRICE

As shown in Figure 18.4, the United States and Canada are divided into North American Electric Reliability Council (NERC) "regions." These regions may also be depicted as three separate "interconnections"—the Western interconnection (shown as WECC in green), the ERCOT or Texas Reliability Entity (shown as TRE in yellow), and the Eastern interconnection, which includes all of the remaining areas. Within an interconnection, all electric companies and all electric generators operate on a synchronized frequency. Frequency refers to the number of cycles per second in

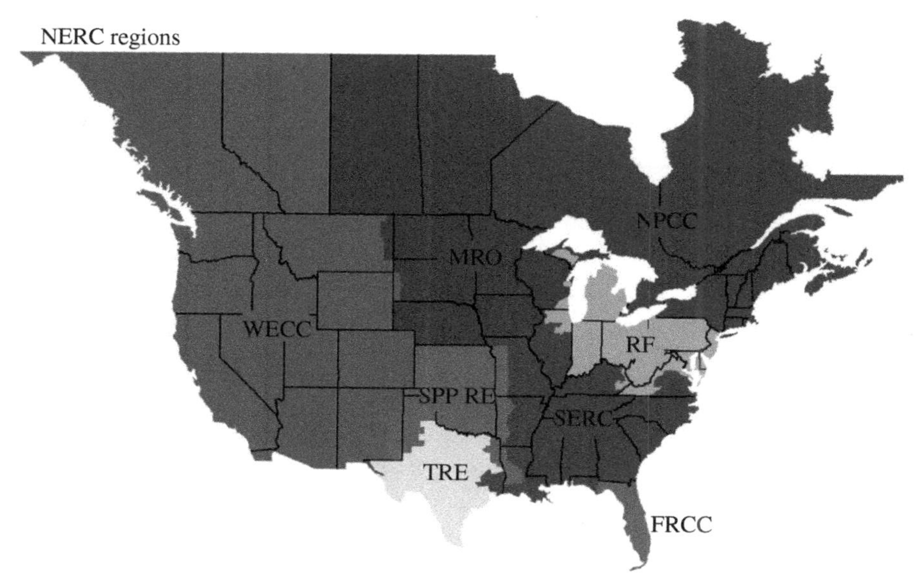

FIGURE 18.4 North American Electric Reliability Corporation (NERC) regions in the United States and Canada (Source: This image from the North American Electric Reliability Corporation's website is the property of the North American Electric Reliability Corporation and is available at http://www.nerc.com/AboutNERC/keyplayers/Documents/NERC_Regions_Color.jpg. This content may not be reproduced in whole or any part without the prior express written permission of the North American Electric Reliability Corporation [5]).

which alternating current (AC) electricity changes direction or "alternates." In the United States and Canada, electric systems are designed to operate at a frequency of 60 Hz, or 60 cycles/s. Electric systems in other countries may operate at frequencies of 50 Hz.

Electric line connections between one NERC interconnection and another NERC interconnection are limited and utilize asynchronous facilities such as high-voltage direct current (DC) stations which convert AC power to DC power and back to AC to prevent the need for synchronism between the two interconnections, with variable frequency transformers, or with DC transmission lines.

Within a given electricity region or market, the average price as well as the marginal price of electricity will be impacted by the mix of generation resources within that region. Some regions may have a higher percentage of coal generation, others a higher percentage of nuclear capacity, and others such as the US Northwest may have large amounts of hydropower. As shown in Figure 18.5, electric generators in Texas generate a larger percentage of electricity from natural gas than the remainder of the United States. The ERCOT market makes up the majority of the state and accounts for about 85% of the state's electricity usage. Due to the volatility of natural gas prices, electricity prices within ERCOT can also be volatile. In periods when natural gas prices are high, the demand for cheaper wind energy increases, which is one of the reasons that Texas leads the nation in installed wind capacity.

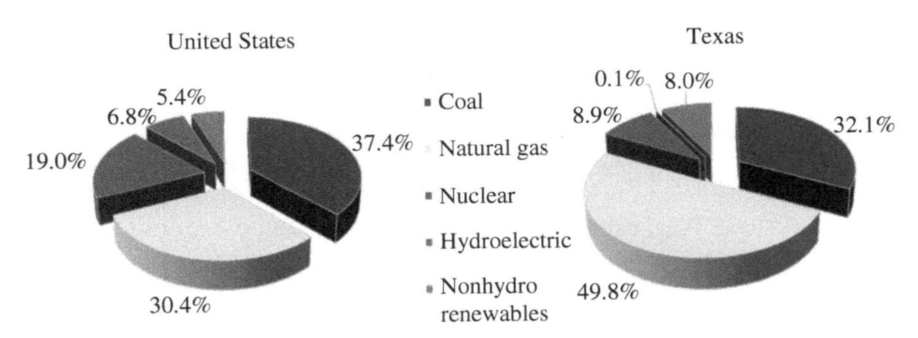

FIGURE 18.5 Comparison of fuel sources used to produce electricity in Texas and the United States during 2012 (Source: R. Walker based on data from US EIA [1]).

Diversification of generation resources is generally beneficial, just as it is considered beneficial to diversity one's investment portfolio. Resource diversity can stabilize the fuel cost component of electricity prices, and when renewable energy sources are used to add diversity, this can serve as a hedge against the potential state or federal imposition of a carbon tax or additional restrictions on emissions from the burning of fossil fuels.

18.5 CAPACITY VALUE OF ELECTRIC GENERATION

In addition to valuing electricity based on fuel prices and power plant efficiency, electric utilities or electric system operators attribute some value to a generator based on the likelihood of being able to use it to meet their customers' needs for electricity at times of peak demand. A highly reliable and fully dispatchable resource such as a well-maintained coal, nuclear, or gas-fired power plant would thus have a higher "capacity value" to the company than an intermittent resource relying on weather conditions such as wind energy. "Dispatchable" refers to the ability to turn on, turn off, ramp-up, or ramp-down the generator as needed to match electricity production with electricity usage.

Wind energy does have capacity value to grid operators, but as a percentage of its rated capacity, an electric system operator may only attribute 10–30% of a wind plant's rated capacity as its "capacity value." Renewable energy resources which may have higher percentages of capacity value include geothermal plants, biomass plants, landfill gas generation facilities, hydroelectric plants, and even solar energy projects. Since the capacity value of a resource is a function of its likelihood of being available during peak demand periods, solar energy production may align well with the peak demand of a utility during hot summer days when peak demand is driven by air conditioning load.

Factors that will impact the capacity value of wind plants include the following:

- Locational diversity of wind plants: Just like diversification in investment portfolios helps to reduce the portfolio's volatility, diversification in the location of wind plants reduces the volatility of the combined output of these wind plants.

Large concentrations of wind plants in a single area (such as the 6000+ MW found around Sweetwater, Texas) tend to result in a wind fleet with little diversity in production profiles, whereas a scattering of wind projects over a wide region can increase the average capacity value attributed to all wind projects.

- Locational differences in timing of wind resources: The timing of wind resources varies across the United States as does the timing of peak periods of electric load. Wind can be caused or affected by heating or cooling of deserts and oceans and by topographic features such as mountains, mesas, or passes between such features. In locations where the timing of the wind resources aligns better with the timing of the peak demand for electricity, wind plants will have higher capacity value as measured in terms of their percentage of rated (or nameplate) generating capacity. For example, this occurs along the Texas Gulf Coast, where onshore afternoon winds coincide well with peak electricity demands.
- Wind forecasting: Accurate forecasting of wind plant output allows electric system operators to prepare for significant increases or decreases in wind generation by changing their dispatch plans for other types of generation resources such as natural gas, coal, or hydroelectric generators. Much work continues in this area as it has become clear that the more accurate the wind forecasting methodology used, the more wind power can be effectively integrated into the grid.

18.6 EFFECT OF CAPACITY FACTOR ON THE ECONOMICS OF ELECTRICITY GENERATION

The capacity factor of a power plant is the ratio of the actual output of a power plant over a period of time to its output if it had operated at full rated capacity the entire time. To determine the capacity factor of a generator, divide the total energy the plant produced during a period of time (such as a year) by the energy the plant would have produced if it ran at its full capacity over that same period, as shown in the following example representative of a typical coal-fired power plant:

Given :	Power plant rated capacity : 500 MW
	Actual production during the year : 3,250,000 MWh
Maximum potential output	$= 500\text{MW} \times 24\,\text{h/day} \times 365\,\text{days/year}$
	$= 4,380,000\text{MWh}$
Capacity factor	$= 3,250,000\text{MWh} / 4,380,000\text{MWh} = 0.742 = 74.2\%$

One may also come across references to "gross" capacity factor (GCF) or "net" capacity factor (NCF). GCF would be based on the amount of energy produced before accounting for items such as electrical losses within the plant site or internal uses, whereas NCF is based on energy that is injected into the electric grid and that can then be sold to a utility company or grid operator.

Capacity factors vary greatly depending on the type of fuel that is used and the design of the plant. **Capacity factor should not be confused with capacity value, availability factor, or efficiency.** These are all separate and distinct concepts. Table 18.1 shows average capacity factors of power plants in the United States by fuel source during 2012. High capacity factors indicate the use of the resource as "baseload" generation, while low capacity factors may indicate the use of the resource for peaking purposes or that the resource is an intermittent source of generation such as wind energy and solar energy.

In the aforementioned table, gas generation averages out to have 33.3% capacity factor. However, newer and more efficient combined cycle units have a substantially higher capacity factor than this, and in periods of low gas prices, may be operated as baseload generation units. However, the average capacity factor of all gas-fired plants is reduced by the continued existence of many old gas-fired steam generation units that may only be operated during periods of exceptionally high customer demand for electricity.

Plant types with high capacity factors have the advantage of being able to spread the recovery of their fixed costs over more megawatt-hours. Fixed costs include things like return on investment, interest on debt, depreciation, property taxes, or insurance that do not vary with the amount of energy produced by the plant.

The following example shows the effect of a generator's capacity factor on the cost of electricity from it:

	Plant A	Plant B
Rated capacity	100 MW	100 MW
Annual fixed costs	$20,000,000	$20,000,000
Variable costs, including fuel	$30/MWh	$30/MWh
Annual capacity factor	**80%**	**40%**
Annual generation	700,800 MWh	350,400 MWh
Total annual variable costs	$21,024,000	$10,512,000
Total fixed and variable costs	$41,024,000	$30,512,000
Average annual cost of energy	**$58.54/MWh**	**$87.08/MWh**

TABLE 18.1 Average Capacity Factors of US Electric Generators By Fuel Type in 2012

Fuel type	Average capacity factor (%)
Nuclear	86.2
Geothermal	70.9
Coal	55.0
Biomass	54.8
Hydroelectric	40.0
Natural gas	33.3
Wind	30.6
Solar	21.0

Source: R. Walker based on data from the US-EIA Electric Power Annual 2012 [6].

The best wind plants in the United States may have average annual NCFs in the range of 40–50%, not that much different from natural gas–fired plants. However, coal and nuclear power plants generally have capacity factors well in excess of this range, giving them a cost advantage in terms of spreading their fixed costs over more megawatt-hours of generation. Of course, some of this is offset by the fact that wind plants do not have fuel costs associated with them.

18.7 EFFECT OF EXPECTED USEFUL LIFE OR ECONOMIC LIFE OF POWER PLANTS ON THE ECONOMICS OF ELECTRICITY GENERATION

Another factor affecting the economic competitiveness of various generation technologies is the expected useful life of the power plant. Sometimes this may also be referred to as the economic life of an asset. As shown in Table 18.2, some power plants built 80, 90, or even more than 100 years ago are still in operation today. The average age of coal plants in the United States, weighted by their rated capacity, was 37 years as of 2011, and some of those built several decades ago may still have high capacity factors as long as their fuel costs remain low, although it is possible that increasing restrictions on air emissions may have necessitated the addition of pollution control equipment at some point in recent years.

While the average age of natural gas–fired power plants in the United States is almost 20 years, again weighted by rated capacity, many of the older natural gas–fired plants have either been "mothballed" or are only used infrequently due to their inefficiency, as new natural gas–fired technology tends to be much more efficient.

Having a longer expected useful lifetime for a power plant allows its initial cost to be spread over more years and, correspondingly, more megawatt-hours (MWh) produced. By spreading these costs over more MWh, the average price of energy over the life of the project (or levelized cost of energy (LCOE)) is reduced as compared with a power plant costing the same amount to construct but not expected to last as many years.

TABLE 18.2 Age of Power Plants in the United States By Fuel Source

	Capacity weighted	
Fuel source	Oldest (years)	Average age (years)
Coal	90	37
Hydroelectric	120	47
Natural gas	96	19
Nuclear	42	30
Petroleum	88	36
Wind energy	36	5

Source: R. Walker based on data from US-EIA data on the age of power plants in the United States [7].

Investors in or lenders to wind energy projects typically require the project developer to provide proof of Project Certification or a Site Suitability Certificate from a recognized third-party firm certifying that all major components of the turbines being used for the project are expected to have a useful life of at least 20 years given the site's wind regime and the developer's proposed spacing and location of wind turbines. Thus, 20 years is often used as the expected useful life of wind energy projects, although turbines or wind projects could last well beyond 20 years assuming good O&M practices are followed, no extreme wind events occur with winds in excess of that customarily anticipated during 20-year periods, and the overall wind regime is not overly harsh.

Based on the age of the existing fleet of power plants, natural gas–fired plants might be expected to have a useful life of 30–40 years, while coal-fired and nuclear power plants might be expected to have a useful life of 40–50 years. The average age of existing hydroelectric power plants is almost 50 years, but many of these may last for several more decades. Thus, it would appear that traditional sources of electricity have some economic advantage over wind energy projects due to the ability to spread construction costs over a longer period of time. Of course, wind energy has a cost advantage over nuclear and fossil fuel power plants in the sense that there is no direct cost of fuel for a wind plant other than perhaps the royalty payments to the landowners where turbines are located.

18.8 BOOK DEPRECIATION VERSUS DEPRECIATION FOR TAX PURPOSES

Depreciation entered into a company's general ledger and reported on the company's financial statements is referred to as "book depreciation." Book depreciation is based on generally accepted accounting principles and is a function of the assumed life of the asset. The federal government has enacted tax legislation designed to encourage investment in certain types of assets by allowing accelerated depreciation for tax purposes, resulting in a much faster write-off of the asset for tax purposes as compared with write-off for book purposes. Accelerated depreciation reduces a project's federal income tax liability in the early years, but increases the tax liability in later years. Therefore, due to the "time value of money," accelerated depreciation for tax purposes will increase the net present value (NPV) of a project, assuming all other things remain constant. However, in competitive markets, the ability to use accelerated depreciation for tax purposes may result in a reduction to the energy price paid by the consumer rather than higher returns to the investor, since project developers would be inclined to reduce their proposed energy sales price to potential customers in situations where they may be bidding against several other projects.

As part of the Tax Reform Act of 1986, Congress enacted a method called the Modified Accelerated Cost-Recovery System (MACRS) for tax depreciation of assets by businesses. This system sets various time frames over which property types may be depreciated, ranging from 3 to 50 years. Wind generators, solar energy systems, and some geothermal equipment are technologies included as 5-year depreciable property, while some biomass facilities have 7-year depreciation periods. Depreciation periods for traditional sources of electric generation include 15 years for gas combustion turbines and nuclear plants, and 20 years for traditional steam generators such as coal-fired plants. Since wind, solar, geothermal, and biomass equipment can be depreciated over a shorter period for tax purposes as compared with traditional generation equipment, this helps to increase the competitiveness of these renewable energy resources.

Due to the on-again, off-again cycle of the Federal production tax credit (PTC), it seems that the majority of wind projects enter commercial operation either in the third quarter or the fourth quarter of a given year. The timing of commercial operation will affect the amount of depreciation for income tax purposes that the project qualifies for during the first year and in subsequent years. As shown in Table 18.3, 5-year property such as wind generators placed in service in the third quarter of the year can be depreciated for tax purposes by 15% in year one, 34% in year two, 20.40% in year three, 12.24% in year four, 11.30% in year five, and 7.06% in year six, which adds up to 100% of the asset, even though the asset may well last an additional 15–20 years.

However, if the project takes a little longer to complete and does not enter commercial operation until the fourth quarter of the year, 5-year property is depreciated by only 5% in year one (see Table 18.4), followed by 38% in year two, 22.80%

TABLE 18.3 MACRS Depreciation Rates For 3-, 5-, 7-, 10-, 15-, or 20-year Property, Mid-quarter Convention, Placed in Service in Third Quarter

	Depreciation rate for recovery period					
Year	3-year	5-year	7-year	10-year	15-year	20-year
1	25.00%	15.00%	10.71%	7.50%	3.75%	2.813%
2	50.00	34.00	25.51	18.50	9.63	7.289
3	16.67	20.40	18.22	14.80	8.66	6.742
4	8.33	12.24	13.02	11.84	7.80	6.237
5		11.30	9.30	9.47	7.02	5.769
6		7.06	8.85	7.58	6.31	5.336
7			8.86	6.55	5.90	4.936
8			5.53	6.55	5.90	4.566
9				6.56	5.91	4.460
10				6.55	5.90	4.460
11				4.10	5.91	4.460
12					5.90	4.460
13					5.91	4.461
14					5.90	4.460
15					5.91	4.461
16					3.69	4.460
17						4.461
18						4.460
19						4.461
20						4.460
21						2.788

Source: US IRS [8].

TABLE 18.4 MACRS Depreciation Rates For 3-, 5-, 7-, 10-, 15-, or 20-year Property, Mid-quarter Convention, Placed in Service in Fourth Quarter

	Depreciation rate for recovery period					
Year	3-year	5-year	7-year	10-year	15-year	20-year
1	8.33%	5.00%	3.57%	2.50%	1.25%	0.938%
2	61.11	38.00	27.55	19.50	9.88	7.430
3	20.37	22.80	19.68	15.60	8.89	6.872
4	10.19	13.68	14.06	12.48	8.00	6.357
5		10.94	10.04	9.98	7.20	5.880
6		9.58	8.73	7.99	6.48	5.439
7			8.73	6.55	5.90	5.031
8			7.64	6.55	5.90	4.654
9				6.56	5.90	4.458
10				6.55	5.91	4.458
11				5.74	5.90	4.458
12					5.91	4.458
13					5.90	4.458
14					5.91	4.458
15					5.90	4.458
16					5.17	4.458
17						4.458
18						4.459
19						4.458
20						4.459
21						3.901

Source: US IRS [8].

in year three, 13.68% in year four, 10.94% in year five, and 9.58% in year six, which again adds up to 100% of the asset.

Accelerated depreciation for tax purposes can have a huge impact on the economics of a project, so completing construction of a project and beginning to take accelerated depreciation in an earlier quarter of the year rather than a later quarter can significantly change the expected NPV of a project. Since the allowable percentages change depending upon the quarter of the year in which the property begins commercial operation, there are different U.S. Internal Revenue Service (IRS) depreciation schedule tables for each quarter of the year, so make sure you are using the correct MACRS table.

Occasionally, Congress will attempt to stimulate the economy by encouraging additional investment in capital equipment by enacting "bonus" depreciation provisions. The Economic Stimulus Act of 2008 included a provision allowing an additional one-half or 50% of certain assets to be depreciated in the first year of operation. The American Recovery and Reinvestment Act of 2009 extended 50% bonus depreciation through 2010, while the Tax Relief, Unemployment Insurance Reauthorization, and Job Creation Act of 2010 authorized 100% first-year bonus depreciation of eligible property placed in service between September 8, 2010 and the end of 2011. During calendar year 2012, bonus depreciation was still available, but reverted back to 50% bonus depreciation in the first year of operation instead of 100%. Wind energy equipment has been included as eligible equipment for bonus depreciation in each of the aforementioned acts, thus helping to reduce the cost of wind energy for utilities and their customers.

18.9 AD VALOREM OR PROPERTY TAXES

Ad valorem taxes, or property taxes, can be a significant component of operating expenses for any electric generation plant. Ad valorem tax rates vary significantly by location of the project. States which have little or no state income tax may collect more funds for schools, roads, or other functions with higher ad valorem states as compared with those states that do have income taxes. In addition, ad valorem tax rates will vary by municipality (if the project is within the city limits of a municipality), by county, and by school district. In addition, some counties may fund water districts, community colleges, or hospitals through ad valorem taxes, and some of these taxes can be quite substantial and may even influence whether or not a power plant is located within a given taxing jurisdiction. Wind plants, in particular, can be quite spread out, with potential turbine locations in multiple counties and multiple school districts. Where one county may have substantially higher ad valorem tax rates due to community college taxing districts or hospital taxing districts as compared with an adjoining county, and if all other things such as wind speed are relatively equal, there is strong incentive for the wind developer to shift planned turbine locations from the county with the higher tax rate to the one with the lower tax rate.

In order to attract businesses (including traditional power plants and wind energy projects) to a community or county, government officials may offer tax abatements or appraised value limitations to these businesses reasoning that some tax revenue is better than no tax revenue and that new jobs in the region can improve the region's economic condition. Where annual ad valorem taxes are near or above 2% of a wind

FIGURE 18.6 New school in Blackwell, Texas, near the Sweetwater Wind Farm (Source: R. Walker).

plant's appraised value, such taxes can be one of the project's largest cost components, along with investor returns, interest expense, and O&M expense. School taxes collected near large power plants often result in areas near such projects being able to construct new schools. In several rural areas of west Texas where most of the nation's largest wind energy projects have been constructed, new schools have been built or old schools have been updated or modernized, whereas without these wind projects, it would have been highly unlikely that these areas, many with declining populations, would have been able to fund the improvements to schools (see Fig. 18.6).

Ad valorem tax rates can also impact the competitiveness of one type of generation resource to another type. For instance, natural gas–fired plants typically cost much less to install per unit of capacity ($/MW) than wind plants, coal plants, or nuclear plants. Thus annual ad valorem taxes paid by a 200 MW gas-fired plant may only be about half that amount paid by a 200 MW wind plant. Assuming both projects (a gas project and a wind energy project) have equal capacity factors or equal amounts of electricity produced annually, then the wind energy project would be paying twice as much tax per unit of electricity generated and sold to consumers.

18.10 FEDERAL SUPPORT AND SUBSIDIES

Governments wanting to boost some sector of the economy (such as farming, housing, energy, or manufacturing) may offer financial incentives to attract additional investment into that economic sector. Examples of such incentives can include investment tax credits, income tax credits or deductions, accelerated depreciation for tax purposes, or ad valorem tax abatements. Governments may also promote industries by allocating funds for research and development that can make those industries more competitive. In the United States, there are literally thousands of financial incentives available for various sectors of the economy that legislators have chosen to support, wind energy being just one of many.

In the United States, virtually every energy technology is supported in one way or another by the federal government. Wind energy is no exception, nor should it be. There are many reasons that governments may want to support the increased use of wind energy and other renewable sources of energy including reducing dependence on energy imports, promoting rural economic development, reducing energy price volatility, improving air quality, and addressing concerns about global climate change. One may quite often hear from opponents of renewable energy that wind or solar energy cannot succeed without federal subsidies, however many such comments are probably coming from other, possibly competing, energy industries that also have a long history of receiving federal support—yet these same individuals or industries rarely mention the subsidies received by the more traditional energy sources.

The following tables (Tables 18.5 and 18.6), which came from the USEIA's report entitled *Direct Federal Financial Interventions and Subsidies in Energy in Fiscal Year 2010* [9], show which energy technologies have received federal support in one form or another. One should point out that on a percentage contribution to energy

TABLE 18.5 Quantified Energy-specific Subsidies and Support By Type, Fiscal Year 2007 (million 2010 Dollars), Table ES2 From *Direct Federal Financial Interventions and Subsidies in Energy in Fiscal Year 2010*

Beneficiary	Direct expenditures	Tax expenditures	Research and development	DOE loan guarantee program	Federal and RUS electricity[a]	Total
2007						
Coal	0	291	582	NA	70	943
Refined coal	0	3,038	0	NA	0	3,038
Natural gas and petroleum liquids	0	1,914	43	NA	53	2,010
Nuclear	0	600	1017	NA	96	1,714
Renewables	110	4,130	717	NA	167	5,124
Biomass	16	5	40	NA	0	61
Geothermal	0	5	9	NA	0	14
Hydro	0	6	0	NA	165	170
Solar	0	8	171	NA	0	179
Wind	0	418	58	NA	1	476
Other	5	6	211	NA	1	224
Biofuels	89	3,682	228	NA	0	3,999
Electricity-smart grid and transmission	0	696	142	NA	243	1,081
Conservation	369	0	0	NA	0	369
End-use	2276	832	509	NA	0	3,618
LIHEAP	2276	0	0	NA	0	2,276
Other	0	832	509	NA	0	1,342
Total	**2755**	**11,501**	**3010**	NA	**629**	**17,895**

Source: U.S. Energy Information Administration [9].

[a] Total will not match Table 24 midpoint of "Estimated Interest Subsidy at Benchmark Interest Rate" because some data cannot be allocated by fuel or activity.

TABLE 18.6 Quantified Energy-specific Subsidies and Support By Type, Fiscal Year 2010 (million 2010 Dollars), Table ES2 From *Direct Federal Financial Interventions and Subsidies in Energy in Fiscal Year 2010*

Beneficiary	Direct expenditures	Tax expenditures	Development	Guarantee program	RUS electricity[a]	Total	ARRA relate
2010							
Coal	42	561	663	0	91	1,358	97
Refined coal	0	0	0	0	0	0	0
Natural gas and petroleum liquids	4	2,690	70	0	56	2,820	0
Nuclear	0	908	1169	265	157	2,499	147
Renewables	4,696	8,168	1409	269	133	14,674	6,193
Biomass	57	523	537	0	0	1,117	10
Geothermal	160	1	100	12	0	273	228
Hydro	17	17	52	0	130	216	16
Solar	496	120	348	178	0	1,134	788
Wind	3,556	1,178	166	85	1	4,986	4,852
Other	95	0	205	0	1	302	130
Biofuels	314	6,330	0	0	0	6,644	169
Electricity-smart grid and transmission	461	58	222	20	211	971	495
Conservation	3,387	3,206	0	4	0	6,597	6,305
End-use	5,705	693	832	1011	0	8,241	1,549
LIHEAP	5,000	0	0	0	0	5,000	0
Other	705	693	832	1011	0	3,241	1,549
Total	**14,295**	**16,284**	**4365**	**1570**	**648**	**37,160**	**14,786**

Source: U.S. Energy Information Administration [9].

[a]Total will not match Table 24 midpoint of "Estimated Interest Subsidy at Benchmark Interest Rate" because some data cannot be allocated by fuel or activity.

supply basis, subsidies to the more traditional energy sources are modest, but that does not refute the point that virtually all sources of energy are subsidized in the United States. However, when one considers that very few coal or nuclear plants have been constructed in recent years, it does seem logical that federal support of renewable energy technologies may make more sense given the rapid growth of wind energy, solar energy, and biofuels.

As shown in Table 18.5, in FY 2007, fossil fuels received more support than all renewables, biofuels were the largest recipient of federal support among the renewable energy sector, while wind energy received only 2.65% of the energy-specific subsidies and support in FY 2007.

And as shown in Table 18.6, in FY 2010, wind energy received a much larger share of federal support, receiving almost $5 billion, or about 13.4% of the energy-specific subsidies and support in FY 2010. Note, however, that the vast majority of this was related to the American Recovery & Reinvestment Act of 2009, which included a provision allowing owners of a wind energy project to take advantage of a 30% Investment Tax Credit or a 30% cash grant from the U.S. Treasury in lieu of receiving the PTC over a 10-year period. Thus, federal support of wind energy other than American Recovery and Reinvestment Act (ARRA)-related items amounted to only about 0.6% of non–ARRA-related energy-specific subsidies and support in 2010.

One of the most beneficial forms of government support for renewable energy in the United States is the federal Renewable Energy PTC. Since the enactment of the Energy Policy Act of 1992 by the U.S. Congress, the PTC has provided an incentive for investors to invest in renewable energy projects. When implemented in 1992, the wind energy PTC provided for a tax credit of 1.5¢/kilowatt-hour (kWh) produced during the first 10 years of a wind project. The annual value of the PTC escalates with the rate of inflation, and stood at 2.3¢/kWh (or $23/MWh) in 2013 for wind energy projects, closed-loop biomass systems, and geothermal energy facilities. Technologies eligible for the PTC include wind energy, landfill gas projects, biomass generation, certain hydroelectric projects, geothermal electricity production, municipal solid waste, anaerobic digestion, tidal energy, wave energy, and ocean thermal energy. Table 18.7 shows the value of the PTC in 2013.

TABLE 18.7 Value of Production Tax Credit For Certain Types of Renewable Energy

Technology type	PTC amount (¢/kWh)
Wind energy	2.3
Closed-loop biomass	2.3
Open-loop biomass	1.1
Geothermal energy	2.3
Solar energy	2.3
Landfill gas projects	1.1
Municipal solid waste projects	1.1
Qualifying hydroelectric projects	1.1

Source: R. Walker based on data from the US IRS [10].

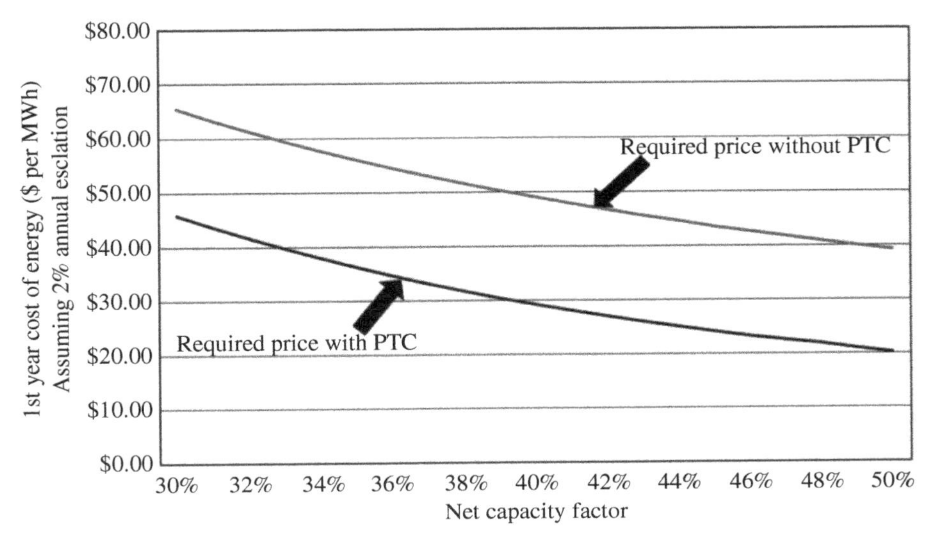

FIGURE 18.7 Estimated break-even price of wind energy in $/MWh as a function of the expected average annual net capacity factor (Source: R. Walker).

Figure 18.7 depicts the impact that the wind energy PTC has on the economics of a wind project. This graph reflects an estimated "breakeven" or minimum acceptable price for sales of wind energy needed for a project to be viable based on various expected annual average NCFs. The black line reflects the break-even price assuming the PTC is available, and the red line reflects the break-even price without the tax benefits of the PTC. Key assumptions used in both scenarios include the following:

- 250-MW project size at an installed cost of $1.61 million/MW of rated capacity
- 2.0% ad valorem tax rate with no tax abatements
- fourth quarter Commercial Operation and corresponding MACRS depreciation, with no bonus depreciation
- 2.0% annual escalation of the energy sales price

While the 2012 average NCF of US wind energy project was just over 30% (see Table 18.1), the large turbine rotors now being offered allows wind developers to select a turbine model that is optimal for their site, so that a 30% NCF is achievable in many parts of the nation. Assuming power prices in a region are $50/MWh, one could expect to have a viable wind project even with a NFC as low as 30%.

In very windy areas of the country, it is likely that NCFs of projects at excellent sites are approaching 50%, particularly now that the CREZ transmission line additions in Texas will greatly reduce the amount of wind project curtailments resulting from transmission congestion. As shown in Figure 18.7, it is possible to get PPA prices beginning well under $30/MWh. This finding is supported by Figure 18.8 from the U.S. Department of Energy (DOE)'s 2012 Wind Technologies Market Report [11] showing levelized wind PPA prices.

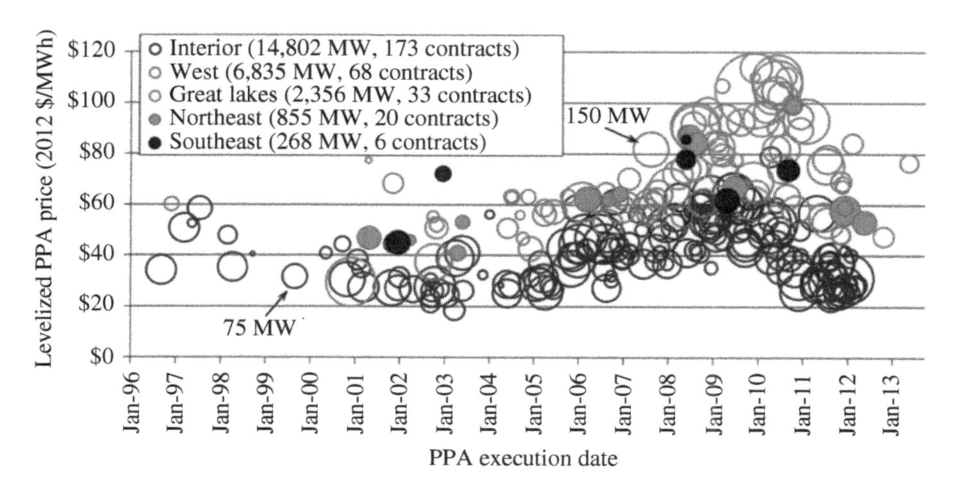

FIGURE 18.8 Levelized wind PPA prices by PPA execution date and region (Source: Figure 32 from 2012 Wind Technologies Market Report [11]).

However, if the PTC is not available, even a project with a 40% NCF requires a PPA price beginning at around $50/MWh, and it would take a phenomenal site, high hub height, and large rotor to be able to sell power for as low as $40/MWh.

The PTC has undoubtedly been a large reason for the rapid growth of wind energy, but it has also created some issues for the wind energy industry. Like many Federal tax credits, Congress must periodically reauthorize or extend them, but in 1999, 2001, 2003, and 2012, it failed to extend the PTC in a timely manner, causing new construction of wind projects to come to a halt until such time as Congress extended the PTC. While Congress did extend the credit retroactively each time to the date on which the PTC expired, wind project developers were hesitant to commit to the projects without knowing for certain that Congress would do so. In 2012, Congress waited until the day the tax credit was scheduled to expire, December 31, 2012, before extending it, and this resulted in few turbine orders for US projects during the second half of 2012 and almost no US projects beginning commercial operation in the first half of 2013. Many energy analysts believe that the fluctuating approval of the PTC has significantly added to the cost of wind energy.

Figure 18.9 clearly shows the impact that lapses in the Federal PTC can have on growth of the wind energy industry. One can clearly see the "down years" of 2000, 2002, 2004, and 2012 when wind project construction in the United States came to a virtual standstill due to lapses or near-lapses in the PTC.

18.11 ENVIRONMENTAL EXTERNALITIES

In addition to direct subsidies or incentives offered by governments, there are other societal costs of electricity generation that may not be reflected in the cost of electricity paid by the consumer. For example, owners of fossil-fueled power plants rarely are required to pay for the release of emissions into the atmosphere provided they stay below any

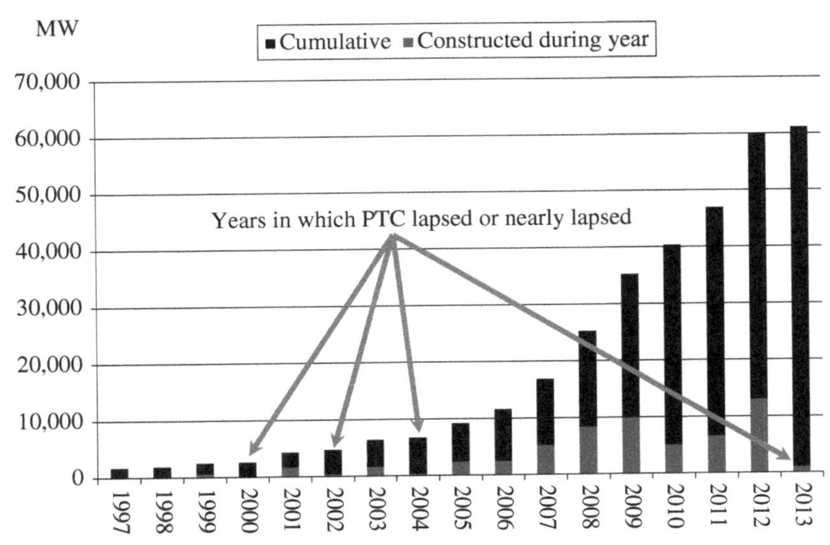

FIGURE 18.9 Growth of wind energy capacity in the United States as affected by the availability of the PTC (Source: R. Walker based on data from AWEA Annual Market Reports).

government-mandated rate of emissions. These emissions, however, can impose a cost on society. Particulate matter from a coal-fired power plant goes out the smokestack and into the air, but eventually comes back down near the Earth's surface where it may be inhaled by an asthmatic person, further complicating his or her medical condition. Thus, there are societal costs or "environmental externalities" (such as increased medical expenses) associated with releases of emissions into the atmosphere. In the vast majority of cases, the company releasing the emissions does not have to pay for the right to emit. Use of any energy source generates externalities (defined as societal costs that are not reflected in market transactions) including sources of renewable energy considered to be environmentally benign compared with traditional sources of electric generation.

For example, wind energy projects emit some level of sound into the environs, and may also be visually displeasing to some people. These would be considered environmental externalities, although by almost any standard, they would be very small compared with the externalities associated with a coal-fired power plant. Some states require that environmental externalities be considered during the process of permitting proposed power plants and as part of economic comparisons of various generation technologies, in effect giving those resources with fewer environmental externalities a better chance of being selected.

18.12 USING LEVELIZED COST OF ENERGY TO COMPARE COMPETING TECHNOLOGIES

Comparing the cost of energy produced by various generation technologies can be quite complicated due to differences in expected useful life, efficiency, fuel costs, capital costs, governmental incentives, and a myriad of other factors. A rather simplistic way to make a ballpark estimate of the average cost of energy over the lifetime of a generation project is to calculate its Levelized Cost of Energy (LCOE) by use of the following method:

$$\text{Levelized Cost of Energy} = \text{LCOE} = \frac{(\text{FCR} * (\text{TCC} + \text{BOS})) + \text{O \& M} + \text{FUEL}}{\text{AEP}}$$

where FCR = fixed charge rate as a percent (the components of FCR are provided later)
TCC = total capital cost of generating unit including balance of station
BOS = balance of station costs (roads, substation, transmission lines, etc.)
O&M = annual operations and maintenance cost (not including fuel)
FUEL = annual cost of fuel
AEP = annual energy production in MWh.

Components of FCR include the following:

Weighted average cost of capital	Assume 8–10%
Ad valorem taxes	Assume 1–2%
Income taxes on return	Assume 2–3%
Levelized replacement cost	100%/Expected life

An example of how this method can be used to compare the cost of energy from a coal-fired plant with that of a wind energy plant is shown as follows:

18.12.1 LCOE Example Comparing a Coal-fired Plant with a Wind Plant

Project 1: 600-MW coal-fired power plant (with scrubber for SO_2 control plus carbon capture and sequestration)

- Total capital cost = \$3,000,000/MW = \$1,800,000,000
- 3,153,600 MWh produced/year = 60% capacity factor
- Assumed nonfuel O&M cost = \$15/MWh produced = \$41,304,000
- 40-year assumed project life
- Heat rate or efficiency = 8600 Btu/kWh (i.e., 8600 Btu needed to produce 1 kWh)
- Fuel cost = \$2.50/million Btu (MMBtu)

$$\text{Annual fuel cost} = 3{,}153{,}600\text{ MWh} * \frac{8{,}600\text{ Btu}*}{\text{kWh}}\frac{1{,}000\text{ kWh}*}{1\text{ Mwh}}\frac{\$2.50*}{\text{MMBtu}}\frac{\text{MMBtu}}{1{,}000{,}000\text{ Btu}}$$

$$\text{Annual fuel cost} = \$67{,}802{,}400$$

$$\text{FCR} = 8\% + 2\% + 2.5\% + (100\% / 40\text{ year life}) = 15\%$$

$$\text{LCOE} = \frac{(\$1{,}800{,}000{,}000 * 0.15) + \$41{,}304{,}000 + \$67{,}802{,}400}{3{,}153{,}600\text{MWh}} = \$117.36\text{ perMWh}$$

Project 2: 250-MW wind energy plant

- Total capital cost = $1,610,000/MW = $402,500,000
- 919,800 MWh produced/year = 42% capacity factor
- Assumed O&M cost = $10/MWh produced = $9,198,000
- 25-year assumed project life
- Fuel cost = $0

$$\text{FCR} = 8\% + 2\% + 2.5\% + (100\% / 25\,\text{year life}) = 16.5\%$$

$$\text{LCOE} = \frac{(\$402,500,000 * 0.165) + \$9,198,000 + \$0}{919,800\text{MWh}} = \$82.20\,\text{per MWh}$$

Note that this simplified method of comparing two different types of plants does not incorporate the effects of things like accelerated depreciation for income tax purposes or the Federal renewable energy PTC; thus, the aforementioned LCOE for wind is far above that reflected in Figure 18.7.

18.13 CONCLUSIONS

Thoroughly describing all factors involved in determining or understanding the economics of electric generation would probably require an entire textbook, rather than the one chapter devoted to this subject within this volume. As one can readily see, there are many different factors to be considered and many of them are in constant flux, such as the cost of natural gas. Trying to compare the average cost of energy produced by one type of generation technology with that of another technology can be extremely tedious and complicated due to differences in factors such as useful life, efficiency, capacity factor, depreciation rates for tax purposes, and numerous other considerations.

Federal, state, and local governmental entities may offer incentives to encourage construction of some types of electric generation and may require quantification of environmental externalities in order to discourage construction of other types of generation. These incentives or disincentives will vary from state to state, county to county, and city to city, thus adding to the complexity of any comparison.

Demand for electric energy in the United States and other nations will continue to grow, and concerns about the environmental impacts of traditional sources of generation will remain strong. Global events such as the Fukushima Daiichi nuclear power plant incident in Japan can impact energy policy in many other nations. This one event probably delayed permitting and construction of new nuclear units being proposed in the United States by several years. Growing demand for fossil fuels such as natural gas may eventually outpace the ability to identify new reserves, which may cause dramatic spikes in prices for these commodities. Concern over global climate change may lead to the imposition of carbon taxes or emission caps, significantly reducing the competitive position of coal-fired generation and making it even harder to permit such plants than it currently is.

In summary, one of the best strategies for meeting the future demand for electricity in the United States and other countries is to diversify the types of resources used to produce electricity. One cannot rely solely on natural gas, on coal, or on nuclear power. Sources of renewable energy must become a significant part of this solution if we are going to stabilize emissions, and doing so can also stabilize the price of electricity.

REFERENCES

[1] U.S. Energy Information Administration website. Total energy. Available at http://www.eia.gov/totalenergy/. Accessed May 23, 2014.

[2] U.S. Energy Information Administration website. Monthly average prices of natural gas for electric power. Available at http://www.eia.gov. Accessed August 8, 2012.

[3] ERCOT website data. Balancing energy services market clearing prices for energy annual report archives. Available at http://www.ercot.com/mktinfo/prices/mcpe. Accessed January 3, 2014.

[4] U.S. Energy Information Administration website. Electricity annual. Available at http://www.eia.gov/electricity/annual/. Accessed December 18, 2013.

[5] North American Electric Reliability Corporation (NERC) website. Available at http://www.nerc.com. Accessed August 8, 2012.

[6] U.S. Energy Information Administration website. *Electric Power Annual 2012*. Washington (DC): U.S. Energy Information Administration; 2013.

[7] U.S. Energy Information Administration. Age of power plants in the United States by fuel source, based on Form EIA-860 annual electric generator report and Form EIA-860M. Washington (DC): U.S. Energy Information Administration; December 4, 2013.

[8] U.S. Internal Revenue Service. *IRS Publication 946, How to Depreciate Property,* Appendix A, MACRS Depreciation Rates for 3, 5, 7, 10, 15, or 20-Year Property, Half-Year Convention. Washington (DC): U.S. Internal Revenue Service; 2010.

[9] U.S. Energy Information Administration. (2011). Direct federal financial interventions and subsidies in energy in Fiscal Year 2010, Washington (DC): U.S. Energy Information Administration.

[10] U.S. Internal Revenue Service. Form 8835. Washington (DC): U.S. Internal Revenue Service; 2013.

[11] U.S. Department of Energy. Figure 32 from *2012 Wind Technologies Market Report*. Washington (DC): U.S. Department of Energy, Energy Efficiency & Renewable Energy; 2013.

19

ECONOMIC IMPACTS AND BENEFITS OF WIND ENERGY PROJECTS AND OTHER SOURCES OF ELECTRIC GENERATION

19.1 INTRODUCTION

As the old saying goes, "there are two sides to every story," whenever a company proposes a new electric generation project in or near a community, there are those who perceive that the project will be beneficial to the community, while others perceive that the project will be harmful to the community. This goes for almost any kind of generation project including wind energy, solar energy, nuclear energy, natural gas–fired generation, hydroelectric generation, or coal-fired generation. The positive attributes and potential negative elements of each type of project have been discussed in previous chapters, but in this chapter we will focus on economic impacts, both positive and negative.

19.2 POTENTIAL ECONOMIC BENEFITS OF ELECTRIC GENERATION PROJECTS

When a new project of any kind is announced in an area, the first thing that comes to many peoples' minds is the job opportunities that come along with new businesses. While jobs at the facility are obviously a key benefit, there are many other types of beneficial economic impacts that such a facility might have. With regard to electric generation projects, economic benefits may include the following.

Wind Energy Essentials: Societal, Economic, and Environmental Impacts, First Edition.
Richard P. Walker and Andrew Swift.

19.2.1 Direct Jobs

There will be ongoing or permanent jobs at the power plant after it has been constructed and commenced operation. Large generation projects or concentrations of projects (such as the numerous wind generation facilities located near Sweetwater, Texas) often spawn start-up of new businesses within the power industry to provide services to the power plant(s). Manufacturing companies may also be attracted to a region where large amounts of its product are being used.

19.2.2 Construction Jobs

While some outside companies may be hired to manage or oversee construction, almost all new power plants will provide opportunities for local individuals or companies in the construction industry while the project is being built.

19.2.3 Local Purchases of Goods and Services

During the construction phase of a large project, business may substantially increase for local restaurants, hotels, cement companies, surveyors, automotive service companies, hardware stores, and many others. Even after construction is complete and operations have commenced, people employed at the power plant and their families will be purchasing groceries, clothes, vehicles, housing, and many other goods and services within the local community.

19.2.4 Indirect Job Creation

As a result of the goods and services purchased by those constructing or operating the power plant, new jobs may also be created at local grocery stores, hardware stores, lumber yards, dry cleaners, hotels, restaurants, car dealers, and other businesses. This is sometimes referred to as a "multiplier effect" or a "ripple effect" because expenditures in one sector of the region's economy lead to additional positive impacts throughout many other sectors.

19.2.5 State and Local Tax Payments

The addition of new tax base in the community can help fund schools, roads, and other state and local infrastructure and help to keep tax rates lower for residents of the community.

19.2.6 Increased Reliability of Electric Supply in the Region

If a region must rely on imports of electricity, particularly if it is located remotely from urbanized areas, its electricity supply is contingent upon the reliability of the electric transmission lines supplying the area.

19.2.7 Reduced or Stabilized Cost of Electricity

If communities or regions rely on transmission lines already at or near their capacity for moving power, they may be paying a high price for their electricity since it may have to be produced from old, inefficient power plants. Of course, there is always the possibility that adding a new power plant may result in increased electricity rates associated with cost recovery of the new facility by the utility company, but in the long run, new power plants are generally justified based on the best choice for providing economic, reliable, and environmentally acceptable power. Addition of renewable energy sources by the utility serving the region can help mitigate fluctuations in the cost of fossil fuels and serve as a hedge against the potential imposition of carbon taxes.

19.2.8 Increased Property Values

Adding new jobs in a community or region can result in increased demand for housing in the area and thus increased property values, particularly where employment at the power plant is very large compared with the population of the area. Wind projects can also substantially increase the value of tracts where turbines are or can be installed, and biomass plants fueled by locally grown crops may increase the value of farmland.

19.2.9 Use of Local or Indigenous Resources

Power plants that rely on local natural resources to produce electricity can have a much larger economic impact on the local area than those that import their fuel. This can include wind plants, solar projects, biomass facilities, hydroelectric projects, or even coal plants using locally mined coal or lignite. Some regions, and even some countries, have very little indigenous oil, gas, coal, or nuclear fuel and have to import virtually all of their energy sources, sometimes at great cost. Island nations, or island states such as Hawaii, often have some of the most expensive electricity rates due to the cost of importing the fuel.

19.2.10 New Revenue Sources for Landowners

Landowners can benefit from the construction of large power projects in the region in a variety of ways. Such projects may help the landowner retain his or her property or allow them to continue farming or ranching as opposed to having to seek employment in some other field. Landowner revenues can include the following:

- Sale of property for power plants
- Royalties from wind turbines on their property
- Sales of water used for cooling at thermal generation plants
- Mining of oil, coal, or gas on their property
- Sales of crops used as fuel in biomass plants
- Sales of rock, gravel, or caliche used for roads and foundations in the project

19.3 POTENTIALLY DETRIMENTAL ECONOMIC IMPACTS OF ELECTRIC GENERATION PROJECTS

As alluded to in the introduction to this chapter, an announcement of a new electric generation project in or near a community is likely to draw opposition from people who perceive that the project will be harmful to the community, whether the project will be using wind power, solar energy, nuclear energy, natural gas, water, biomass, municipal solid waste, landfill gas, or coal to produce electricity. Issues specific to each of these resources are discussed in detail in previous chapters, but in this section impacts potentially detrimental to the region's economy are discussed. Some of the economic impacts that developers of any type of electric generation project must consider or address during the project design phase can include the following.

19.3.1 Environmental Impacts That May Translate to Reduced Property Values

Large coal or nuclear plants and tall wind turbines can be seen for many miles, and there are those who contend that such visual impacts can reduce the value of properties surrounding such projects. Air emissions and odors produced by some generation projects may also impact property values of adjacent properties. Others contend that audible noise produced by wind turbines and industrial facilities has been responsible for reduced property values. Even traffic associated with a power plant may have an effect on the value of adjacent or nearby properties.

19.3.2 Environmental Impacts That May Translate to Reduced Tourism in the Area

Areas known for their scenic views attract tourists and may in fact have a large portion of their local economic activity derived from tourism. Community leaders in areas with forests, canyons, hiking trails, rivers, or lakes may wish to discourage industrial development in areas near these resources and thus may oppose projects that have an impact on the area's scenery. Tall objects, such as wind turbines, smokestacks, or cooling towers, would likely fall into this category. In addition, air emissions (including steam from cooling towers) may also be considered detrimental to tourism. Following are examples of such impacts.

- **Big Bend and Guadalupe Mountains National Parks:** A report produced by Texas Natural Resource Conservation Commission (now the Texas Commission on Environmental Quality) titled *Blurry Big Bend* [1] states that "visibility impairment at Big Bend and Guadalupe Mountains National Parks can be attributed to regional haze, which involves the long-range transport of fine particles from a group of sources over a broad geographic area. Recent scientific evidence indicates that there are contributing sources in the U.S. as well as Mexico. Carbón I and Carbón II, for example, which are enormous coal-burning power plants near Piedras Negras in the Mexican border state of Coahuila, are thought to be regular contributors to the haze that sometimes obscures parts of West Texas. Although both plants meet Mexican air quality standards, neither is equipped with controls for sulfur dioxide emissions that are required in the United States. It is also probable that there are occasional air invasions from heavily industrialized areas and petrochemical operations on Texas' Gulf Coast."

- **Great Smoky Mountains:** The U.S. National Park Service's website [2] discusses in detail the effect that air pollution is having on the Great Smoky Mountains, including the following statements:
  - Air pollution is shrinking scenic views, damaging plants, and degrading high-elevation streams and soils in the Great Smoky Mountains. Even human health is at risk. Most pollution originates outside the park and is created by power plants, industry, and automobiles.
  - Research and monitoring conducted in the park has shown that airborne pollutants emitted from mostly outside the Smokies are degrading park resources and visitor enjoyment. The burning of fossil fuels (such as coal, oil, gasoline, or natural gas) causes most of the pollution. Inadequate pollution-control equipment in power plants, factories, and automobiles is the primary problem.
  - Wind currents moving toward the southern Appalachians transport pollutants from urban areas, industrial sites, and power plants located both near and far. The height and physical structure of the mountains, combined with predominant weather patterns, tend to trap and concentrate human-made pollutants in and around the national park.
  - Views from scenic overlooks at the Great Smoky Mountains National Park have been seriously degraded over the last 50 years by human-made pollution. Since 1948, based on regional airport records, average visibility in the southern Appalachians has decreased 40% in winter and 80% in summer.
  - The burning of fossil fuels produces tiny airborne sulfate particles that scatter light and degrade visibility. Increasingly, visitors no longer see distant mountain ridges because of this haze. Annual average visibility at Great Smoky Mountains National Park is 25 miles, compared with natural conditions of 93 miles.

19.3.3 Potential Disruption of Housing Market and School System Due to Temporary Influx of Construction Workers and Their Families

Construction of very large electric generation projects can result in a large influx of construction workers into an area. Since wind farms, coal plants, and nuclear power plants are typically located in rural areas, such an influx of workers can potentially overwhelm the local housing market and school system. High demand on housing may drive up the cost of building or purchasing a new home in the area, and after construction is completed, property values may return to more normal levels. Thus someone moving to the area when the project is being built, even though they may be unaffiliated with the project, could have paid more for their housing than otherwise would have been the case. Rural school systems may have to expend additional funds to bring in portable classrooms or hire additional staff. However, in the case of school systems, these impacts are probably temporary since the potential increase in *ad valorem* tax revenues associated with a large power plant would far exceed any additional expense incurred during the construction period.

19.3.4 Impacts on the Local Job Market

While new jobs and economic activity in a community are generally considered to be good for an area, a large power plant can reduce the availability of qualified workers for other local businesses in the area and may require such businesses to increase the wages they pay in order to attract or retain their workforce. A second consideration involves the average salary of permanent workers once construction is completed and operations have commenced. If the new jobs being added are on average lower-paying jobs than the regional average, this may be viewed by some people as having a detrimental effect on the area's economy.

19.3.5 Impacts on Local Roads, Utilities, and Emergency Services

Another consideration of adding a large electric generation project in or near a small-to-midsize community is the potential impact on roads, utilities, and emergency services. Construction of such projects can be particularly hard on roads, and county commissioners will generally require the company building the project to restore the local roads to the condition which existed prior to construction of the project. Additional traffic both during and after construction also should be a consideration. Facilities requiring large volumes of water or producing large volumes of wastewater can create significant problems for local utility systems as well. Impacts on emergency services also needs to be addressed to ensure that emergency responders are not overwhelmed by a sudden influx of construction in the area and that construction traffic does not present a barrier to effective emergency response. While it is likely that additional tax revenues and economic activity associated with the project will exceed the cost of such impacts, community leaders must evaluate the timing of such revenues in relation to the demand for these services. The additional cost of providing these services also needs to be considered in the event the developer of the generation project will be requesting tax abatements.

19.3.6 Cost of Health Impacts

As discussed in Chapter 8 of this book, some people have alleged that wind energy projects can be detrimental to the health of individuals living near wind turbines, usually related to audible noise produced by the project. Large gas, nuclear, or coal power plants also generate audible noise and thus may receive similar allegations.

A much larger health impact associated with fossil-fueled plants is the health impacts of emissions they produce, as discussed in Chapter 14.

19.4 TAX ABATEMENTS OR OTHER INCENTIVES TO ATTRACT GENERATION PROJECTS

Local government officials wishing to attract new businesses to their city, county, or region may offer to reduce *ad valorem* taxes, or property taxes, for the business during the first few years of the business' operation. They may also offer such tax abatements to expansions of existing businesses. Reasons often cited for granting tax abatements to businesses may include the following:

- Creation of new jobs in the area.
- Increasing the tax base in the area, which may help hold tax rates down in the future.
- County tax abatements may attract businesses that pay full school property taxes, thus allowing for significant improvements to facilities or services at the region's schools.
- Encouraging construction of public facilities or infrastructure (e.g., streets and roads).
- Redevelopment of economically depressed or blighted areas within the community.
- Providing access to services for residents, such as housing or retail shopping.

Since taxes do affect many business location decisions and because tax abatements allow local government officials to be proactive or action-oriented in their economic development efforts, the use of tax abatement offers to business is a widespread practice. However, not everyone is in agreement that the benefits of such agreements always exceed the cost to the community. Those who argue against tax abatements may include some of the following arguments:

- Taxes are not the only factor considered in business location decisions.
- The businesses being attracted to an area may place a costly burden on local services such as roads, emergency services, water, or sewer systems.
- The selective use of abatements raises questions of equity.
- An abatement offered to a new business may place an existing, competing business in the community at a disadvantage.
- Abatements pull public dollars away from local expenditures that benefit business.
- Local property taxes are a small portion of overall costs faced by a firm.
- Local property taxes are deductible against federal income taxes.
- Tax abatements may cause a rise in local land prices.

West Texas was the "hotbed" of wind development from 2005 to 2010, and several West Texas communities have been particularly aggressive in pursuing wind energy projects and associated businesses, although it is possible that they may offer similar incentives to other types of electricity generation projects due to the associated jobs and tax base. The Texas House Research Organization's Focus Report titled *Capturing the Wind* [3] evaluated the merits of tax abatements and lists the following pros and cons of offering tax abatements to wind energy projects:

Pros

- Incentives such as tax abatements are an important tool for attracting and keeping wind developers in Texas.
- Texas has a high property tax rate on capital-intensive projects like wind energy.
- While Texas has many resources that are attractive to wind energy developers seeking to locate in the state, the high tax rate is a disincentive when location decisions are made.
- While opponents argue that offering tax abatements costs revenue, that argument assumes that those companies were going to locate in Texas even without the incentive of a tax abatement.
- In the rural economically distressed areas of West Texas and the Panhandle, where wind projects and turbine-manufacturing plants would consider locating, the creation of 10 or more good jobs is a boon to the local economy.

Cons

- Tax abatements are an unnecessary loss of revenue to the state, especially tax abatements for school districts.
- School districts are guaranteed a certain amount of property tax revenue from the state. If a school district abates those taxes, the state will have to absorb the cost through the school finance system and make up the difference with other types of funding, including general revenue.
- With superior wind resources, a trained work force, an open and unregulated market, a closed electric grid that prevents competition from wind energy imported from other states, and abundant natural gas reserves serving as a complement to wind energy, Texas provides multiple inducements to locate in the state without the need to forego taxing potential and revenue.

19.5 SUMMARY OF STUDIES ASSESSING THE ECONOMIC IMPACTS OF WIND PLANTS

Several studies have been conducted over the past few years to analyze the economic impacts of wind facilities. Following are brief summaries of some of them.

19.5.1 Renewable Northwest Project Economic Impact Study of the 24-MW Klondike Wind Project in Oregon

This study [4] examined local business activity, employment, landowner revenue, and tax revenue for the county, and its conclusions included the following points:

- Benefits of the wind project were widespread throughout the county and the surrounding region.
- Employment from development, construction, and operations stimulated regional businesses and continue to provide personal income in the county.
- Sherman County as a whole receives substantial tax revenues.
- Individual farmers receive additional income from royalty payments while continuing farming operations.
- All of these benefits will drastically increase as the Klondike project is expanded in the coming years.
- In a county where farming is a way of life, wind has provided value and stability to the economy.

19.5.2 National Wind Coordinating Committee's (NWCC) *Assessing the Economic Development Impacts of Wind Power*

The National Wind Coordinating Committee contracted with Northwest Economic Associates to evaluate the economic impact of wind projects ranging between 25 and 107 MW in three separate regions of the country. In the February 2003 report titled "Assessing the Economic Development Impacts of Wind Power, Final Report" [5] the impacts of the Lake Benton I Project in Minnesota, the Vansycle Project in Oregon, and the Delaware Mountain Project in Texas were described. The findings from this study included the following:

- In all of the case studies, there was a modest to moderate boost in economic activity that was attributed to the construction phase.
- All of the case study areas also benefited from continuing operation and maintenance (O&M) activities, with the amount of activity very much related to the size of the development and previous development of wind power and planned future development of wind power.
- These cumulative effects of multiple projects can be important in the decision to perform O&M work with a local workforce rather than import these workers from outside the area.
- Tax effects, particularly property taxes that support local entities, were important in all cases. If the entities' budgets do not increase as a result of the project, the assessed value of the tax base increases, and there is a redistribution of the local tax burden. This, in effect, shows up as an increase in household income, which can directly affect the local trade and services sectors and to a lesser extent other local economic sectors.

- The annual revenue from leases and easements received by households in the areas was a significant source of household income and had a significant total effect on the economies.
- The cost of foregone opportunities from farming and livestock grazing was small compared with the revenues obtained.

The authors of this study did receive anecdotal reports of negative impacts such as the killing of birds or damages to existing roads from construction and O&M activities.

19.5.3 Nolan County: Case Study of Wind Energy Economic Impacts in Texas [6]

The area surrounding Sweetwater, Texas is the most concentrated area of wind energy development in the United States, if not the world. Within 75 miles of Sweetwater, approximately 6000 MW of wind generation has been installed, beginning with the Trent Mesa Wind Farm in 2001. This study found that benefits to Nolan County alone included the following:

- An estimated local economic impact of over $315 million in 2008 and almost $400 million in 2009 based on an assumed sevenfold multiplier principle to the primary direct wind energy base payroll
- In 2008, 1124 direct wind jobs with payroll in excess of $45 million, increasing to 1330 direct wind jobs in 2009 with payroll in excess of $56.6 million (Fig. 19.1)

FIGURE 19.1 E.ON Climate & Renewables' Roscoe Wind Farm Operations Center in Nolan County (Source R. Walker).

FIGURE 19.2 New school building in Trent, Texas (Source: R. Walker).

- Estimated annual landowner royalties of over $12 million in 2008 based on 2500 MW of installed wind capacity, increasing to over $17 million in 2009 based on expected capacity of 3600 MW
- A huge increase in total taxable property values in Nolan County from $500 million in 1999 to $2.4 billion in 2008 and a projected $3.5 billion by 2010
- Cumulative wind energy project property taxes paid in Nolan County (including Trent Independent School District (ISD)) from 2002 to 2007 of over $30 million
- Between 2004 and 2010, $24 million in new school construction in Nolan County school districts between 2004 and 2010 (Fig. 19.2)

19.5.4 Economic Impact of a Proposed Wind Farm in Kittitas County, Washington [7]

As part of the planning process for two wind projects (165 and 225 MW), the Phoenix Economic Development Group hired ECONorthwest to evaluate the potential economic impacts of constructing and operating the wind plants in Kittitas County. Specifically, ECONorthwest was asked to analyze and help quantify impacts in three key areas of interest:

1. Property values: Local residents have voiced concern that constructing numerous wind turbines in the valley will detract from views and ultimately reduce property values.
2. Economic impacts: The wind plants will create jobs and increase spending in the economy during the construction phase and during plant operations.
3. Tax revenues: The increase in jobs and local spending will also increase tax revenues for Kittitas County.

19.5.5 Lawrence Berkeley National Laboratory Study of *The Impact of Wind Power Projects on Residential Property Values in the United States* [8]

In 2009, researchers at the Lawrence Berkeley National Laboratory released the results of their study which investigated a fairly commonly expressed concern that property values can be adversely affected by wind energy facilities. The study collected data on almost 7500 sales of single family homes situated within 10 miles of 24 existing wind facilities in nine different states, and used eight different pricing models to determine if impacts on property values could be conclusively demonstrated.

The study indicated that concerns about the possible impact of wind power facilities on residential property values can be divided into the following categories:

- Area stigma: A concern that the general area surrounding a wind energy facility will appear more developed, which may adversely affect home values in the local community regardless of whether any individual home has a view of the wind turbines.
- Scenic vista stigma: A concern that a home may be devalued because of the view of a wind energy facility and the potential impact of that view on an otherwise-scenic vista.
- Nuisance stigma: A concern that factors that may occur in close proximity to wind turbines, such as sound and shadow flicker, will have a unique adverse influence on home values.

The conclusions of the report included the following statements:

- The result is the most comprehensive and data-rich analysis to date on the potential impacts of wind projects on nearby property values.
- Although each of the analysis techniques used in this report has strengths and weaknesses, the results are strongly consistent in that each model fails to uncover conclusive evidence of the presence of any of the three property value stigmas.
- Based on the data and analysis presented in this report, no evidence is found that home prices surrounding wind facilities are consistently, measurably, and significantly affected by either the view of wind facilities or the distance of the home to those facilities.
- Although the analysis cannot dismiss the possibility that individual or small numbers of homes have been or could be negatively impacted, if these impacts do exist, they are either too small and/or too infrequent to result in any widespread and consistent statistically observable impact.
- Moreover, to the degree that homes in the present sample are similar to homes in other areas where wind development is occurring, the results herein are expected to be transferable.

19.6 STUDIES ON ECONOMIC BENEFITS OF NON-WIND ELECTRIC GENERATION PROJECTS

Traditional electric generation plants using nuclear energy, coal, or natural gas as fuel also provide economic impacts within the region where they are located. As with wind energy, some of these impacts may be perceived as negative, but large electric generation projects can also provide significant amounts of economic benefits to communities. Following are summaries of a couple of studies in this regard.

19.6.1 Nuclear Energy Institute Study of the *Economic Benefits of PPL Susquehanna Nuclear Power Plant: An Economic Impact Study* [9]

Following are some of the economic benefits suggested by the Nuclear Energy Institute related to the construction and operation of PPL's Susquehanna Nuclear Power Plant in Luzerne County, Pennsylvania:

- Jobs
  - The operation of PPL Susquehanna and the secondary effects of the plant account for more than 380 jobs in Luzerne County and almost 4200 jobs in Pennsylvania.
  - These jobs result in $39.5 million in earnings to workers in Luzerne County and $293.3 million in Pennsylvania.
  - PPL Susquehanna employs 1528 people, including corporate support and peak outage and other supplemental PPL craft employees.
  - Twenty percent, or 305 employees, live in Luzerne County.
  - An estimated 196 full-time employees live in the Luzerne County cities of Nescopeck, Shickshinny, Hazleton, Wilkes-Barre, and Sugarloaf.
  - These jobs pay substantially higher salaries than the average salary in Luzerne County.
- Taxes
  - The plant pays $35.3 million in state and local taxes each year.
  - Adding the economic activity generated by PPL Susquehanna through increased business, corporate, payroll, and personal taxes results in a total state and local tax impact of $50.1 million.
- Economic activity
  - In 2005, PPL Susquehanna paid $37.3 million in compensation to employees living in the county and an additional $159.4 million to employees residing elsewhere in Pennsylvania.
  - Economic activity created by the plant accounted for an additional $2.2 million in employee compensation in Luzerne County and an additional $94.4 million in other areas of the state.
  - PPL Susquehanna makes substantial purchases in Luzerne County, where the plant spent nearly $1 million in 2005 (not including employee compensation).

  - Purchases totaled over $56 million in Pennsylvania and nearly $188 million in the United States.
  - Economic activity generated by PPL Susquehanna's purchases and operation also led to $6.1 million in increased output in the county and $251.7 million in the state.
  - In 2005, operation of PPL Susquehanna increased Pennsylvania's economic output by $251.7 million, including $6.1 million in Luzerne County.
  - Adding the direct value of the plant's electricity generation brings the economic output attributable to PPL Susquehanna to $1.16 billion in Pennsylvania and $915 million in Luzerne County.
- Intangible benefits; such as charitable giving; recreational opportunities; cleaner air; community involvement; environmental stewardship; and stable, affordable electricity prices.

19.7 OPPORTUNITIES CREATED BY A TRANSITION TO A "CLEAN ENERGY ECONOMY"

As third-world nations strive to increase the electrification of their nations and as concerns mount about global climate change, depletion of fossil fuel resources, and emissions associated with burning of fossil fuels, many companies are trying to establish a foothold in the emerging market for clean energy technologies, such as the manufacturing of wind turbines or solar modules. Many communities, states, and nations are aggressively trying to attract such businesses to locate there. Within the United States, the state of Arkansas has successfully attracted several wind energy-related industries in recent years despite having limited wind resources. China and South Korea have added several very large wind turbine manufacturing facilities, with China being home to four of the top 10 wind turbine manufacturing companies in terms of megawatts of sales in 2010, while the United States is home to only one of the top 10 companies. China and Taiwan are home to most of the largest solar cell and solar module manufacturers.

A report produced in 2010 by Billy Hamilton Consulting for the Cynthia and George Mitchell Foundation titled *Texas' Clean Energy Economy: Where We Are. Where We're Going. What We Need to Succeed* [10] found that policies promoting clean energy within the state could have substantial economic benefits for the state. Findings within the report included the following:

- Increased development of clean energy holds enormous potential to create jobs and to provide significant tax revenues to state and local governments, while also reducing airborne pollutants from traditional power plants.
- Extending and expanding the state's renewable portfolio standard (RPS) is essential to achieve a greater share of its total electricity generation from renewable sources.

- A modest investment in wind and solar energy would create 6000 jobs/year from 2010 to 2020 and increase the state's Gross State Product (GSP) by $802 million annually. State and local governments also would collect an additional $177 million/year in tax revenues.
- Making a stronger commitment to renewable energy would produce results nothing short of spectacular. By raising the state's RPS to accommodate another 13,000 MW of power, including 3500 MW in new solar photovoltaic energy, Texas' economic gains would more than triple, with job gains of up to 22,900/year, an additional $2.7 billion/year in GSP, and roughly $279 million more/year in state and local taxes.

A 2009 report titled *The Clean Energy Economy: Repowering Jobs, Businesses and Investments Across America* [11] was released by the Pew Charitable Trusts. According to the report, "a clean energy economy generates jobs, businesses and investments while expanding clean energy production, increasing energy efficiency, reducing greenhouse gas emissions, waste and pollution, and conserving water and other natural resources." Included in the report were the following conclusions:

- The number of jobs in America's emerging clean energy economy grew nearly two and a half times faster than overall jobs between 1998 and 2007.
- Jobs in the clean energy economy grew at a national rate of 9.1%, while traditional jobs grew by only 3.7% between 1998 and 2007.
- There was a similar pattern at the state level, where job growth in the clean energy economy outperformed overall job growth in 38 states and the District of Columbia during the same period.
- By 2007, more than 68,200 businesses across all 50 states and the District of Columbia accounted for about 770,000 jobs.
- The emerging clean energy economy is creating well-paying jobs in every state for people of all skill levels and educational backgrounds.
- Venture capital investment in clean technology crossed the $1 billion threshold in 2005 and continued to grow substantially, reaching a total of about $12.6 billion by the end of 2008.
- In 2008 alone, investors directed $5.9 billion into American businesses in the clean energy economy, a figure that represents a 48% increase over 2007 investment totals and accounts for 15% of all global venture capital investments.
- Federal and state lawmakers also see the sector as helping to spur America's economic recovery and protect the environment.
- Every US state offers some form of financial incentive to drive its clean energy economy.
  - Twenty-three states have adopted regional initiatives to reduce the global warming pollution from power plants.

- Forty-six states offer some form of tax incentive to encourage residents and corporations to use renewable energy or adopt energy efficiency systems and equipment.
- Twenty-nine states and the District of Columbia have established RPSs, which require electricity providers to supply a minimum amount of power from renewable energy sources.

19.8 CONCLUSIONS

Industrial development, including construction of new electric generation projects of any technology or fuel source, will have both positive and negative impacts on the region surrounding it. While environmental impacts often get the most attention, particularly from those who may be opposed to the project, there will be economic impacts on the region as well. Many of these will be beneficial to the local economy (such as the creation of new employment opportunities) although others (such as impacts on human health) can have detrimental economic impacts. A critical step for developers of any such project is effectively making the case that the economic benefits associated with a new electric generation project will far exceed any associated environmental impacts and detrimental economic effects.

REFERENCES

[1] Texas Natural Resource Conservation Commission. Blurry Big Bend, Austin; March 12, 1999.

[2] U.S. National Park Service website. Nature and science/air quality page. Available at http://www.nps.gov/grsm/naturescience/air-quality.htm. Accessed August 10, 2012.

[3] House Research Organization, Texas House of Representatives. *Capturing the Wind: The Challenges of a New Energy Source in Texas*. Austin: House Research Organization, Texas House of Representatives; 2008. Focus Report.

[4] Ouderkirk B, Pedden M. *Windfall from the Wind Farm, Sherman County, Oregon*. Portland: Renewable Northwest Project; 2004.

[5] Northwest Economic Associates. *Assessing the Economic Development Impacts of Wind Power, Final Report*. Washington, DC: National Wind Coordinating Committee; 2003.

[6] New Amsterdam Wind Source LLC. *Nolan County: Case Study of Wind Energy Economic Impacts in Texas*. Sweetwater: West Texas Wind Energy Consortium; 2008.

[7] ECO Northwest. *Economic Impacts of Wind Power in Kittitas County: A Report for the Phoenix Economic Development Group*. Portland: ECO Northwest; 2002.

[8] Hoen B, Wiser R, Cappers P, Thayer M, Sethi G. *The Impact of Wind Power Projects on Residential Property Values in the United States: A Multi-Site Hedonic Analysis*. Berkeley: Ernest Orlando Lawrence Berkeley National Laboratory, Environmental Energy Technologies Division; 2009.

[9] Nuclear Energy Institute in cooperation with PPL Corporation. Economic Benefits of PPL Susquehanna Nuclear Power Plant: An Economic Impact Study. Nuclear Energy Institute, Washington, DC; 2006.

[10] Billy Hamilton Consulting. *Texas' Clean Energy Economy: Where We Are. Where We're Going. What We Need to Succeed.* Austin: Cynthia and George Mitchell Foundation; 2010.

[11] Pew Charitable Trusts. *The Clean Energy Economy: Repowering Jobs, Businesses and Investments Across America.* Washington, DC: Pew Charitable Trusts; 2009.

20

ENVIRONMENTAL IMPACTS AND ECONOMICS OF OFFSHORE WIND ENERGY

20.1 OFFSHORE WIND RESOURCE AREAS IN THE UNITED STATES

We begin this section with a discussion about offshore wind farms, what they are, and how they could impact the development of wind energy both in the United States and worldwide. The term *offshore wind*, as its name implies, refers to wind farms that are placed near the coastline but in a marine system. They are usually within several miles of the shore and in relatively shallow waters, typically less than 60 m in depth (Fig. 20.1), although research and development of floating turbine platforms for use in deeper locations is being undertaken by several companies or institutions.

Some of the attributes of offshore wind installations include the following:

- A more stable wind resource with higher wind speeds
- The ability to sustain larger wind turbines as compared with land-based systems
- The potential to locate wind farms near large business and population centers, not only in the United States but worldwide. This is due to the fact that for most countries with access to the ocean, populations tend to migrate there because of business opportunities or tourist and recreational opportunities
- The ability to locate turbines far from occupied buildings, eliminating any issues related to turbine noise
- This industry can be built on concepts already developed for offshore oil and gas drilling using many of the same technologies and techniques

Wind Energy Essentials: Societal, Economic, and Environmental Impacts, First Edition.
Richard P. Walker and Andrew Swift.

Presently, offshore wind projects are located mostly off the coast of Europe and the United Kingdom with only planned projects off the coast of the United States. According to RenewableUK (formerly the British Wind Energy Association), the United Kingdom "is the world leader in offshore wind with as much capacity already installed as the rest of the world put together." Their website [1] further indicates there are 22 operational wind farms in UK waters totaling 3653 MW, 3.8 GW under construction, and 7.8 GW in planning stages. The 630-MW project called the London Array was completed in 2012, as well as the 504-MW Greater Gabbard project.

As of the end of 2012, the European Wind Energy Association estimated that almost 5000 MW of offshore wind farms were in operation in Europe, as shown in the following chart (Fig. 20.2).

FIGURE 20.1 Middelgruden Offshore Wind Farm (Photo Credit: Kim Hansen, http://commons.wikimedia.org/wiki/File:Middelgrunden_wind_farm_2009-07-01_edit_filtered.jpg).

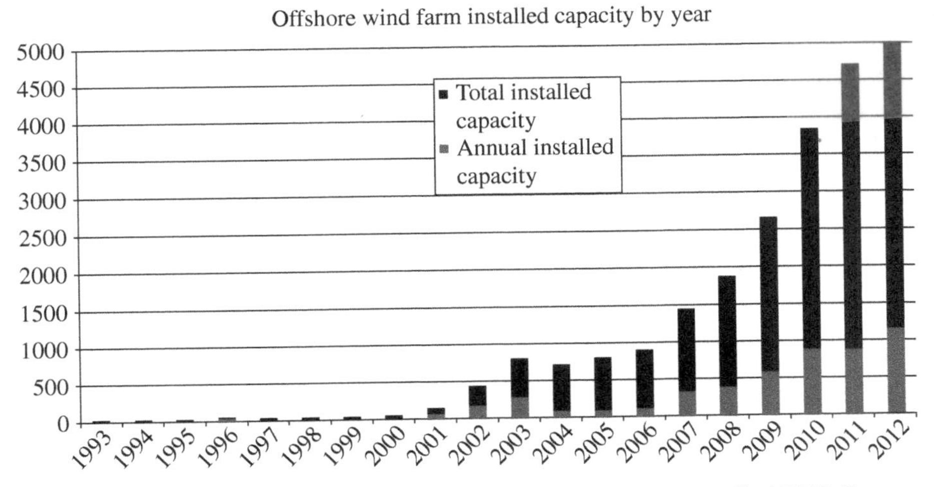

FIGURE 20.2 Offshore Wind Capacity in European Waters as of Year-End 2012 (Source: Reproduced by permission of K. Jay [2]).

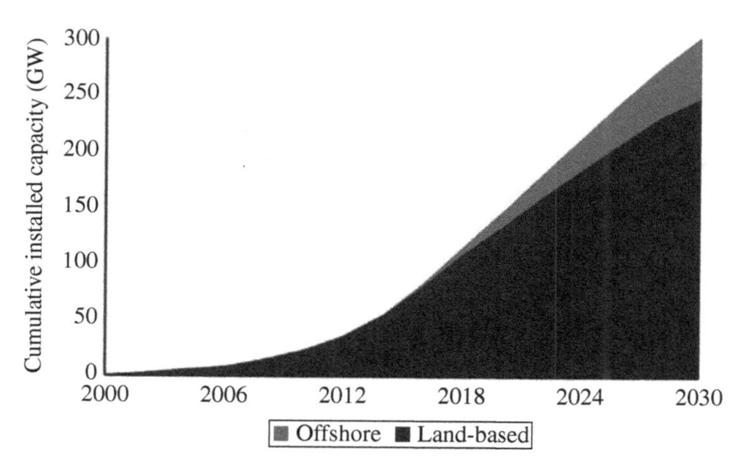

FIGURE 20.3 DOE's projected wind power growth need to achieve 20% wind by 2030 (Source: *20% Wind Energy by 2030* [3]).

In Asia, both China and Japan have operational offshore wind farms. While several planned US offshore wind projects have been announced, only a few have made it through the permitting process including the Cape Wind project and the Delaware Offshore Wind Farm. Permitting for offshore wind projects in the United States can be costly and time-consuming, as shown by the Cape Wind project, which first applied for federal permits in 2001 before finally receiving all necessary state and federal approvals near the end of 2011.

The U.S. Department of Energy's (DOE's) *20% Wind Energy by 2030* [3] report shows a significant offshore component beginning around 2015 and growing to a total of approximately 50 GW of development or about one-sixth of the planned 300 GW by 2030, as shown in Figure 20.3.

To understand and evaluate offshore wind resources the following factors are considered:

- The annual wind speed at hub height
- Sea level and water depth to understand the feasibility of offshore platforms
- Legal definitions and regulations of state and federal jurisdictions
- Site-specific information to include distance from mainland to identify transmission and transportation costs

The U.S. DOE and the National Renewable Energy Laboratory have together developed a database of wind speeds in coastal areas according to state boundaries and jurisdictions. A resulting wind speed map is shown in Figure 20.4.

Wind speed data are shown at 90 m height above sea level (wind turbine hub height). The importance of these data is to find the potential offshore sites for wind farms with a steady and high wind resource. These maps were produced using a physics-based computer model.

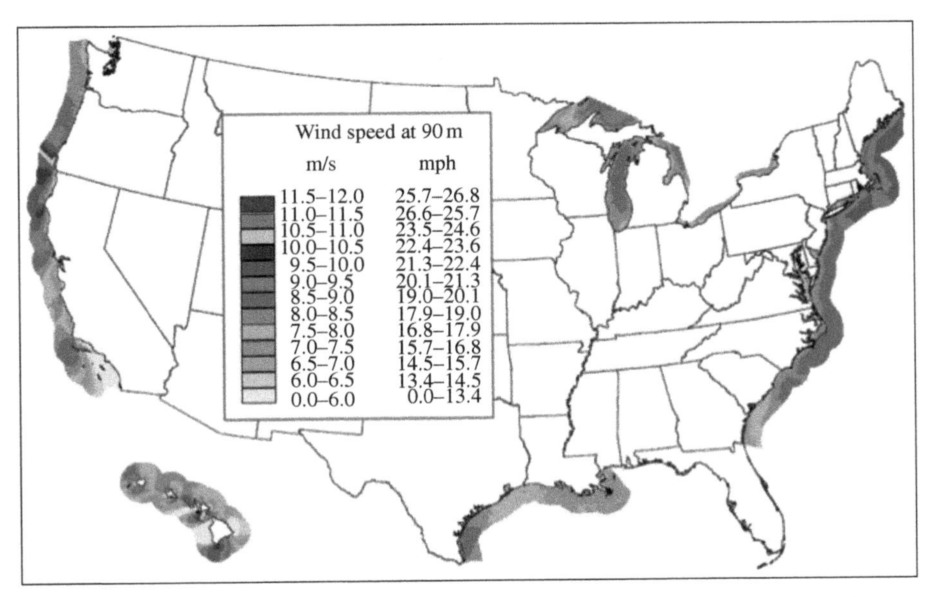

FIGURE 20.4 Wind Resource Map of the United States (Source: DOE/NREL [4]).

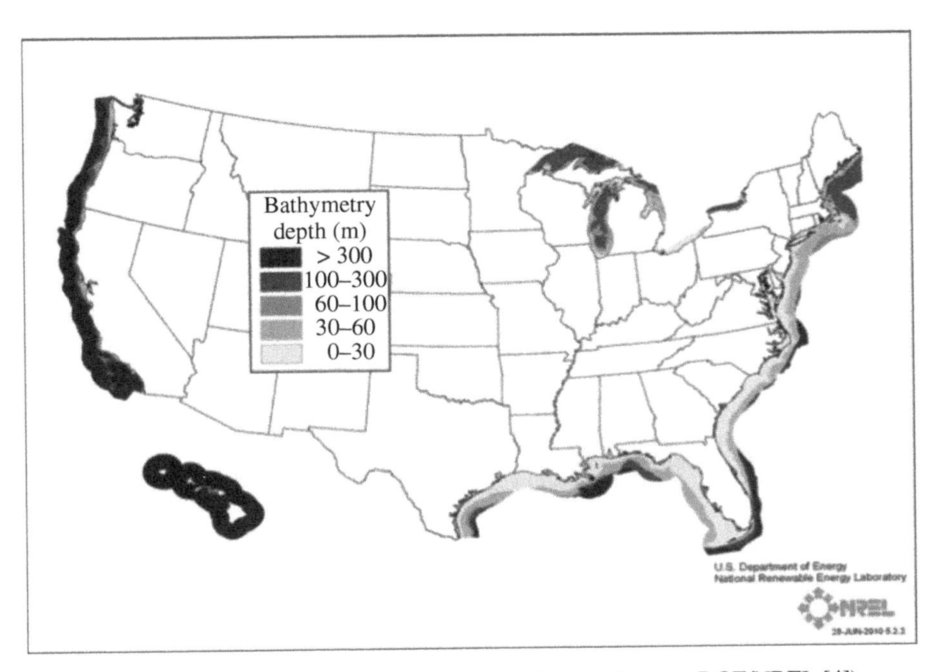

FIGURE 20.5 Bathymetry Map of the United States (Source: DOE/NREL [4]).

The most energetic areas are off the Atlantic coast of the northeastern United States with other areas identified on the Pacific coast, the Gulf of Mexico, and the Great Lakes region. Given the highly regulated environment of coastal marine use, numerous state and federal agencies are involved with offshore development and at a level of involvement much greater than for land-based wind energy development. Figure 20.5 shows the Bathymetry index, which is essentially water depth that has to be compared with the available wind resource map and then related to possible turbine locations and the types of foundation technology required to support the wind turbine.

The Bathymetry map for the United States shows the distribution of water depths around the coastal region, that is, the Pacific and Atlantic oceans. The importance of these data is to know the type of foundation technology that is required to support the wind turbine.

20.2 ECONOMICS OF OFFSHORE WIND VS. ONSHORE WIND

The first variable in determining the economics of both offshore and onshore wind is to estimate the annual energy production from the wind power plant. In some respects, wind resource measurement and turbine output estimation is simpler offshore because of the uniform topography over the sea surface. Surface roughness changes with wave height but other than that factor, the wind shear variations are relatively easily estimated using a log law profile. Typically, offshore turbines will be larger—for example, the GE 3.6-MW offshore turbine with 104-meter rotor diameter. These wind turbines will require significant spacing, typically in the order of 10 rotor diameters in the prevailing wind direction.

Next, a market assessment must be completed to assess the electricity markets and the proximity of the wind farm and the transmission infrastructure by which the electricity would be transmitted to those markets. Being offshore this would require underground cabling with associated regulation and permitting. Using estimates of capacity factor and the size of the wind farm to estimate annual output, along with the interannual and seasonal variability, negotiations for utility pricing agreements can be entered into either through a fixed price power purchase agreement or spot market pricing structure.

Integration of offshore wind farms may be easier than onshore wind farms, especially in the United States, depending on how the timing of the output of the turbine coincides with peak load requirements of the regional utility. In Texas, for example, there is a good match between coastal winds and the timing of peak loads—especially when loads are high on hot summer afternoons due to air-conditioning loads. Next the *pro forma* business plan would be laid out and if the project meets acceptable criteria, it would be negotiated for financing with banks and other financial institutions.

In general, offshore wind farms tend to be significantly more expensive—raising the cost of energy to about twice the cost of onshore wind farms. Transportation and maintenance requirements are a more difficult issue in some respects, given that it is a maritime environment and that one has to consider the corrosive aspects of a salt water atmosphere. On the other hand, offshore wind farms can accommodate larger

TABLE 20.1 Potential Socioeconomic Impacts

Positive impacts	Negative impacts
Raw material and supply chain availability in coastal areas	Noise and visual impacts
Improved workforce availability in coastal areas	Conflicts with fishing and recreational regions
Related economic development	Potential for increased accidents and collisions affecting ship traffic
Noise impacts normally would be less than that from onshore wind projects	Opposition from coastal residents

wind turbines, which will help reduce the cost of energy due to economies of scale. Additionally, because of barge transportation, larger components are feasible.

Reduced wind shear offshore will allow for larger rotors with less variability of the wind across the rotor diameter; thus, there will be reduced constraints on turbine size and component size as compared with land-based systems for which components must be transported to the site by truck or rail, which have both weight and size limitations. For offshore wind turbines, the cost of the turbine itself is only approximately 30% of the cost of the power plant. The remainder is the balance of station and installation costs, which are much higher for offshore than onshore. However, due to the location close to large coastal population centers, offshore wind installation should have good access to a ready workforce. Table 20.1 summarizes the socioeconomic impacts with some being positive and others negative.

20.3 ENVIRONMENTAL IMPACTS UNIQUE TO OFFSHORE WIND ENERGY

There are a number of environmental impacts that have to be considered for offshore wind turbine installations. These are outlined as follows:

- **Marine life:** Offshore wind turbine foundations can support new artificial reefs and consequently create new habitats for sea species.
- **Birds:** Offshore projects may increase bird strike mortality as well as cause navigational disorientation to birds.
- **Electromagnetic field effects:** Electromagnetic fields that are produced by electric transmission cables may be harmful to sea biosystems.
- **Emissions:** There would a significant reduction in the carbon and other power plant emissions.
- **Marine traffic and recreation:** The wind farms are likely to have a negative effect on ship traffic, human recreational and tourist activities, and sea space used by the defense forces for training.
- **Visual impacts:** Include impacts from towers, rotating turbine blades, and navigation and aerial warning lights.

- **Noise emissions:** May cause impairment of hearing to sea mammals. The large wind turbine rotor blade tip noise transmits low-frequency underwater noise (Fig. 20.6).
- **Vibration:** Vibrations transferred from the wind turbine foundation to the water may induce different wave patterns affecting the habitat of a wide variety of sea species from whales to lobsters.

While offshore wind development will have several siting issues similar to the development of onshore wind projects, it also has some issues unique to only offshore projects. Table 20.2 summarizes both environmental and human constraints.

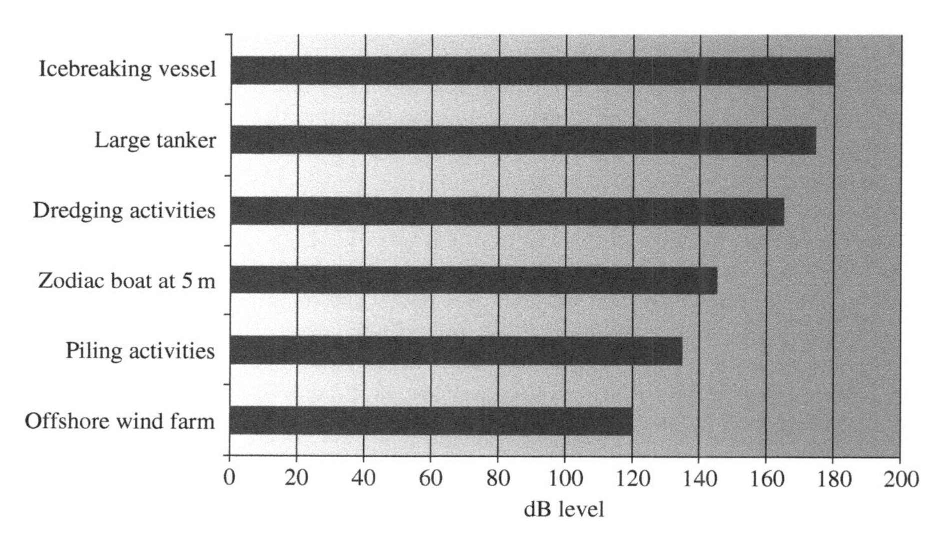

FIGURE 20.6 Sound pressure levels in decibels for various sea traffic and activities. Notice that the wind turbine is actually one of the lower frequency, lower decibel activities listed (Source: Reproduced by permission of K. Jay [5]).

TABLE 20.2 Offshore Wind Development Constraints

Environmental constraints	Human use constraints
Natural obstructions	Military, commercial, and private aviation areas
Submerged cultural resources	Department of Defense (DOD) restricted areas
Reefs—natural and artificial	Long-range military radar zones
Sensitive/critical habitat or biological resource areas	Military practice areas
	Unexploded ordnance
Avian resource areas	Undersea cable and pipeline routes
Bat resource areas	Dumping grounds
Scenic locations	Mining areas
National Register Of Historic Places	Oil and gas exploration/excavation areas
National Wildlife Refuges	Commercial and recreational fishing areas
Wildlife management areas	

20.4 POTENTIAL IMPACT OF HURRICANES

Due to their coastal location, offshore wind farms will be exposed to the risk of hurricanes. In many spots (such as the Gulf of Mexico as shown in Fig. 20.4), there is significant wind resource availability; however, there is also a high risk of damage to wind power plants due to the high winds associated with severe hurricanes. It should be pointed out that not all hurricanes are of the same severity. For example, a Category 1 storm would not exceed the wind speed design criteria laid out in current design codes for wind turbines and thus probably inflict very little if any damage to the wind farm. However, beginning with Category 2 and increasingly as the categories grow to the severe winds of a Category 5 storm, the potential damage to a wind power plant becomes severe. The highest risks in US Coastal waters are in the Gulf of Mexico and the southern Atlantic coast, based on historical hurricane frequency and intensity data. However, the entire Eastern coast of the United States is at some risk, while the West coast has minimal risks of hurricane impact. Recent research from Carnegie Mellon University [6] estimates that up to 50% of wind turbines may be destroyed over a 20-year lifetime for offshore turbines deployed in the areas of the Texas and Louisiana Gulf Coast, based on historical records of hurricane frequency and severity.

The report also points out that there are strategies to mitigate these effects and significantly improve the survivability of turbines in more severe hurricanes. First, insuring that the turbines can maintain the capability to yaw with the wind direction will increase survivability substantially. This would involve a backup battery system to ensure that the yaw motors and wind direction sensors are robust and can continue to operate even if the utility power is not available. This will make sure the turbine is aligned with the wind and can avoid excessive side loads on the blades. Other strategies include increasing the maximum wind design criteria to higher values than currently used and develop stronger towers and blades by adding additional steel and fiberglass. All of these strategies will increase turbine costs, possibly up to 30%, adding to the cost of energy, which as pointed out earlier, is already higher for offshore wind turbines. Currently in the United States, there is significant interest in pursuing offshore wind power development. The DOE has recently put in place a $43 million research program to lower the cost of offshore wind energy and deal with some of the problems of offshore wind development in line with the more than 50 GW goal for offshore wind in the 2030 DOE 20% wind scenario.

REFERENCES

[1] Renewable UK website. 2014. Offshore wind. http://www.renewableuk.com/en/renewable-energy/wind-energy/offshore-wind. Cited December 21, 2013.

[2] European Wind Energy Association. Deep Water: The Next Step for Offshore Wind Energy, A Report by the European Wind Energy Association, Brussels, Belgium; 2013.

[3] U.S. Department of Energy, Energy Efficiency and Renewable Energy. 20% Wind Energy by 2030: Increasing Wind Energy's Contribution to U.S. Electricity Supply (DOE/GO-102008-2567). U.S. Government Printing Office, Washington, DC; 2008.

[4] Schwartz, M., Heimiller, D., Haymes, S., Musial, W. (2010). Assessment of Offshore Wind Energy Resources for the United States (Technical Report NREL/TP-500-45889). Golden: National Renewable Energy Laboratory.

[5] Vella G. *The Environmental Implications of Offshore Wind Generation*. Liverpool: Centre for Marine and Coastal Studies, University of Liverpool; 1981.

[6] Rose S, Jaramillo P, Small MJ, Grossmann I, Apt J. Quantifying the Hurricane Risk to Offshore Wind Turbines (CEIC paper). Pittsburg, PA: Carnegie Mellon University; 2011.

21

STATE AND NATIONAL ENERGY POLICIES

21.1 INTRODUCTION

The energy policy of the United States has evolved over many decades, several presidential administrations, and numerous congressional sessions. Policies are molded over time by many entities including the President, Congress, judicial system, and branches of the Administration such as the Department of Energy (DOE) and the Environmental Protection Agency (EPA). Policies are often driven by threats to the nation's energy security (such as the OPEC oil embargo of 1973 or Iraq's invasion of Kuwait in 1990) or by environmental issues (such as acid rain, global climate change, or urban air quality). The OPEC oil embargo resulted in US government policies promoting the development and use of renewable energy technologies and raised related concerns about reliance on foreign energy imports and their contribution to trade imbalances.

Energy supply shortages and associated price disruptions also influence policy-making, such as the spikes in natural gas prices that occurred in 2000, 2005, and 2008, shown previously in Figure 14.7. Other supply issues leading to energy policy changes have included electricity shortages in California during 2001 and a blackout of large areas of the US Northeast and Midwest regions in 2003.

In addition to Federal government entities, states and even municipalities are involved in setting energy policy, adding to the diversity and complexity of an overall national energy policy. As a result, US energy policy might best be described as a

Wind Energy Essentials: Societal, Economic, and Environmental Impacts, First Edition.
Richard P. Walker and Andrew Swift.

"hodgepodge" of initiatives rather than a thoughtful, comprehensive, and cohesive policy. Some policies or regulations may be enacted in reaction to a major supply issue or to a significant environmental disaster that is receiving extensive media attention, without really thinking of the long-term consequences of the change in policy or regulation.

A recent example is the 2010 *Deepwater Horizon* oil spill in the Gulf of Mexico. Eleven men were killed in an explosion and the resulting oil spill was the largest marine oil spill in the history of the petroleum industry. About one month before this disaster, the Obama Administration announced plans to allow more offshore oil and gas exploration in the eastern Gulf of Mexico and parts of the Atlantic Ocean. But soon after the disaster, they reversed their position and announced that offshore oil drilling would not be allowed in the eastern Gulf of Mexico or off the Atlantic and Pacific coasts as part of the next 5-year drilling plan, reversing previously announced policy changes. While this decision may or may not have been a reasonable action, policy decisions made in reaction to a single event or circumstance that is the focus of media attention at that point in time quite often will turn out to be short-sighted or to have significant unintended consequences.

Other nations have experienced similar direction-changing events to their energy policies, such as Japan's policy regarding the use of nuclear power following the Fukushima Daiichi nuclear disaster in March 2011, which was triggered by an undersea earthquake and a subsequent tsunami. Since Japan has to import the vast majority of its energy, it relied heavily on nuclear power for electricity, but following the Fukushima disaster, it has significantly scaled back its plans for nuclear power in the future.

21.2 HISTORY OF US ENERGY POLICY

The U.S. Department of the Interior was created in 1849 to oversee federally owned lands and natural resources. This has evolved into responsibility for oil, natural gas, and coal policy. In the nineteenth century, coal supplanted wood and whale oil as the primary fuel source and continues to be used in abundance, mainly for electricity generation. During World War II, the first atomic bomb was developed under the Manhattan Project, which was overseen by the US Army Corps of Engineers. In 1947, after the end of World War II, the Atomic Energy Commission was formed to oversee the development of nuclear energy.

In response to the OPEC Oil Embargo of 1973, President Nixon proposed a goal of energy independence for the United States within a decade and established the Federal Energy Office. In 1974, the Nuclear Regulatory Commission was established to replace the Atomic Energy Commission's oversight of the use of nuclear energy for electricity production. In 1975, President Gerald Ford also proposed a plan focused on energy independence that led to passage of the Energy Policy and Conservation Act, which included establishment of a Strategic Petroleum Reserve as a defense against future disruptions of oil supplies. That same year, the Energy

Research and Development Administration (ERDA) was created to focus the federal government's energy research development activities into one unified agency. ERDA's scope included development of renewable energy projects, and it provided funding for some of the early wind and solar demonstration projects. In 1977, President Jimmy Carter created a Cabinet-level DOE by merging functions of several Federal agencies, including ERDA, and charged DOE with developing a comprehensive national energy policy.

One of the most significant policy changes affecting electric power production in the United States occurred in 1978 when Congress passed the Public Utility Regulatory Policies Act (PURPA) in order to create a market for power produced by nonutility generators or independent power producers (IPPs), including renewable energy facilities. Essentially, PURPA forced vertically integrated electric utilities to purchase energy produced from certain types of generators called "qualifying facilities" ("QFs"). Cogeneration facilities were the earliest beneficiaries of PURPA, but many wind generation plants have also been able to sell their output under PURPA contracts. Cogeneration refers to the production of both electricity and thermal energy (such as steam) with the steam then being used in an industrial process. Use of the thermal energy produced during power generation is more efficient than just wasting the thermal energy byproduct as has been a common practice in the past. IPP-owned generation now makes up over one-third of all electric generating capacity in the United States. In recent years, the deregulation of electricity markets and open access of the transmission grid have reduced the significance of PURPA, but in some locations, it remains a viable option for selling electricity from a QF.

Three Energy Policy Acts have been passed, in 1992, 2005, and 2007, which have included many provisions for conservation, such as the Energy Star program, and energy development, with grants and tax incentives for both renewable energy and nonrenewable energy development. Provisions within these acts have helped to spur growth in the wind and solar energy sectors, but most other renewables have shown very little growth.

As shown in Table 21.1, while the amount of energy consumed in the United States has grown significantly since 1952, the mix of resources used to produce this energy has not changed substantially since 1960 with the exception of the increased use of nuclear energy. The percentage of energy coming from renewable sources increased only slightly over the past six decades, rising from 8.0% in 1952 to 9.3% in 2012.

A significant factor in this is that hydroelectric generation, which accounted for almost one-half of renewable energy used in 1952 and still accounts for almost one-third, has seen virtually no growth in the past four decades. Yet despite rapid growth of wind energy, solar energy, and biofuels in recent years, the percentage of energy used in the nation coming from nonhydroelectric renewable sources only amounted to about 6.46% of total energy consumed in the United States during 2012. Thus, if the United States is going to markedly increase the proportion of its energy coming from renewable sources, it will require significant changes to the nation's energy policy.

TABLE 21.1 US Energy Consumption (1952–2012) in Quadrillions of Btu[a]

U.S. Primary energy consumption by source										
	Quadrillion Btu					Percentage of Total Consumption				
Year	Coal	Natural gas	Petroleum	Nuclear electric	Renewable energy	Coal (%)	Natural gas (%)	Petroleum (%)	Nuclear electric (%)	Renewable energy (%)
1952	11.31	7.55	14.96	0.00	2.94	30.8	20.5	40.7	0.0	8.0
1962	9.91	13.73	21.05	0.03	3.12	20.7	28.7	44.0	0.1	6.5
1972	12.08	22.70	32.95	0.58	4.38	16.6	31.2	45.3	0.8	6.0
1982	15.32	18.36	30.23	3.13	5.98	21.0	25.1	41.4	4.3	8.2
1992	19.12	20.71	33.52	6.48	5.82	22.3	24.1	39.1	7.6	6.8
2002	21.90	23.51	38.22	8.15	5.73	22.4	24.1	39.1	8.3	5.9
2012	17.36	26.00	34.58	8.05	8.82	18.3	27.4	36.4	8.5	9.3

[a]Source: R. Walker; based on data from US-EIA Annual Energy Review 2012 [1].

21.3 KEYS ISSUES WITHIN ENERGY POLICY

There are numerous issues that the energy policy of a nation, state, or city will try to address. Following are some of the issues that can have the most impact on the use of wind energy or other renewables:

- Taxation policies, such as the Federal renewable energy production tax credit (PTC), or the use of accelerated depreciation for income tax purposes, or the use of tax credits for purchase of renewable energy equipment or projects
- State or national mandates regarding the use of renewable energy, such as renewable portfolio standards (RPSs) or renewable electricity standards (RESs)
- National policies regarding dependence on imported energy sources or placing tariffs on imported energy sources
- Mandated purchases of renewable electricity by government facilities
- Mandated utility purchases of renewable energy from renewable energy generation facilities including customer-owned facilities that at times may produce more energy than being used by the customer (net metering or feed-in tariffs (FITs))
- State or national goals regarding the use of renewable energy, which may be achieved through mechanisms other than mandated purposes, such as tax policies, siting policies, electric transmission policies, net metering policies, or grid operations policies
- State or federal funding of renewable energy technology research and development (R&D)
- Development of state or regional plans for electricity production and transmission
- Establishing or tightening emissions limits or placing a price or tax on emissions, both of which would tend to make most forms of renewable energy more economically competitive
- Entering regional or international agreements or treaties limiting the amount of emissions that can be released into the atmosphere (such as the Kyoto Protocol)

Other energy policy issues that would have a lesser impact on the use of wind energy or other renewables include the following:

- Building codes addressing the energy efficiency of homes or businesses
- Appliance energy efficiency standards
- Vehicle mileage standards
- Policies promoting use of mass transportation
- Policies promoting the use of electric or hybrid-electric vehicles
- State or federal funding of nonrenewable energy technology R&D

During the remainder of this chapter, several of these key policy issues will be discussed in more detail.

21.4 ENERGY IMPORTS

21.4.1 US Petroleum and Natural Gas Imports

Heavy reliance on energy imports can expose a nation to national security risks as well as contribute to trade imbalance, although some may argue that importing energy preserves a nation's own fossil fuel resources, which does have some merit. As shown in Figure 21.1, after the OPEC oil embargo of 1973, the United States made some significant gains toward reducing the amount of petroleum it imported, but beginning in 1986, imports rose fairly steadily until 2006. Since 2006, imports have dropped and domestic production has increased, although as of 2012, over one-half of the petroleum consumed in the United States was being imported. Table 21.2 shows the top 10 nations that provide petroleum to the United States and the breakdown between OPEC versus Non-OPEC countries.

As depicted in Figure 21.2, the United States produced the vast majority of the natural gas consumed in the country, although imports rose steadily from 1986 until 2007, when the domestic shale gas production began to alleviate the need for imports.

21.4.2 Nations Providing Energy Imports to the United States

Another important aspect of energy imports is the country of origin and the ongoing status of diplomatic relationships with that country. In 2012, more than 57% of petroleum products consumed in the United States during 2012 were imported, with more than 28% of the total coming from OPEC nations. Table 21.1 shows the top 10 nations that provide petroleum to the United States and the breakdown between OPEC versus non-OPEC countries. While some countries are close allies, others on the list would not quickly come to the minds of most Americans as countries that the United States should rely heavily upon for our energy supply, and some of them would make a list of countries that the United States has strained relationships with.

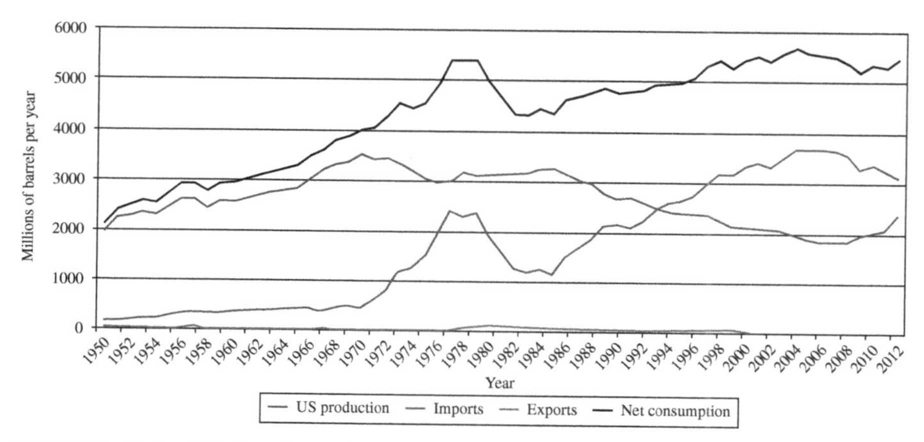

FIGURE 21.1 US Petroleum Production, Imports, and Exports (millions of barrels) (Source: R. Walker based on data from US-EIA website [2]).

TABLE 21.2 US Petroleum Imports During 2012 and Country of Origin (thousands of Barrels)[a]

Source	2012 imports (1000 barrels)	Percentage of 2012 imports (%)
Total imports	3,878,852	
Non-OPEC countries	2,315,579	59.7
OPEC countries	1,563,273	40.3
Canada	1,078,412	27.8
Saudi Arabia	499,595	12.9
Mexico	378,692	9.8
Venezuela	351,220	9.1
Russia	174,612	4.5
Iraq	174,080	4.5
Nigeria	161,558	4.2
Colombia	158,586	4.1
Kuwait	111,586	2.9
Algeria	88,487	2.3

[a]Source: R. Walker based on data from US-EIA Annual Energy Review 2012 [2].

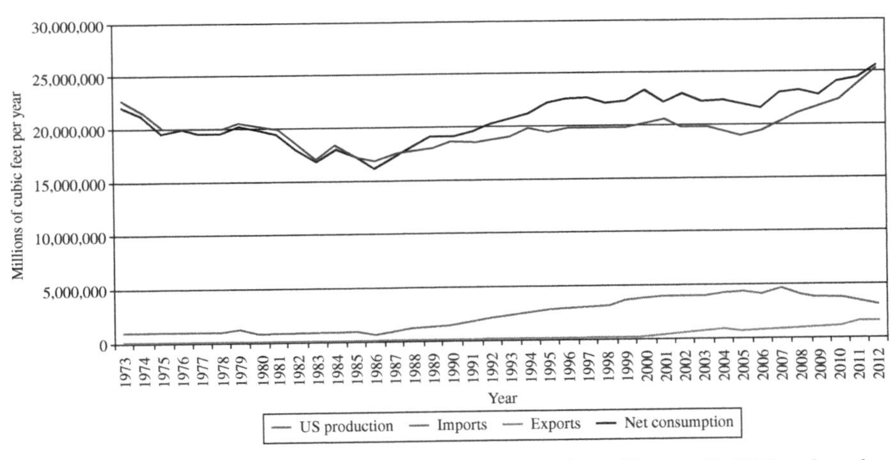

FIGURE 21.2 US Natural Gas Imports (million cubic feet) (Source: R. Walker based on data from US-EIA website [2]).

Strained relationships with nations providing natural gas to the United States is not much of an issue since almost all of the natural gas consumed in the United States comes from either domestic production or from Canada.

21.4.3 Can Renewable Energy Reduce Energy Imports?

One argument frequently made in support of the use of renewable energy is that its use can reduce energy imports and thus reduce the US trade deficit. In response to such an argument, opponents of renewable energy will often point out that the major use of petroleum in the United States is for transportation, with only about 1% of the nation's electricity being produced using petroleum; and since petroleum is by far the largest energy import, increased use of renewable energy cannot significantly reduce US energy imports or improve on the trade deficit.

While these points may have some validity based on the current use of energy and state of technology, technological advances and policy changes can alter this situation. Figure 21.3 shows what types of fuels are being used to displace petroleum for transportation purposes, and a discussion of some of these follows.

- Increasing use of electric and hybrid-electric vehicles while also increasing the amount of electricity produced from renewable resources is one way to reduce petroleum used for transportation while also reducing emissions from the transportation sector. Most car manufacturers are now offering hybrid-electric vehicles, and electric vehicle–charging stations are becoming more common. In many areas, consumers are able to choose to purchase electricity produced from sources of renewable energy and thus would be able to demonstrate that renewable energy fuels their electric or hybrid-electric vehicles.
- The use of biofuels for transportation purposes has been rising rapidly in the United States and other nations. As shown in Figure 21.4, US production of ethanol has been increasing rapidly since the year 2000, increasing at an average annual rate of almost 20%, although production declined slightly between 2011 and 2012.
- Natural gas that would be used for electricity production could be diverted to transportation purposes, thus reducing petroleum imports while using renewable energy sources to replace the electricity that would have been produced from natural gas. One of the pillars of T. Boone Pickens' *Pickens' Plan* is to "use America's natural gas to replace imported oil as a transportation fuel in addition to its other uses in power generation, chemicals, etc...." [4]
- Ethanol derived from corn, sugarcane, and soy is increasingly being used to produce biofuels, and R&D of other promising biofuel sources is also being

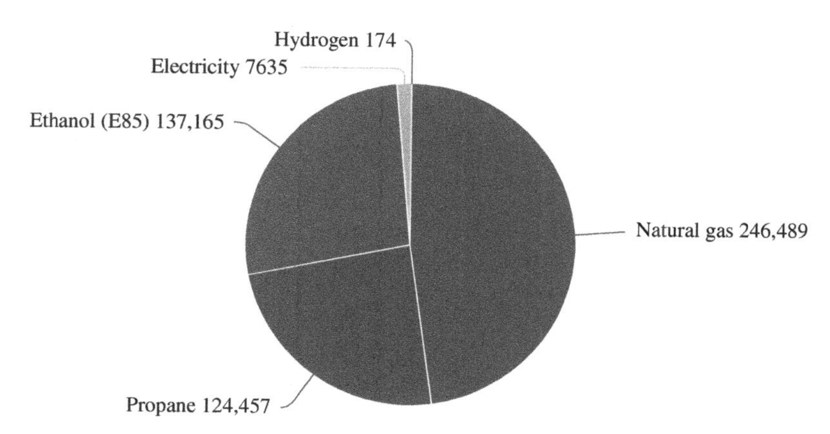

FIGURE 21.3 US Consumption of Alternative Fuels by Vehicles in 2011 (1000 gasoline-equivalent gallons) (Source: US-EIA [2]).

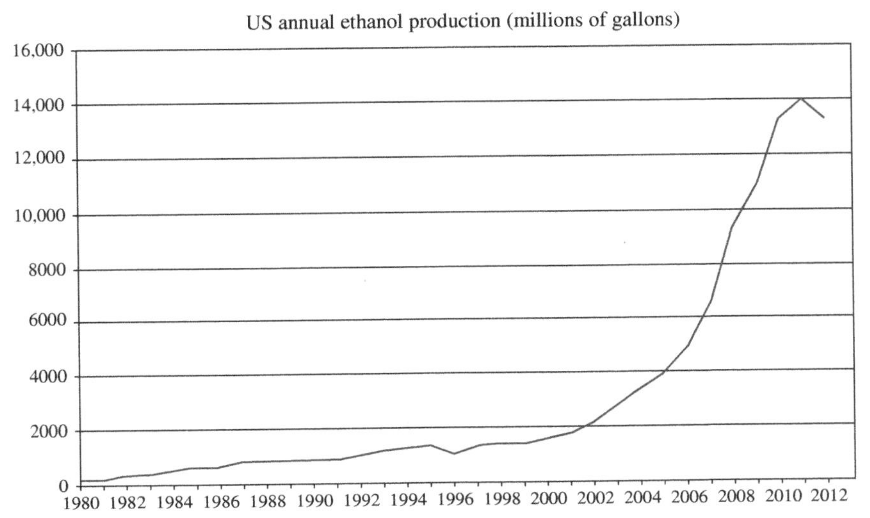

FIGURE 21.4 US Annual Ethanol Production (millions of gallons) (Source: R. Walker based on data from the Renewables Fuels Association website [3]).

conducted. One of the potential sources of significant interest is algae. Algae grows naturally throughout the world and since algae grows in water, it would not compete with traditional uses of land. Lipid oil, or fat, can make up one-half of the weight of algae and can be used for ethanol production.

21.5 GOVERNMENTAL MANDATES, TARGETS, OR GOALS

21.5.1 Renewable Portfolio Standards

A RPS is a government mandate that requires electric utilities or retail electricity providers to obtain some percentage or a specified amount of their electricity supply from qualifying sources of renewable energy. Such a mandate may also be referred to as a RES or perhaps an Alternative Energy Standard.

The US Congress has considered a national RPS in the past, and in some congressional sessions, either the Senate or House has passed a bill to enact such a mandate, but the House and Senate have never been able to agree upon one that would be passed on to the President. On the other hand, as shown in Figure 21.5, most of the states in the United States have implemented their own RPS or RES, while several others have enacted renewable energy goals or alternative energy goals. In addition to state RPS policies, several cities within the United States have established their own RPS policies.

Generally, electric utilities can meet their required supply of renewable energy by owning and operating renewable energy projects, by purchasing energy and its associated environmental attributes from Independent Power Producers (IPPs) producing renewable energy, by purchasing energy from "behind-the-meter projects" (such as homes or businesses that have solar modules or small wind turbines) when more energy is produced than the owner uses, or by purchasing renewable energy

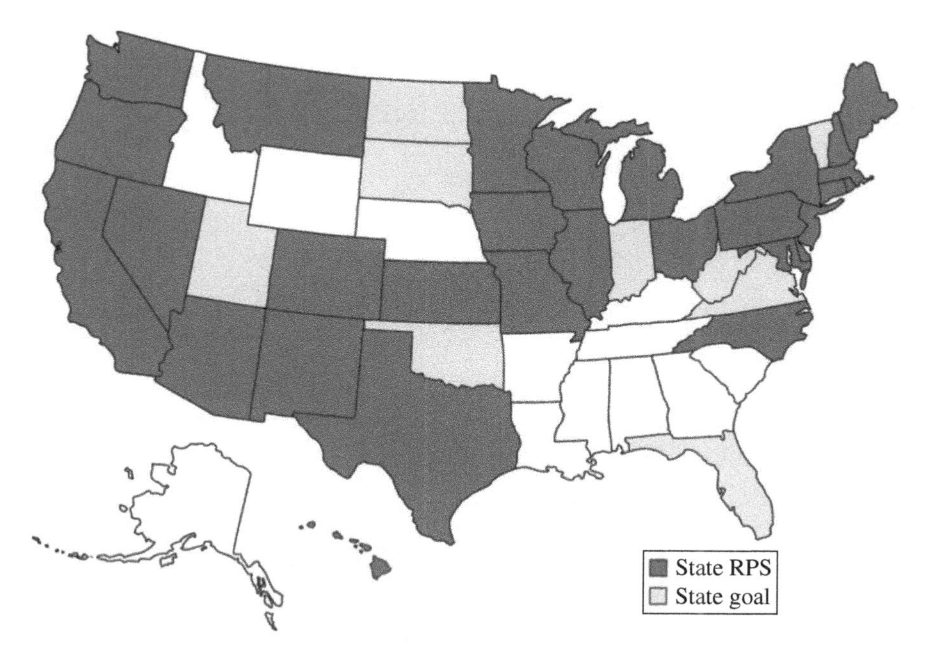

FIGURE 21.5 States with Renewable Portfolio Standards, Renewable Electricity Standards, or Renewable Energy Goals as of December 31, 2013 (Source: Reproduced by permission of Amy Marcotte [5]).

credits (RECs). A REC essentially represents the environmental attributes associated with a block of 1000 kilowatt-hours (kWh). So it is possible to purchase the environmental attributes without purchasing the actual electrical energy at the same time.

States may or may not require that the renewable energy used to meet an RPS be produced within the state. For example, Texas requires that the energy being produced in the state or "capable of being metered in the state" which would imply that a project just outside Texas could qualify if energy is transmitted directly into Texas without intermingling of nonrenewable energy.

States also have varying definitions of what types of resources qualify as renewable energy and which ones do not. Generation types that qualify in all states with an RPS include the following:

- Biofuels
- Biomass
- Hydroelectric (although some states would not include new hydroelectric generation unless it is a "run-of-the-river" facility as opposed to hydroelectric generation relying on a large dam)
- Landfill gas–fueled electric generation
- Solar photovoltaics
- Wind energy

Tidal energy, wave energy, and ocean thermal energy would presumably qualify in all states other than the fact that some states do adjoin an ocean. A few states exclude solar thermal generation, although the majority does include it as a qualifying resource. Most include geothermal energy as a qualifying resource, and perhaps the reason that some do not list it is because of the lack of any viable resources in the state. About one-half of states with an RPS include municipal solid waste (MSW) as a renewable resource, and a few even include waste tires used to fuel electric generation as a qualifying resource.

21.5.2 Goals for Renewable Energy Use

While some states may choose to mandate the increased use of renewable energy, others may choose instead to establish incentives that promote greater use of renewable energy rather than mandate it and often set a goal for the state to achieve. Incentives that states may offer can include the following:

- State income tax credits or deductions
- Franchise tax exemptions or deductions
- Property tax abatements or exemptions
- State sales tax exemptions or incentives
- Funding of R&D
- Expedited permitting and siting approval
- Regulatory authorization or approval of new transmission lines needed to move wind energy from windy areas to urban areas using large amounts of electricity

21.5.3 Goals or Mandates for Renewable Energy Use at Government Facilities or by Government Agencies

Local, state, or federal governments may implement goals or mandates to increase the use of renewable energy resources at their facilities or by their agencies. For example, the federal Energy Policy Act of 2005 requires government facilities to obtain 7.5% of their energy from renewable resources by the year 2013. A Presidential Memorandum [6] by President Obama in December 2013 addressing federal leadership on energy management increased the target to 20% by the year 2020. This can be in the form of on-site installations, purchases of electricity from green power providers, or via the purchase of RECs.

21.6 FEDERAL TAX INCENTIVES

21.6.1 Production Tax Credits

In 1992, one of the greatest contributors to the rapid growth of wind energy in the United States was enacted by Congress as part of the Energy Policy Act of 1992. This was the Renewable Electricity PTC. It allowed a tax credit of 1.5 ¢ for each kilowatt-hour (kWh) produced by a wind energy or closed-loop biomass systems (closed loop refers to organic matter (crops) planted and grown for the exclusive purpose of providing fuel for electric generation, as opposed to the use of wood wastes or agricultural wastes). The PTC is indexed for inflation, reaching 2.3¢ per kWh in 2013, and only applies to energy actually produced during the first 10 years of the project's operation. It has been modified, lapsed, expanded, and/or renewed several times since 1992 and now includes poultry waste, solar energy, open-loop biomass, landfill gas, combustion of MSW, marine and hydrokinetic production, and certain types of small hydroelectric generators, although open-loop biomass, landfill gas, MSW, marine and hydrokinetic generation, and qualified hydroelectric plants only receive 1.1¢ per kWh produced.

The fact that Congress has allowed the PTC to expire or lapse in the past has been problematic for the renewable energy industry, particularly for the wind energy industry. As discussed previously in Chapter 18 and as shown in Figure 21.6, installations of new wind generation capacity came to a virtual halt immediately after the PTC was allowed to lapse in 2000, 2002, and 2004 or very nearly lapsed, as in 2012 when Congress waited until the day it was scheduled to expire before extending it. While those who oppose wind energy may point to this as proof that wind energy is not competitive without this incentive, it is very likely that similar consequences would result if a tax credit for oil and gas drilling or the construction of coal or nuclear plants were available and then lapsed—those companies undertaking such activities would stop and wait to see if Congress will extend the credit once again.

As of the writing of this book, the PTC applicable to wind energy expired on December 31, 2013, other than for projects under construction as of that date.

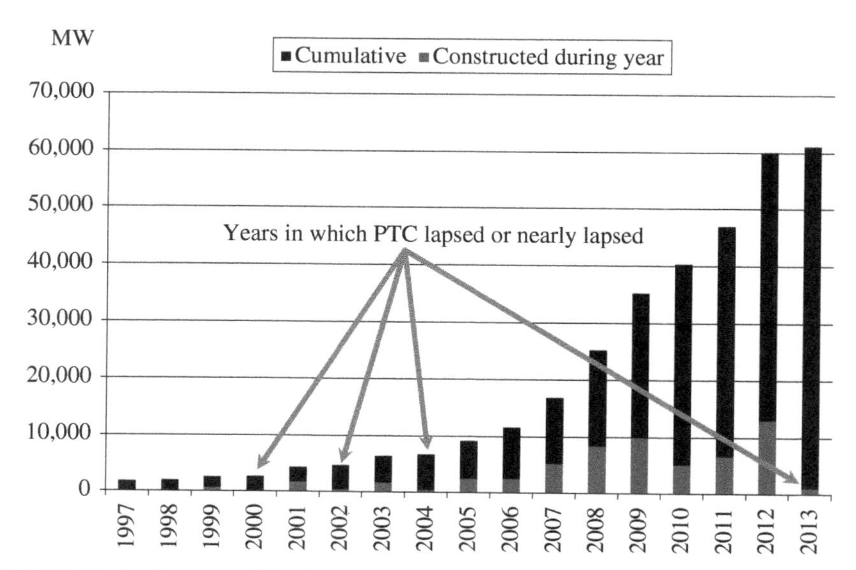

FIGURE 21.6 Growth of Wind Energy Capacity in the United States as affected by the availability of the PTC (Source: R. Walker based on data from AWEA Annual Market Reports).

21.6.2 Renewable Energy Production Incentive

In order to receive and directly benefit from a tax credit, such as the PTC, an individual, entity, or corporation must first be a taxpayer. Thus, state and local governments or nonprofit utility systems such as rural electric cooperatives would not qualify for PTCs from renewable energy projects they own. Therefore, Congress included a provision in the Energy Policy Act of 1992 setting up a Renewable Energy Production Incentive (REPI) to encourage investment in renewable energy technologies by entities that do not pay Federal income taxes. REPI established an incentive payment of 1.5 ¢ per kWh beginning in 1993 and indexed for inflation for electricity produced by solar, wind, ocean, wave, biomass (excluding MSW), or geothermal energy. However, this incentive is funded by appropriations to the US DOE, which have generally been less than $5 million per fiscal year, leaving many otherwise eligible projects without REPI payments. Thus, it may be difficult or unwise to use the potential receipt of REPI funds as economic justification for moving forward with a project.

21.6.3 Investment Tax Credits

As the names imply, investment tax credits (ITCs) differ from PTCs. To qualify for PTCs, one must *produce* electricity from certain renewable energy technologies, whereas to qualify for ITCs, one must *invest* in certain renewable energy technologies.

Much of the early growth in US wind energy capacity occurred in California during the early 1980s due to the combination of a 25% state ITC, an equivalent Federal ITC, and the enactment of PURPA. Unfortunately, some of the California wind projects utilized poorly made equipment and quickly fell into a state of disrepair, so ITCs are sometimes viewed skeptically by policy-makers. The federal Tax Reform Act of 1986 repealed such incentives for wind energy, resulting in a prolonged period of little to no investment in wind energy.

Due to the significant economic downturn in the United States during 2008–2009, companies that had invested in wind plants (or that were considering investments in wind plants) found themselves unable to utilize all of the PTCs that a large wind plant might produce, and this quickly became a barrier to investment in new projects. Subsequently, the American Recovery and Reinvestment Act of 2009 allowed companies that qualified for the Federal PTC to elect to receive either an ITC or a 30% grant from the US Treasury Department in lieu of the PTC. If one calculates the net present value (NPV) of 10 years of PTCs to the owner of a wind plant, it approximately equals 30% of the original construction cost of the wind plant, which is the basis of how the 30% cash grant amount was determined. When Congress extended the PTC for another year on December 31, 2012, it also extended the option of the 30% ITC, but it did not reinstate the option for the 30% cash grant.

Federal business energy ITCs are also available to some renewable energy technologies under the Energy Improvement and Extension Act of 2008. These ITCs are available for solar projects, fuel cells, small wind turbines, certain geothermal systems, microturbines, and combined heat and power projects placed in service prior to December 31, 2016. In the case of solar energy projects, since they typically have high capital costs and low net capacity factors, a 30% ITC is a much better option for solar project owners than the PTC.

21.6.4 Accelerated Depreciation for Tax Purposes/Bonus Depreciation

As discussed in Chapter 18, the US government has enacted tax legislation designed to encourage investment in certain types of assets by allowing accelerated depreciation for tax purposes. This reduces the federal income tax liability of a project or its owner in the early years, but increases the tax liability in later years. Due to the "time value of money," accelerated depreciation for tax purposes will increase the NPV of a project, assuming all other things remained constant. Deferring income tax payments for several years allows investors to increase their cash flow early in the project, and these funds can be reinvested into other types of projects.

As shown in Tables 18.4 and 18.5 earlier in the book, the Modified Accelerated Cost-Recovery System for tax depreciation of assets by businesses sets various time frames over which property types may be depreciated, ranging from 3 to 50 years. Wind generators, solar energy systems, and some geothermal equipment are technologies included as 5-year depreciable property, while some biomass facilities have 7-year depreciation periods. Depreciation periods for traditional sources of electric generation include 15 years for gas combustion turbines and nuclear plants and 20 years for traditional steam generators such as coal-fired plants. Since wind, solar, geothermal, and biomass equipment can be depreciated over a shorter period for tax purposes as compared with traditional generation equipment, this helps to increase the competitiveness of these renewable energy resources. Those same tables also illustrate how depreciation rates can vary depending upon which quarter of the calendar year they enter commercial operation.

Occasionally, Congress will attempt to stimulate the economy by encouraging additional investment in capital equipment by enacting "bonus" depreciation provisions. The Economic Stimulus Act of 2008 included a provision allowing an additional 50% of certain assets to be depreciated in the first year of their lives. The American Recovery and Reinvestment Act of 2009 extended 50% bonus depreciation through 2010, while the Tax Relief, Unemployment Insurance Reauthorization, and Job Creation Act of 2010 authorized 100% first-year bonus depreciation of eligible property placed in service between September 8, 2010, and the end of 2011. During calendar year 2012 and 2013, bonus depreciation will still be available but reverts back to 50% bonus depreciation in the first year of operation instead of 100%. Wind energy equipment has been included as eligible equipment for bonus depreciation in each of the aforementioned acts, thus helping to reduce the cost of wind energy for utilities and their customers.

21.7 PUBLIC BENEFIT FUNDS

Several states have chosen to implement Public Benefit Funds for the purpose of increasing the use of renewable energy, funding energy efficiency programs, providing assistance to low-income customers of electric utilities, or supporting R&D of cleaner

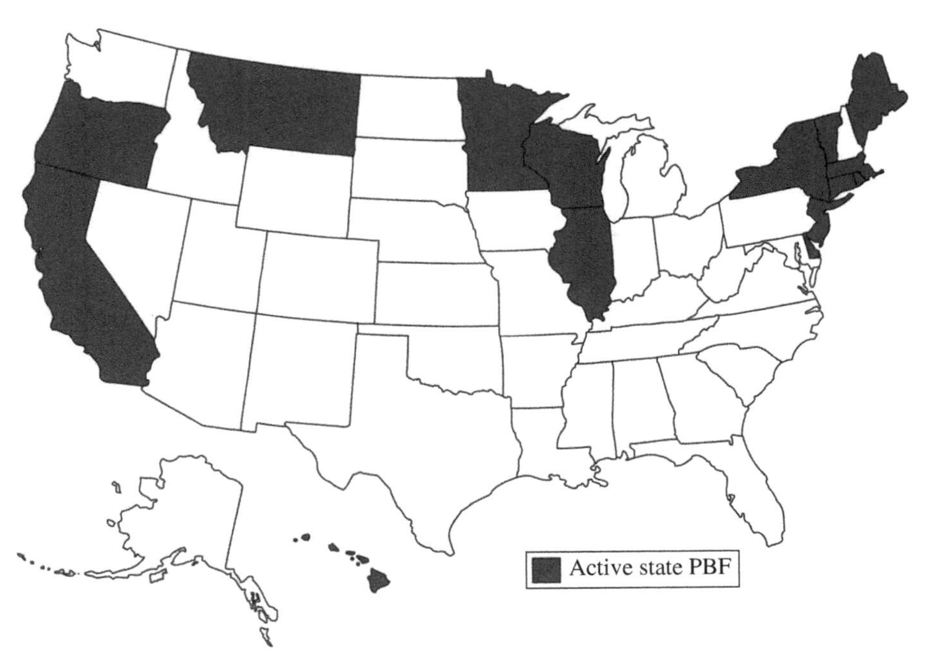

FIGURE 21.7 States with Public Benefit Funds that Support use of Renewable Energy as of December 31, 2013 (Source: Reproduced by permission of Amy Marcotte [5]).

or more efficient energy technologies. Most of these funds rely on very small surcharges to the bills of all customers of regulated public utilities or by voluntary contributions of utility customers. The funds are then accumulated by either the utility itself or a state agency and allocated to various purposes. As shown in Figure 21.7, 15 states have Public Benefit Funds, as do Washington, DC and Puerto Rico, for the purpose of increasing the use of renewable energy, while several others have implemented this policy in the past but no longer are actively collecting such funds. Some of these state Public Benefit Funds can accumulate several million dollars, such as the one in California that collects about $400 million per year, much of which is being devoted to support of solar energy systems in the State. California recently exceeded 1000 MW, or 1 GW, of solar energy installations.

21.8 FEED-IN TARIFFS

One policy that several European nations have implemented to promote the growth of renewable energy is the use of an FIT. An FIT promotes development of new sources of renewable energy by mandating a specified rate of payment for all energy produced by a project for some specified period of time. Usually, the utility that the project is interconnected with is required to purchase the project's energy as well as being required to allow the project to interconnect with the electric grid. The "Renewables 2011 Global

Status Report" [7] indicates that more than 50 nations have implemented a national FIT. Germany and Spain are both noted for their aggressive FITs, and both are among the top five nations in terms of installed wind energy capacity, as well as being the global leaders in installed solar energy capacity.

The success of a FIT may depend largely on the rate that is paid for the electricity and the length of time that the payment is guaranteed. If the payment is based on the levelized cost of energy from the renewable resource and is assured for 10 or more years, the project owner is more likely to be able to obtain financing for the system and to earn a reasonable return on investment. On the other hand, if the utility only pays its "avoided cost of energy," roughly equivalent to the value of fuel not burned as a result of purchasing the renewable energy, this is probably not enough to justify installing the renewable energy system. A middle ground between these two examples would be to include the value of "environmental attributes," similar to a REC, and some value for reducing the demand for electricity on the utility's system near the location of the renewable resource.

FITs are not in widespread use in the United States, although one could consider that PURPA's mandated purchase of energy by utilities from qualifying renewable energy facilities at the utility's' avoided cost a form of FIT. The US wind energy industry has its roots in California, which began to take off under California's Standard Offer Contract No. 4, which established a fairly high rate for purchases of renewable energy.

21.9 NET METERING

Net metering refers to laws or regulations that require electric utilities to allow homes or businesses to install "behind-the-meter" renewable energy systems such as solar photovoltaic or small wind generators and to use electric meters that keep track of any excess energy produced by the renewable energy system that is fed back into the electric grid. Most of the states in the United States have some form of net metering laws, and some utilities have implemented net metering policies on their own without being mandated to do so by state regulators. Most of these laws or regulations limit the size of the renewable energy facility that the utility is required to purchase the excess energy from, with typical size limits being 100kW or less, although a few states have much higher limits.

Some of these policies allow the owner of the renewable energy system to get paid for the excess energy produced while others allow a "netting" of energy for some period of time, typically up to 1 year. For example, in some months, the renewable energy system may produce more electricity than is consumed by the owner's facility, but in other months the facility may consume more electricity than is produced. The utility would allow excess energy produced in prior months to reduce or completely offset the customer's bill in months when the system produces less electricity than is consumed at the facility. However, if the renewable energy system consistently produces more than is consumed on site, excess amounts "banked" more than 1 year earlier may be kept by the utility.

Smaller electricity consumers, such as homes or small retail businesses, typically pay a rate based on their monthly electricity usage, measured in kilowatt-hours or kWh. So as long as the output of their behind-the-meter renewable energy system remains below their own usage (either on a monthly basis or an annual basis), net metering allows them to offset their full retail rate, which may be on the order of 8–20 ¢/kWh, depending upon where they live. However, larger-utility customers, such as commercial, industrial, or municipal businesses, may be billed on rates that have a "demand" component. This means part of their monthly bill is based on the peak instantaneous usage of power at any time during the month, measured in kilowatts or kW. Since their renewable energy system may not be generating at the same time that they are using the most power during the month, they may not offset any of this demand component in their electric bill, meaning the value of net metering to them would be diminished. And in either case, when production from the renewable energy system exceeds the owner's own electricity usage (either on a monthly basis or an annual basis), in many locations they are likely to receive little to no payment for the excess electricity, so it usually would not pay off to size the renewable energy system larger than needed to meet the owner's own usage.

21.10 RESEARCH AND DEVELOPMENT

In the United States, the DOE is the lead agency for overseeing the allocation of federal funds supporting energy-related R&D. Several national laboratories falling under the purview of DOE perform much of our nation's energy technology R&D, including the National Renewable Energy Laboratory, the National Energy Technology Laboratory, Sandia National Laboratory, Lawrence Berkeley National Laboratory, Los Alamos National Laboratory, and more than a dozen others.

DOE's Strategic Plan [8], released in May 2011, states that President Obama has set the following goals:

- Reduce energy-related greenhouse gas emissions by 17% by 2030 and 83% by 2050, as compared with a baseline based on 2005 emissions.
- By 2035, produce 80% of America's electricity from clean energy resources, defined as renewables, nuclear power, combined-cycle gas generation, and fossil energy with carbon capture and storage.
- Put 1 million electric vehicles on US roads by 2015.

Some of DOE's initiatives set out in its Strategic Plan to accomplish the aforementioned goals include the following:

- Drive energy efficiency to reduce demand growth
- Demonstrate and deploy clean energy technologies
- Modernize the electric grid
- Enable prudent development of our natural resources
- Make the federal government a leader in sustainability

A February 2011 newsletter produced by DOE's Office of Energy Efficiency and Renewable Energy [9] indicated that DOE requested $3.2 billion for renewable energy and energy efficiency purposes in fiscal year 2012, representing a 44% increase over 2010 levels. According to the article, "the proposed budget aims to strengthen renewable energy sources, boost clean energy research, and cut expenses as the United States pursues the President's vision of generating 80% of its electricity from clean sources by 2035."

21.11 INCREASING RESTRICTIONS ON FOSSIL-FUELED POWER PLANT EMISSIONS

Since the enactment of the US Clean Air Act in 1970, fossil-fueled power plants have had to deal with increasing regulation of air emissions. The Clean Air Act defines the EPA's responsibilities for protecting and improving the nation's air quality and the stratospheric ozone layer. Major changes in the law occurred in 1990 with enactment of the Clean Air Act Amendments, and several minor changes have been made since that time. The Clean Air Act is probably the most significant barrier to installation of new fossil-fueled power plants, coal plants in particular.

As shown in Figures 21.8 and 21.9, emissions from electricity generation have declined significantly. However, additional or more stringent regulations have been proposed, including restrictions on mercury, particulate matter, and carbon dioxide. Any of these new restrictions that come to pass can cause the cost of power from

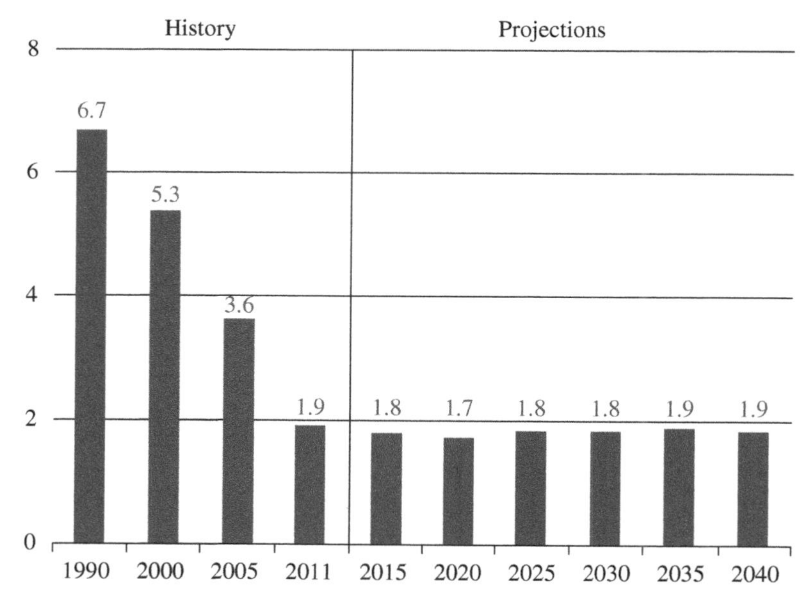

FIGURE 21.8 Nitrogen oxide (NO_X) emissions from US electricity generation (million short tons) (Source: US EIA [1]).

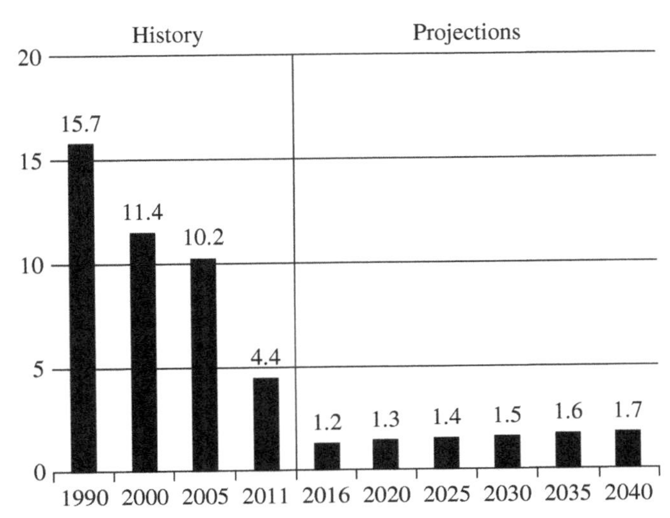

FIGURE 21.9 Sulfur dioxide (SO_2) emissions from US electricity generation (million short tons) (Source: US EIA [1]).

many fossil-fueled plants to increase, sometimes substantially, and some power plants may even be forced to shut down.

One of the more recent regulations in this regard is the US EPA's Cross-State Air Pollution Rule (CSAPR) [10], released in July 2011, although adjustments to the rule are still being finalized as of the writing of this book. This rule requires states to significantly improve air quality by reducing power plant emissions that contribute to ozone and/or fine particle pollution in other states. CSAPR requires a total of 28 states to reduce annual SO_2 emissions, annual NO_X emissions, and/or ozone season NO_X emissions to assist in attaining the 1997 ozone and fine particle and 2006 fine particle National Ambient Air Quality Standards. Several utilities have already announced plans to shut down power plants as a result of this new rule.

As can be seen in Figure 21.10, many existing power plants do not have flue-gas desulfurization (FGD) equipment, sometimes known as "scrubbers," used to remove SO_2. Some owners of these plants plan to install FGD systems, but for others, the most cost-effective solution may be to shut the plant down.

Removal of mercury can also be quite expensive. A May 2005 report [11] from the US Government Accountability Office estimated that installation of mercury removal equipment for a 975-MW coal-fired power plant could cost as much as $36 million in capital with annual operating costs of $12.8 million (in 2003 US dollars).

Capturing and sequestering carbon dioxide is also being considered in response to growing concerns about greenhouse gas emissions. As previously shown in Table 14.4, adding carbon capture and sequestration to existing pulverized coal power plants could substantially increase their levelized cost of energy. Some estimates suggest that requiring carbon capture and storage for fossil-fueled plants could increase the cost of electricity from such plants by about 20–50% [12].

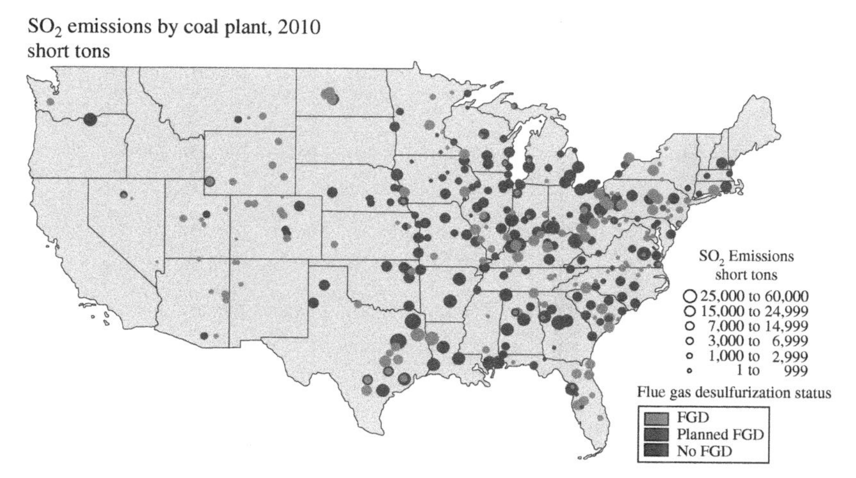

FIGURE 21.10 US Coal-fired power plants with and without flue-gas desulfurization systems (Source: US EIA [11]).

In summary, emissions from fossil-fueled power plants are a significant issue, and local, state, and federal government regulators are continuing to force emission reductions. Such reductions will be costly and sometimes will force closure of older power plants. These factors combine to increase the economic competiveness of and the need for sources of renewable energy.

21.12 CARBON TAXES/CAP-AND-TRADE PROGRAMS

As discussed in more detail in Chapter 16, the potential for climate change due to the emission of greenhouse gases is one of the major issues we face. Carbon emissions related to our use of energy are the largest source of man-made greenhouse gas emissions. As shown in Figure 21.11, US energy-related emissions of carbon in 2012 were about 12% lower than the peak emissions year of 2007. Reasons for this decline may include improved energy efficiency, increased use of renewable energy sources, greater use of natural gas in place of coal for electricity generation, and an economic slowdown beginning in December 2007.

Two potential ways to force reductions in carbon dioxide emissions are the enactment of a carbon tax or the implementation of a carbon cap-and-trade system. Both are fairly controversial and may cause significant economic disruptions. A carbon tax would levy a tax on emissions of carbon resulting from the combustion of fossil fuels. Essentially, this would increase the cost of using fossil fuels, again increasing the cost-effectiveness of using renewable energy. Proponents of a carbon tax argue that it is a good way to recognize the social costs associated with burning fossil fuels, such as increased health-care costs. Several nations have already enacted carbon taxes including Great Britain, Sweden, Finland, and New Zealand [14]. In the United States, a form of carbon tax was proposed by the Clinton Administration in 1993

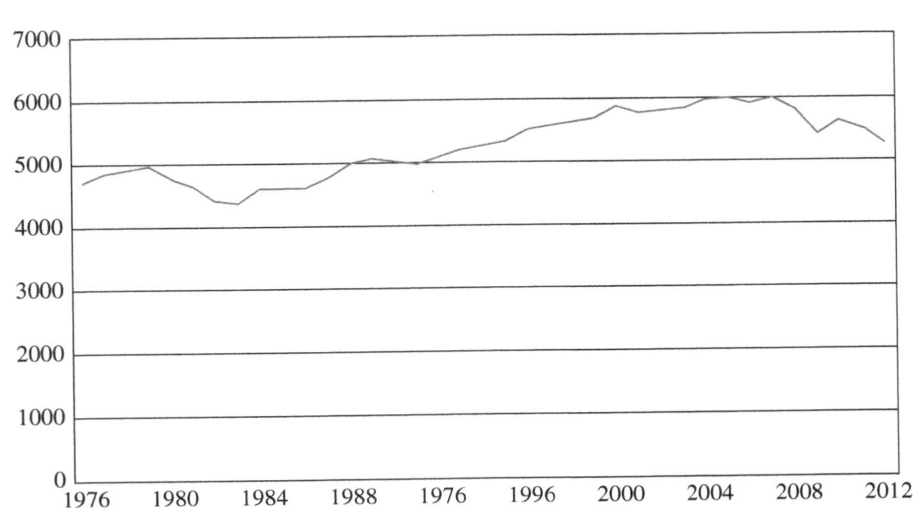

FIGURE 21.11 US Energy-Related Carbon Emissions (million metric tons) (Source: R. Walker based on data from US-EIA [13]).

when it proposed an energy tax on coal, natural gas, liquefied petroleum gases, and gasoline. However, there was significant, broad-based opposition to this proposal, and it was instead replaced with a Transportation Fuels Tax on gasoline, diesel, and special motor fuels.

A cap-and-trade system would impose government restrictions or "caps" on carbon emissions. The government would issue carbon credits to companies allowing them to emit some specified amount of carbon. Presumably, the initial amount of credits a company receives would be in some proportion to the amount of carbon emissions it has historically released. Companies needing to or desiring to emit greater amounts due to expansions or additions of new facilities would have to purchase the additional carbon credits from the market for such credits. Those who support such a program argue that it is an economically efficient way to reduce emissions since some companies can reduce emissions more easily or more cost-effectively than other companies.

21.13 CONCLUSIONS

Energy represents a significant portion of the budget for most individuals, corporations, cities, states, or nations. In addition to the economic issues surrounding energy, the environmental consequences of how we use energy are enormous. The security of a nation can hinge upon its energy supply. Therefore, energy is a very important issue to most people, yet elected officials often focus on short-term issues such as the cost of gasoline, the economy, or jobs and thus may be ignoring longer-term consequences or issues such as global climate change, environmental impacts, trade imbalances, and national energy security.

In order to preserve our environment, to strengthen the security of our nation's energy supply, and to ensure that adequate energy resources exist for future generations, it is essential that a comprehensive, cohesive, long-term national energy policy is developed. Such a policy should be focused on achieving long-term goals rather than being radically altered any time a highly publicized environmental disaster occurs or a new president is elected. Certainly, energy policy must incorporate new information and new concerns, but to radically change directions every presidential administration or every time a new issue arises is a highly inefficient and ineffective process.

REFERENCES

[1] U.S. Energy Information Administration website. Annual Energy Review; Primary Energy Consumption Estimates by Source, 1949–2012. Available at http://www.eia.gov/totalenergy/. Cited December 14, 2013.

[2] U.S. Energy Information Administration website. Available at http://www.eia.gov. Cited August 8, 2012.

[3] The Renewable Fuels Association website. Available at http://www.ethanolrfa.org/pages/ethanol-facts-energy-security. Cited August 10, 2012.

[4] The Pickens' Plan website. Available at http://www.pickensplan.com/. Cited August 10, 2012.

[5] Database of State Renewable Incentives for Renewables & Efficiency (DSIRE™). Available at http://www.dsireusa.org/. Cited August 10, 2012.

[6] Obama B. Presidential Memorandum—Federal Leadership on Energy Management. Washington (DC): The White House Office of the Press Secretary; 2013.

[7] Swain J, Martinot E. Renewables 2011 Global Status Report; Renewable Energy Policy Network for the 21st Century (REN21), Paris, France; 2011.

[8] U.S. Department of Energy. 2011 Strategic Plan (document DOE/CF-0067). U.S. Department of Energy, Washington (DC); 2011.

[9] U.S. Department of Energy. EERE Network News, February 2011, electronic newsletter, Washington (DC); 2011.

[10] U.S. Environmental Protection Agency website. Cross-State Air Pollution Rule. Available at http://www.epa.gov/airtransport/. Cited August 10, 2012.

[11] U.S. Energy Information Administration. Today in Energy, December 2011 Issue, Washington (DC); 2011.

[12] U.S. Government Accountability Office. Report to Congressional Requesters on Clean Air Act and Emerging Mercury Control Technologies (GAO-05-612). Washington (DC); 2005.

[13] U.S. Energy Information Administration. Today in Energy, May 2013 Issue, Washington (DC); 2013.

[14] Carbon Tax Center website. Available at http://www.carbontax.org/progress/where-carbon-is-taxed/. Cited August 10, 2012.

22

GLOBAL WIND ENERGY POLICY AND DEVELOPMENT

22.1 INTRODUCTION

This chapter explores worldwide policies and development of wind energy over the next several decades. Wind energy is the leading renewable energy technology for new sources for electric power generation worldwide. This chapter will first explore the global thrust to develop renewable energy sources. Next we will examine how and why wind energy is a major component of these renewable energy development initiatives by considering how the technology, national policies, and public opinion contribute to meet growing worldwide electric demand.

22.2 RENEWABLE ENERGY DEVELOPMENT—A GLOBAL PERSPECTIVE

As discussed previously in this text, public opinion can be a strong and determining factor in future energy development. Recent global public opinion data demonstrate strong support for renewable energy technology and development. A 2012 Global Consumer Wind Study [1] conducted by TNS Gallup shows that 85% of global consumers want more renewable energy, while 49% are even willing to pay more for products made using renewable energy. The survey polled 24,000 people from over 20 countries.

Wind Energy Essentials: Societal, Economic, and Environmental Impacts, First Edition.
Richard P. Walker and Andrew Swift.

In the United States, recent polls show similar support for renewable development. A 2013 poll of Americans showed strong support for renewable energy development as compared with more traditional forms of energy production. The poll asked which source of energy production should the United States "place more emphasis" in the future. Table 22.1 shows the strong support for renewable energy [2].

In addition to public opinion, state and national energy policies strongly affect energy development. In the United States, it was shown in Chapter 21 that state renewable portfolio standard (RPS) mandates and the Public Utility Regulatory Policy Act of 1978 were important precursors to US renewable energy development. Recent international data show strong worldwide support for energy policies favoring renewable energy development. The data show that 138 countries have defined renewable energy targets and 127 countries have renewable energy support policies in place—two-thirds of which are developing countries [3].

The international business community has also shown support for renewable energy development through growing investments in the technology. For example, recent net investment in renewable power capacity outpaced net investments in fossil fuel generation [3], and the fraction of global financial investment in renewable power is expected to rise substantially, as compared with traditional fossil and nuclear power capacity [4].

Recent studies have also shown that large contributions to the electrical energy supply mix from renewable energy sources are both technically and economically reasonable. The U.S. National Renewable Energy Laboratory recently completed an 80% renewable study for the United States. Key findings showed that renewable electricity generation from technologies commercially available today, in combination with a more flexible electric system, are more than adequate to supply 80% of total US electricity generation by 2050 [5].

The convergence of public opinion, national and state policies, and commercially available renewable energy technologies supported by business and financial markets are strong indicators of continued worldwide growth potential in the renewable electric sector. Arthouros Zervoc, the Chairman of REN21: Renewable Energy Policy

TABLE 22.1 Survey Results in the United States Showing Support For Renewable Energy[a]

Energy source	Percent of respondents who want more emphasis on renewable energy
Solar energy	76
Wind energy	71
Natural gas	65
Oil	46
Nuclear energy	37
Coal	31

[a] Source: A. Swift: created from Gallup Poll data [2].

Network for the 21st Century, a large, international renewable energy policy group, summed it up well:

> We stand on the cusp of renewables becoming a central part of the world's energy mix. [3]

As we will see in the next section, wind energy is presently the leading renewable energy technology worldwide, in new renewable power installations.

22.3 THE WIND ENERGY INDUSTRY'S ROLE IN THE GLOBAL ENERGY MARKET

22.3.1 Recent Growth

In the previous section, we discussed the worldwide growth of renewable power. Figure 22.1 shows the global exponential growth in wind energy capacity from 1996 to 2013, while Table 22.2 shows that although solar electric power installations are also growing rapidly, wind energy is currently the dominant leader in worldwide renewable power additions. Figure 22.2 shows the current worldwide installed capacity of wind power generation and its recent growth, by country and region.

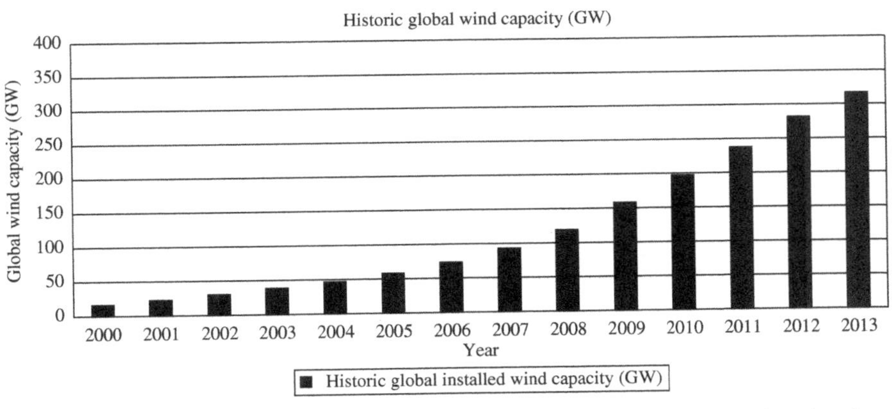

FIGURE 22.1 Global Wind Energy Capacity by year (Source: A. Swift based on data from Global Wind Energy Council [6]).

TABLE 22.2 Total Global Wind and Solar Power Capacity, in GW[a]

Resource	2010	2011	2012
Wind power	198	238	283
Solar/PV	40	71	100

[a] Source: A. Swift based on data from NREL [5].

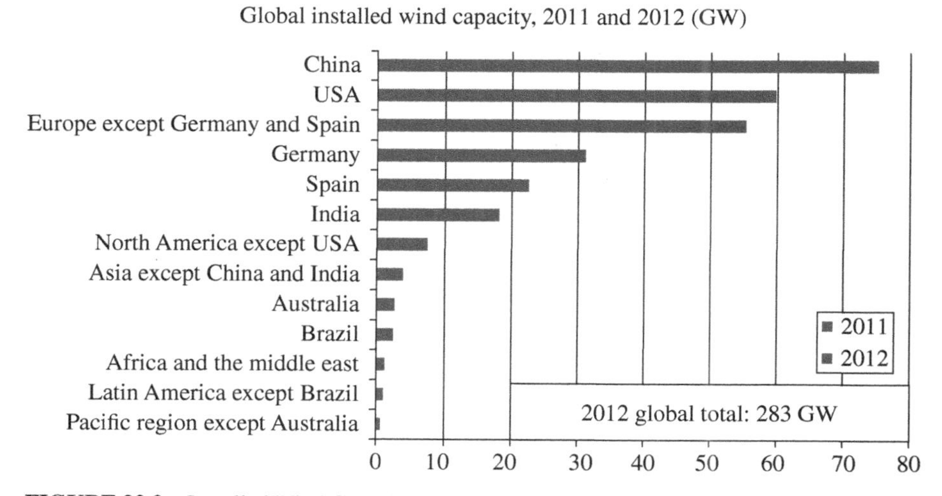

FIGURE 22.2 Installed Wind Capacity by Country/Region, 2011 and 2012 (Source: A. Swift based on data from Global Wind Energy Council [6]).

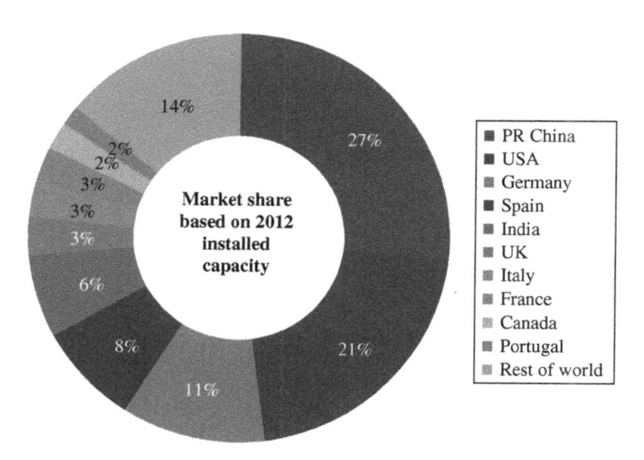

FIGURE 22.3 Top 10 Global Wind Power Markets in 2012 (Source: A. Swift based on data from Global Wind Energy Council [6]).

22.3.2 Global Wind Energy Markets

The global wind-electric power generation industry is a large business with heavy capital investment. There are no fuel costs, thus the cost of wind-generated electricity is related more to capital investment than operations. The top 10 global wind power markets in 2012 are shown in Figure 22.3.

Development is both onshore and offshore, depending on the wind resource, land availability, and access to offshore sites. Most wind development to date has been onshore, but the offshore developable wind resource worldwide is very large, and many nations are considering offshore development. Global manufacturers are developing offshore wind turbine designs in anticipation of that growing sector of the future market. Offshore development is discussed in Chapter 20 with data for European offshore wind development.

22.3.3 Global Demand for Electric Power

As discussed in Chapter 1, access to affordable electric power is a necessity for a modern society and an important element in providing a reasonable quality of life (QOL). As one looks to the future, global electric power consumption is predicted to grow substantially, driven by growing world population, growing demand for electric power in underdeveloped nations, and new technologies that depend on electric power—such as computer data storage centers and electric vehicles. Some energy analysts predict a doubling of electric consumption in the next two decades.

The drivers of global electric demand are outlined as follows:

- Population: World population continues to increase. U.S. Census Bureau projections estimate the world population will grow from about 6 billion in the year 2000 to 9 billion in the year 2050—a 50% increase. Most of this increase will be in emerging economies and developing countries—each of which will be interested in advancing their standard of living through access to electrical power and the services it provides.
- Underdeveloped countries: Presently, according to the World Bank, there are approximately 1.2 billion people without access to electrical power—mostly in Africa and India [7]. The United Nations has implemented programs with the goal of providing access to electrical power for these people—requiring substantial new generating capacity [8].
- Power plant replacement: Power stations generally have relatively long lifetimes. Large coal and nuclear power plants can operate for 50 years. Wind power plants at present have a design life of 20 years, though some may continue to operate beyond their design life and many may be "repowered" with new technology extending their useful life.
- Increased electrification, such as electric vehicles and data centers: There are about 240 million vehicles registered in the United States. Two-thirds, or about 160 million, are light-duty cars and trucks—one-third are heavy-duty trucks, busses, and delivery vehicles. Although the numbers are modest to date, the growth of plug in electric vehicles has been very strong, as shown in Table 22.3. If rapid growth continues, electric vehicles could significantly add to the electric demand both in the United States and globally.

A single plug-in electric vehicle, charged overnight, will about double the electric energy consumption for an average household—depending on the daily driving needs of the vehicle owner. Because the power will be consumed overnight during off-peak hours, additional generation capacity may not be substantially required,

TABLE 22.3 Electric Drive Vehicle Sales in the United States[a]

Year	Hybrids	Plug-in electric and plug-in hybrids	Total electric	Total vehicle sales
2007	352,274			11,777,314
2008	313,673			13,260,747
2009	290,292			10,429,014
2010	274,210	345	274,555	11,588,783
2011	266,329	17,735	284,064	12,734,356
2012	434,645	52,835	487,480	14,439,684
2013	495,530	96,702	592,192	15,531,609

[a] Source: R. Walker based on data from Electric Drive Transportation Association [9].

although additional electric energy will have to be generated from thermal sources, requiring additional fuel use, or from new renewable sources.

For example, at locations where the winds are substantially higher at night, such as the Great Plains of the United States, wind energy could be a primary source of power for these vehicles, while solar covered parking spaces associated with work locations could provide charging energy during the day. The rapid growth in home electronic equipment, global computing power, the Internet, smart phones, and the related need for large data-storage centers, such as *cloud computing*, have all contributed to an increased demand for electrical services.

22.3.4 Estimating the Future Demand for Electricity

In order to better understand and estimate the growing demand for electricity, one can develop a spreadsheet model to forecast future electric demand. Past experience with these models has indicated that it is very difficult to predict future energy demand and costs with reasonable certainty. There are a number of reasons–as changing technologies; difficulties in anticipating future costs, markets, and value; economic growth; changes in public opinion; and unanticipated policy changes. In spite of these difficulties, it is still valuable to develop forecast models and consider various future scenarios—understanding that there are a number of underlying uncertainties.

Consider electric demand and the required generating capacity through the year 2050 for both a world model and a US model, using the following metrics as input data:

- World and US Population projections
- QOL and electrical energy consumption data, as presented in Chapter 1
- Provision for electrical service to those who currently have none
- Historical electric generation data to estimate overall generation capacity factors
- Fifty-year power plant replacement
- New demand, such as electric vehicles

Assumptions used for the model include the following:

- World and US Population: Used Census Bureau projections.
- QOL and electrical energy consumption data as presented in Chapter 1: Assumed global kWh per person per day energy intensity increases approximately 25% from the current value of 9.8 (2010 data not including 1.2 billion without electric access) to 12 based on increasing global QOL. The US average is currently 39 kWh/person/day.
- Provision for electrical service to those who currently have none: It is assumed that almost all (one billion) of the 1.2 billion without electrical service are provided service by 2050.
- Historical electric generation data to estimate overall capacity factors: Used 46% power plant capacity factor based on historical data for 2010.
- Fifty-year power plant replacement: Assumed existing power capacity will be replaced at 20% per decade.
- New demand, such as electric vehicles and data centers, are assumed to grow minimally in comparison with the other factors.

The model results are shown in Figure 22.4 and indicate that under these assumptions demand for electric power capacity will approximately double over the 40-year period—a very large increase, necessitating the addition of about 9 TW (or 9,000,000 MW) of new electric generating capacity.

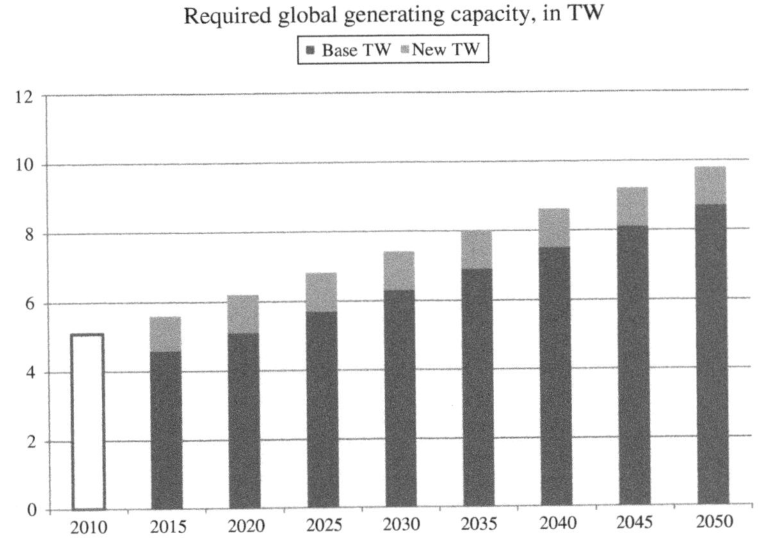

FIGURE 22.4 Estimated Global Electric demand through 2050 showing Base Generation Requirements and Additions each 5 years; 2010 is historical from EIA whereas 2015–2050 are model projections (Source: A. Swift based on data from U.S. EIA).

22.3.5 Meeting the Demand

To meet the global demand for new electric generating capacity over the next decades will require substantial investments in new power plant facilities totaling trillions of dollars. The authors estimate total investments, including replacement generation costs, of approximately one-third trillion dollars per year (in 2013 dollars), or 13 trillion dollars over the 40-year time interval. This is for power plant capacity only and does not include fuel costs. In addition to the monetary investments, choices of generating technology (fuel mix), transmission infrastructure, and computer-based smart grid business models that implement consumer choice and demand management along with improved energy efficiency will also be important elements to limit operating costs and provide reliable electric service to meet the growing electric demand.

These growth numbers are not the same for all regions. The United States, for example, is projected to have modest growth in line with population increases, as shown in Figure 22.5, the US-EIA's (US-Energy Information Administration) electric generation projections through 2040 from the Annual Energy Outlook 2013 [8]. Generation capacity investments in the United States will therefore be needed mostly for replacement of aging equipment or to meet new regulatory requirements. Assuming the average 50-year power plant lifetime criteria, this will result in additions of approximately 30 GW/year representing an average annual investment of approximately $45 billion in new power generation equipment.

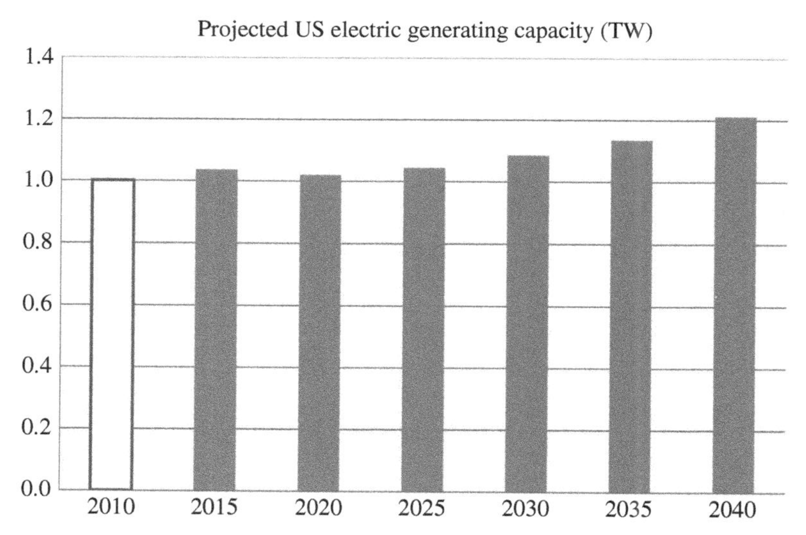

FIGURE 22.5 Projected US Electric Demand through 2040; 2010 is historical data, 2015–2040 are projections (Source: A. Swift based on data from U.S. EIA [8]).

22.3.6 Fuel Choice—The Three Ps

The term "Fuel Choice" describes the difficult decision of choosing generating technology and related infrastructure to meet the future demand. Fuel here has the broader meaning of *Energy Source* for power generation, rather than the more restrictive definition related to only combustible material. Electric generating technologies include coal; natural gas; nuclear; hydropower; and other renewables such as wind, solar, biomass, and so forth. Other technologies are possible, but for the foreseeable future, choices will come from this list. In this book, we have looked at the economics of these technologies and some of their advantages and disadvantages, but what will be the driving criteria of choice?

There are a number of factors that determine fuel choice for electric power production. According to Daniel Yergin, three important factors when considering energy supply choice are the three Ps: *policy, politics, and public opinion* [10]. Additional factors include regional and geographic constraints and attributes, economics, technology readiness and availability, and energy security issues (such as fuel and equipment supply, electric service availability, and reliability). Policy would include public subsidies and incentives; environmental law and regulations; public safety; as well as state, federal, and international law. Political considerations would include the impact on jobs, the local economy, economic growth, and tax revenues generated or received. Public opinion would include technology acceptance, Not In My Back Yard (NIMBY) considerations, concerns about impacts on property values, and cost of service. The list is summarized in Table 22.4.

22.3.7 Greenhouse Gas (GHG) Issues

One constraint on fuel choice that is global in nature is the issue of GHG emissions. As discussed in Chapter 16, this is a global issue, and not by any means resolved. Global climate models evolve, climate impact forecasts change, public opinion and

TABLE 22.4 Fuel Choice: Factors For Choosing Electric Power Generation Technologies[a]

Factors	Description
Public policies and regulations	Public subsidies and incentives; taxes; environmental law and regulations; public safety; as well as state, federal, and international law
Political issues	Impact on jobs, tax issues, the local economy, and economic growth
Public opinion	Technology acceptance, Not In My Back Yard (NIMBY) considerations, availability, and cost of service
Geography	Constraints and attributes of the region
Economics	Cost of plant, infrastructure, and electric service
State of the technology	Maturity of the technology and availability for deployment
Energy security	Fuel and equipment supply, electric service availability, reliability, and pricing

[a]Source: A. Swift; adapted from Daniel Yergin [10].

world politics surrounding the issue change, and so forth. Imposition of a carbon tax by a regulating entity can make the economics of a fossil fuel generating plant completely reverse.

22.3.8 Fuel Choice: Natural Gas and Renewables

As electric power is a growth sector for the global economy, so too is natural gas as a primary energy source. Natural gas emits about one-half the carbon emissions per unit of electrical energy generated as compared with coal-fired electric power generation. Also, over the past decade, there have been significant technological improvements in directional drilling and hydraulic fracking—using water at high pressure to fracture underground shale rock formations releasing large quantities of natural gas. This has resulted in large increases in natural gas reserves and simultaneous increases in projections for natural gas use in the future. Some energy analysts are citing the partnership of natural gas and renewable energy technologies to be the preferred combination of resources for future electric power production. Natural gas–fueled power stations can rapidly change their output to match variable generation sources of power, such as wind and solar. With the increased availability of natural gas supply, world consumption is expected to increase by about 65% from 2010 to 2040 with a significant portion of that increase as fuel for electric power production. Additionally, a number of countries are looking to use natural gas to replace oil as a transportation fuel in the heavy duty transportation sector for large trucks, buses, and delivery vehicles [11].

Although coal is presently the dominant fuel choice for electric power generation on a global basis, as shown in Figure 22.6, some energy analysts project that natural gas and renewable power, including hydroelectric, will generate one-half of global electric power by the year 2035 [12], as shown in Figure 22.7. More aggressive renew-

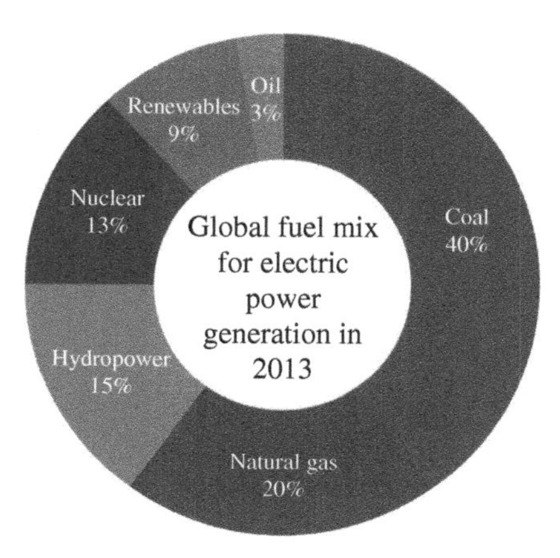

FIGURE 22.6 Estimated global fuel choice mix for electric power generation in 2013 (Source: A. Swift based on data from BDEW [12]).

able energy forecasts indicate that renewables could contribute a substantially larger share of global electric power over the next decades [3].

The future of global wind power development can be derived from the electric demand model of the previous section with some simple assumptions. If one assumes one-half of the new electric generating capacity installed globally is from renewable sources and one-third of that renewable capacity is wind energy technology, by 2050 global wind energy capacity could exceed 1.5 TW, as shown in Figure 22.8.

In the United States, fuel choice for new power plants is also expected to favor natural gas and renewable power sources. Figure 22.9 shows the growth in renewable energy generation technologies for the United States as projected by the US-EIA [8].

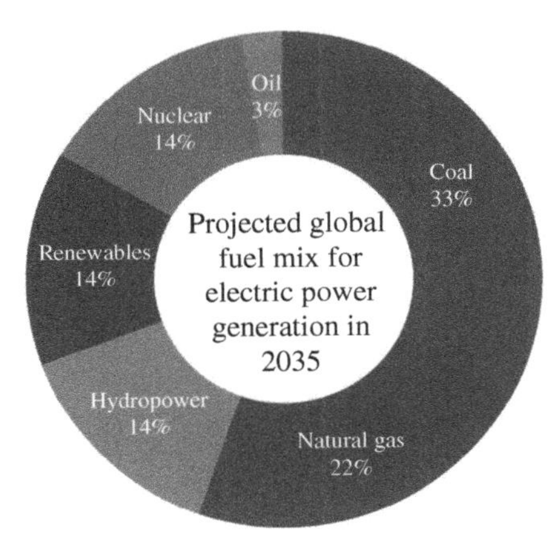

FIGURE 22.7 Projected global fuel choice mix for electric power generation in 2035 (Source: A. Swift based on data from BDEW [12]).

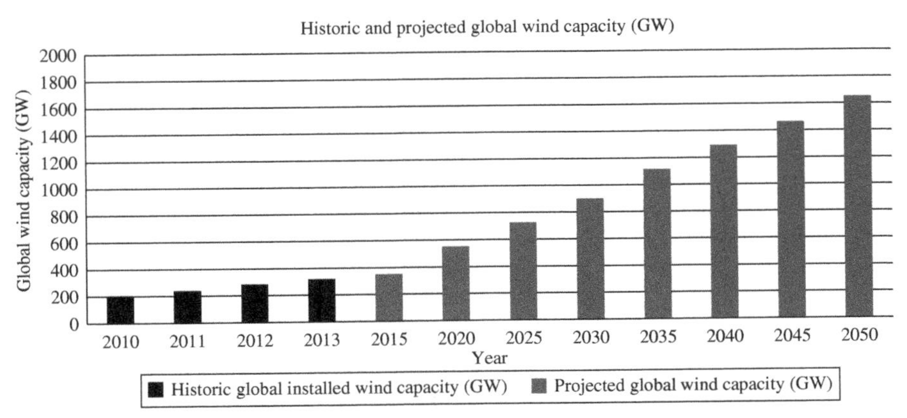

FIGURE 22.8 Estimated global wind generation capacity (gigawatts or GW); 2010–2013 is historical data, 2015–2050 is based on model projections (Source: A. Swift model & assumptions).

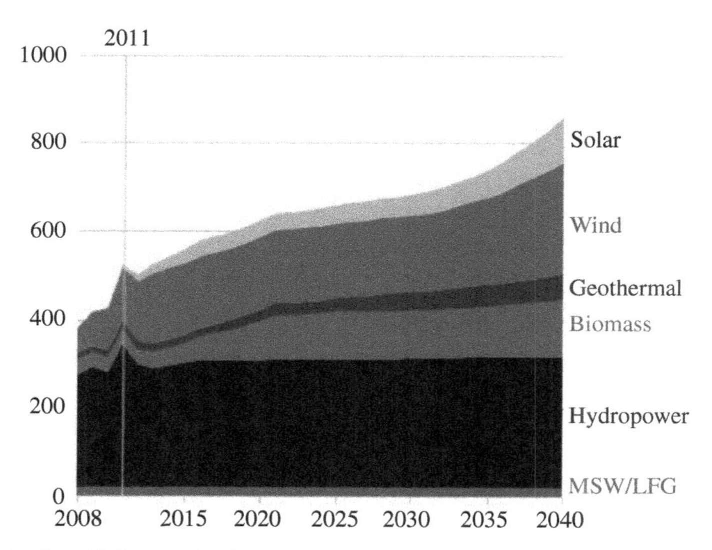

FIGURE 22.9 US Renewable Electricity Generation by type in Terrawatt-hours (TWh) (Source: US-EIA Annual Energy Outlook 2013 [8]).

Thus, although the overall capacity growth is projected to be modest in the United States over the next several decades, significant numbers of new renewable generation and natural gas facilities will be built as aging power plants are retired and the fuel mix is adjusted to accommodate more natural gas and renewable power. As shown in the figure, wind is projected to lead the growth in renewable electric power generation.

22.4 WIND ENERGY BY REGION

Global wind statistics for 2012 compiled by the Global Wind Energy Council [6] show that Europe, Asia, and North America have far more installed wind power capacity than other regions of the world, as shown in Table 22.5. Policies driving wind energy development in these regions is discussed in this section, and additional information is presented for an important nation or state in the region, to include electric consumption data, population, QOL data, wind energy growth rates, and policies in place that support the development of wind energy and renewable power.

22.4.1 Europe

The Europeans are leaders in modern wind power development. With 50 countries and 11% of the world population (~740 million), Europe reported in 2012 an installed wind capacity of over 109 GW, the highest regional wind power installed capacity in the world. Figure 22.10 shows an 80-m wind map of the globe showing good to

TABLE 22.5 Global Wind Power Capacity By Region At Year-end 2012 (MW)[a]

Region	Installed wind capacity at year-end 2012 (MW)	Percentage of world total (%)
World	282,482	
Europe	109,237	38.7
Asia	97,810	34.6
North America	67,576	23.9
Latin America/Caribbean	3,505	1.2
Australia/Pacific Region	3,219	1.1
Africa/Middle East	1,135	0.4

[a] Source: R. Walker based on data from Global Wind Energy Council [6].

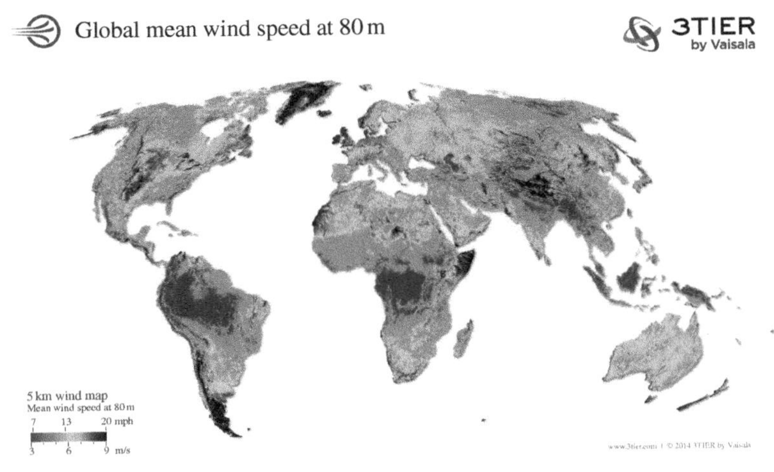

FIGURE 22.10 3TIER global wind map (Source: 3TIER by Vaisala [13]).

excellent wind resource locations as illustrated by the "hotter" colors—yellows and reds; in Europe, these tend to be concentrated in the northwestern areas of the region and along the coasts.

As of year-end 2012, European Wind Energy Association (EWEA) statistics [14] indicate that the leading European countries by installed wind capacity are Germany with 31.3 GW; Spain with 22.8 GW; and United Kingdom with 8.4 GW. Growth rates in the last year averaged about 9%. In terms of wind penetration percentages, or the ratio of wind energy generation to total electricity consumption, Denmark is the leading country at 27%, followed by Portugal at 17% and Spain with 16%. However, since Denmark is a relatively small country, the total installed capacity is relatively modest when compared with the larger countries.

In general, the QOL Index for Germany, Spain, and the United Kingdom is quite high (see QOL table in Chapter 1) with an average ranking of 22 out of 111 countries ranked. Their average per capita per day electric consumption is 19.5 kWh/day, in the midrange for industrialized countries with a high QOL index. EWEA statistics [14] also indicate that the overall electric generation fuel mix for Europe is about 25% coal, 25% nuclear, 25% natural gas, 15% hydropower, 4% wind, and 3% oil. There are exceptions, such as France, which have chosen to use nuclear power as their dominant source of electric power generation, or Germany, which intends to phase out its nuclear power by 2022 after the Fukushima earthquake and resulting nuclear accident in Japan.

22.4.2 Germany

Germany has a very aggressive renewable energy growth strategy and is a leader in both policy and implementation of renewable technology. In the first 6 months of 2011, renewables (led by wind at 7.5%, biomass at 5.6%, solar photovoltaic (PV) at 3.5%, hydropower at 3.3%, and other at 0.8%) contributed almost 21% of the total electrical demand of the country. The German goal of at least 35% renewable electricity by 2020, set in place by the revised Renewable Energy Sources Act in 2012, is one of the most ambitious national renewable energy policies in the world [15]. In support of the stated goal, there are a number of energy policies and financial incentives in place. A generous feed-in tariff (FIT) awards wind a "premium tariff" that provides preferred grid access and a tariff for each kWh produced. The FIT has been the major incentive for the development of the German onshore, offshore, and repower market. Repowering is the term used when aging wind turbine technology is replaced with newer and more efficient wind technology. The German Wind Energy Association estimates that by 2020, Germany could be generating 25% of its electrical needs from a combination of onshore and offshore wind energy plants. In addition to the FIT, Germany has a number of financial incentives, including public financing, and tax incentives that contribute to its renewable energy goals.

22.4.3 Asia

Asia is the second ranking region in terms of installed wind power capacity with 97.6 GW, as of year-end 2012 [6]. The region experienced a 19% growth in capacity additions during 2012. The leading Asian countries by installed wind capacity in 2012 are China with 75 GW and India with 18 GW. Growth rates during 2012 average about 19%.

China and India are the most populous countries on the planet, with 2.4 billion people, a third of the world population, about equally divided between the two countries. The QOL Index for these countries is about 6 (see QOL table in Chapter 1) with an average ranking of 67 out of 111 countries ranked. The average per capita per day electric consumption is very different between the two countries, with China consuming about 7 kWh/person/day and India only using about 2 kWh/person/day—in the low range for this metric. The two countries together consume about 20% of the global electric energy generated with 33% of the global population. As China and

India continue to develop and consume more electrical energy, their need for increased electrical generation capacity will dominate global trends.

The electric generation fuel choice for both China and India has been mostly coal—about 80% for China and 70% for India. Hydropower generation accounts for 16% in China and 13% in India. Natural gas and nuclear power are small fractions in both countries. China is reevaluating its nuclear expansion plans after the earthquake and Fukushima nuclear accident in Japan.

22.4.4 China

China has rapidly emerged as the global leader in renewable power development, and it has implemented a very aggressive renewable energy growth strategy. China's wind power market doubled each year between 2006 and 2009, and in 2010 China became the world's leading nation in terms of installed wind power capacity. Figure 22.11 shows the windiest areas in China, while Table 22.6 shows the government forecast for aggressive growth in wind capacity to the year 2020 in these regions, or bases, assuming that sufficient transmission infrastructure can be developed to accept the generated electric power.

In addition to strong wind energy project development, China's rapid emergence in the wind energy field is also demonstrated by its rapid growth in turbine-manufacturing capabilities. There are numerous domestic turbine manufacturers as well as

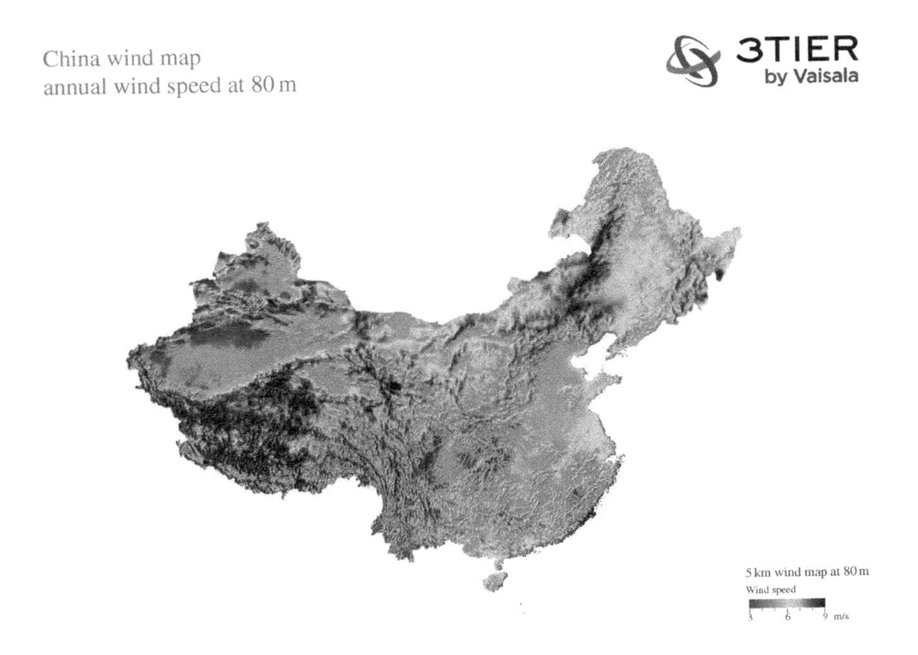

FIGURE 22.11 China Wind Speed Map (Source: 3TIER by Vaisala [13]).

TABLE 22.6 China Wind Power Bases and Growth Forecast[a]

Wind power base/region	2010 (installed MW)	2015 (planned MW)	2020 (planned MW)
Heibei	4,160	8,980	14,130
Inner Mongolia East	4,211	13,211	30,811
Inner Mongolia West	3,460	17,970	38,320
Jilin	3,915	10,115	21,315
Jiangsu	1,800	5,800	10,000
Gansu Jiuquan	5,160	8,000	12,710
Xinjiang Hami	0	5,000	10,800
Total	22,706	69,076	138,086

[a] Source: R. Walker based on data from the Global Wind Energy Council, Wind Energy Statistics 2012 [6].

FIGURE 22.12 Visual impact of air pollution in China (Photo Credit: http://en.wikipedia.org/wiki/File:Beijing_smog_comparison_August_2005.png [17]).

component and parts suppliers so that domestically produced equipment dominates the Chinese market and is also available for export. Of the top 10 global wind turbine manufacturers in 2012, four are Chinese companies.

This rapid expansion in wind energy development in China has been the result of strong, government-planned national and regional renewable power development policies. The key drivers behind these policies are the need to boost energy security, ease the pressure of environmental pollution, improve energy supply in rural areas, and provide an overall boost to national energy security [16]. In 2006, China became the world's number one emitter of CO_2. In 2012, the President of the China Medical Association warned that air pollution, see Figure 22.12, could become the biggest health threat for the nation [16].

China is looking to wind power and other renewable energy technologies to directly address both these national challenges. In 2004, China first presented its national plan to ramp-up renewable energy production at an international conference in Germany and the following year it enacted its Renewable Energy Law, which provided the legal statutes for the country's renewable energy policies. Passed in

2007, the Medium- and Long-Term Development Plan for Renewable Energy established the national guiding principles for development. Key elements of these plans included the following:

- Establishment of national targets for renewable energy development—15% of energy supply from renewables by 2020; mandatory grid access—stating that utility companies had to purchase generated renewable power
- Implementation of feed-in tariff (FIT) pricing with set pricing to be paid for wind power, by region
- Use of price balancing to distribute the costs of renewable power nationally
- Establishment of a Renewable Energy Fund for Development, financed through the government budget
- Authorization of a number of tax exemptions and subsidies for renewable generation equipment

Strong government policies have supported the rapid expansion of wind energy in China but have also led to challenges moving forward. A major obstacle is lack of sufficient grid access—as is occurring in other countries with rapid wind energy development. In China, many of the best wind resources lie in areas that are less economically developed, such as the northeastern provinces and Inner Mongolia. Also, national installed capacity targets have resulted in advances in installation capacity at the expense of energy production in some areas. There is also a lack of national product testing and certification standards leading to equipment-quality issues, to a need to direct more effort to solar and wind resource assessment, and to develop a workforce to support large-scale development of renewable technologies [18].

22.4.5 North America

The North American continent has 23 independent countries and ranks third in continental land mass but has only 8% of the world population—about 530 million. The wind map in Figure 22.10 indicates strong wind resources in the central plains, Alaska, and the far northeastern parts of the continent, including Greenland, which has an exceptional wind resource but virtually no population. Over 85% of the continent's population lives in the United States, Mexico, and Canada. Canada and the United States are very large consumers of electricity. Canada, with abundant hydropower and a relatively small population, cold climate, and large land area, ranks number one in the world at 52 kWh/person daily, while the United States uses about 39 kWh/capita/day; however, Mexico only uses about 6 kWh/person/day. The QOL rankings, as reported in Chapter 1, are 14, 13, and 32, respectively, out of 111 countries ranked.

North American wind development has been led by the United States and Canada, starting with the OPEC oil embargo of the 1970s. Following a pattern of energy concerns similar to Europe—in that there were energy supply concerns as well as a recognition of significant environmental effects of large-scale electricity production—governments began alternative energy research and development programs that included wind and solar energy. Solar energy development, led by photovoltaics and large concentrating collector designs such as the "solar power tower" (see Chapter 13), received significant attention. Wind power development was not as well-funded, but there were start-up programs in the 1970s and 1980s as described previously (see Chapter 1). US Federal initiatives such as the Public Utilities Regulatory Power Act led to state programs for alternative energy development, such as funding for pilot demonstrations, resource assessment, and tax incentives. Canada has been somewhat slower in adopting policies for renewable energy development but is currently ranked 9th in the 2013 global wind power market list, Figure 22.3. The Canadian Wind Energy Association was formed in 1984, supporting the development of wind power in the country. Mexico has more recent activity with development of about 1.4 GW in 2012, supported by modest tax incentives such as accelerated depreciation.

22.4.6 Texas, USA

The discussion of North American wind development will focus on the state of Texas since that state leads the United States in wind power development. With over 12 GW of wind power installed as of the end of 2012, Texas would rank 6th globally in wind capacity installed if it were a country. Texas is rich in wind energy resources with both onshore resources in the western and northwestern parts of the state and offshore opportunities along the southern coastline on the Gulf of Mexico. As discussed in Chapters 12 and 18, three National Electricity Reliability Council (NERC) interconnections cover the entire continental United States—the Eastern Interconnection, the Western Interconnection, and the ERCOT (or Texas Reliability Entity (TRE)) Interconnection. ERCOT stands for the Electric Reliability Council of Texas, and it is the only one of those three entirely within the boundary of a single state, meaning that it only has to deal with one regulatory commission (the Public Utility Commission of Texas), one governor, and one state legislature, generally making it easier to effect policy changes that will benefit the state. In addition, ERCOT is the balancing authority for a majority of the state, making power management more straightforward as compared with more complex arrangements of multiple system operators across state lines. ERCOT's policies and procedures, backed by the support of the Texas legislature, have been conducive to wind energy development.

In the late 1990s, a series of Deliberative Polls® were undertaken by utility companies in Texas, which were required by state regulators to obtain customer input as part of the generation planning process. During these polling processes, several hundred Texas residents learned about various sources of electric power generation and were then asked to provide input on what direction they wanted to see their utility go in order to meet the needs of their customers in the future. The participants in each poll, taken at several different cities in Texas, overwhelmingly favored wind and solar energy development over the more traditional means of producing electricity—natural gas, coal, and nuclear energy.

In 1999, the state legislature passed the first RPS, to be followed in 2005 with a second—both setting forth aggressive renewable energy development targets for utilities in the state. Additionally, a renewable energy credit (REC) market was established that could be used to by the utilities to purchase RECs in order to reach the mandated limits. A few years later, in order to address transmission congestion issues limiting the amount of electricity that could be moved from the windy, but less populous, northwest parts of the state, a $6 billion transmission development plan was proposed for construction. Called the Competitive Renewable Energy Zone (CREZ) transmission plan, it called for identifying areas of the state with excellent renewable energy resources and the construction of transmission lines from those regions to the more populous areas of the state. With a capacity of 18 GW, the new CREZ lines are being paid for by ERCOT customers and are expected to be completed in 2014.

Also within the state are long-standing trade organizations that advocate for renewable energy technologies and lobby the state legislature for policies to hasten their adoption. One such group has been the Texas Renewable Energy Industries Association (TREIA). Organized in 1984, this group has been a strong advocate for renewable energy technologies and policies within the state. In recent testimony to the Texas House State Affairs Committee, TREIA outlined the direct benefits to the State of Texas in furthering the use of renewable energy [19]. The points presented were as follows:

Direct benefits of renewable energy as a complementary energy source include the following:

- Quick to market.
- Competitive and declining costs.
- Low to zero water use.
- Driver of local jobs.
- Increased generation diversity helps maintain grid reliability.
- Renewables can address demand side as well as the supply side.
- Complementary with conventional generation, especially natural gas.

The final point, concerning the integration of renewables with natural gas generation, was addressed in a recent study by the Brattle Group [20]. The in-depth analysis examined combining increased use of natural gas for electric generation and renewable energy concluded that the ERCOT electricity region would experience no technical difficulties obtaining up to 43% of its electricity from renewable resources. Many analysts have expressed concern that large amounts of variable resource generation from wind and solar power could cause significant technical difficulties on a utility system. This study refutes those claims.

For the state of Texas, it is the combination of positive in-depth technical analysis and results, aggressive state policies, a mandate for renewable development by the people of the state, a unique single utility operating entity, open land available for development, and an excellent wind resource that has led to the rapid growth of wind energy in the state.

22.4.7 South America

The continent of South America has 13 countries and a population of 390 million—about 6% of the world's population. Brazil, the largest country, is home to 50% of the population of the continent. Installed wind energy capacity was only about 3 GW at the end of 2012, with most of the development in Brazil, a country known for its aggressive renewable energy development initiatives. The wind resource map in Figure 22.10 indicates strong wind resources in the northern, northeastern, and far southern parts of the continent. Per capita per day electric energy consumption in Brazil is about 7 kWh/capita/day, about the same as China, and the country is ranked 39 out of 111 on the QOL scale as presented in Chapter 1.

22.4.8 Brazil

Electric energy in Brazil is dominated by hydropower. Shared with Paraguay, the Itaipu dam and hydropower plant is the largest in the world—rated at 20 GW. Brazil's total installed electric generating capacity was 113 GW in 2010 as reported by the US-EIA [21], with about 90% of its electricity coming from hydropower sources, with the remainder coming from nuclear, natural gas, and some wind power.

In 2002, the Brazilian government created the *Program for Incentive of Alternative Electric Energy Sources* (Proinfa) to encourage the use of other renewable sources, such as wind power, biomass, and small hydroelectric power stations. One of the major issues with Brazil's dependence on hydropower resources is the issue of low water supply during times of drought and the resulting lack of electrical power. The Proinfa program was initiated in hopes of alleviating the problem. An assessment of wind energy resources in the country indicated a potential of over 140 GW of developable wind power, and since 2002 wind power development has been substantial. Brazil now leads the region in wind power development with 2.5 GW installed, over 80% of total, installed capacity in South America. The Brazilian government has established a wind power installed capacity goal of 10 GW by the year 2020.

22.4.9 The Pacific Region (or Oceania), Including Australia and New Zealand

The Pacific Region, or Oceania—the region centered on islands of the tropical pacific ocean—has 14 countries and about 1% of the global population. The largest country in the region is Australia, which has an active renewable energy program and approximately 2.6 GW of wind energy installed as of 2012. Australia and New Zealand both rank very high on the 111-nation QOL list in Chapter 1, with Australia ranked number 6 and New Zealand ranked number 15. Both also have high relative consumption of electricity per person at 33.5 and 24 kWh/capita/day, respectively.

22.4.10 New Zealand

A rather small land area located in the southern pacific region, New Zealand has a population of 4.5 million, about 13% of the region. Currently, the country has an installed electric generating capacity of approximately 10 GW, with approximately 61% of its electrical energy from hydroelectric power, 25% from natural gas, 13% from geothermal resources, 5% from coal, and 5% from wind energy. New Zealand, due to its location along the "roaring 40s" latitude, a region between the 40° and 50° latitude that is conducive to strong and consistent westerly winds (see Chapter 4), is well suited to wind power development. At the end of 2012, the country had over 600 MW of installed wind power capacity [6], accounting for about 5% of the electricity needs of the country.

New Zealand currently obtains almost 80% of its electricity from renewable energy sources, one of the highest percentages in the world. In 2008, the government released an energy strategy to increase that fraction to 90% by the year 2025 [22]. Additionally, a separate strategy is to increase the generation of wind power to 20% by the year 2030. With a projected increase of about 1% per year in electrical demand, the additions of wind power and renewable energy to the electric mix will have to be substantial.

An additional challenge for the nation is the current "NIMBY" movement within the nation, organized for the purpose of slowing or halting wind power development altogether. In 2009, the New Zealand Environment Court ruled against the development of a 630-MW wind farm that had been recently approved for construction. In its ruling, the court stated "Despite the potentially large contribution of energy to the national grid, it would be inappropriate to put a huge wind farm in such a nationally important natural landscape" [23]. The area chosen for development was an unusually picturesque location and the local inhabitants preferred not to disrupt the landscape—demonstrating the difficulties that can sometimes be associated with large wind project siting and development—in spite of strong national policies for renewable energy development.

22.4.11 Africa

The continent of Africa has 54 countries with a population of approximately 1 billion. Overall, the continent is rich in natural resources, including wind and solar, with both onshore and offshore development opportunities for wind development. The global wind resource map in Figure 22.10 shows excellent wind resources in the northeastern and northwestern areas of the continent as well as in the southern regions. Wind development to date has been modest, with about 1 GW installed on the continent as of 2012 and Egypt and Morocco the leading countries. Being a large continent, there is great diversity in economic development, the per capita per day use of electrical energy and rankings on the QOL index presented in Chapter 1. To the north, for example, Egypt ranks number 80 out of 111 countries ranked on the QOL index scale, using about 4 kWh/person/day. However, much of central and southern Africa is an area of developing nations, with a large indigenous population that lacks access to electrical power. Nigeria, for example, ranks 108 out of 111 on the QOL index and uses less than 0.4 kWh/day/person of electrical energy.

22.4.12 Kenya

Kenya, in the northeastern part of the continent, has approximately 44 million people, about 4.5% of the continent's population, and the most rapidly growing economy in East Africa. The country has recently announced two large wind development projects—a 300-MW project called the Lake Turkana Wind Project, expected to be completed in 2016, and the 61-MW Kinangop Wind Park, expected to come on line in 2015. With a total installed electric capacity of only 1700 MW planned, wind energy will become a significant part of Kenya's future energy mix following the completion of these projects [24]. Kenya, as many of the Sub-Saharan countries in Africa, has been hampered in development due to the lack of electrical power. The government of Kenya has a reported national goal of 2 GW of wind power installed by the year 2030, meeting about 9% of its projected energy needs [25].

22.5 CONCLUSIONS

As we conclude this section on global renewable energy policy and development, it is striking to note the rapid pace and worldwide emphasis on wind energy and other renewable energy technologies for electric power generation across many continents, regions, and countries. A recent research report by Citi Bank [4], referenced earlier in this chapter, discussed the opportunities for global investment in renewable energy

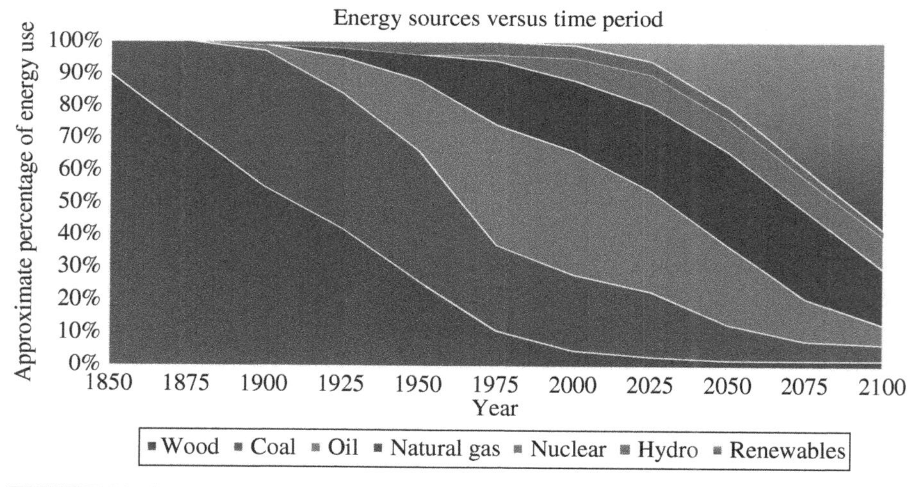

FIGURE 22.13 Global Energy Transitions in Primary Fuel Sources and the Dawn of the Age of Renewables. Note that source data have been approximated and combined to emphasize global energy transition trends over time. "Wood" includes the use of other biofuels such as animal products, for example, whale oil and animal waste, while "Renewables" includes newly derived biomass and biofuels as well as other renewable sources other than hydroelectric power, which is listed separately (Source: A. Swift based on sources from Citi IEA, EIA, Citi Research, the petroleum-economist.com, sbc.slb.com, and the energy collective [4,25–27]).

technologies and at its conclusion proposed that the world may be in the early stages of a long-term energy transition to a renewable energy future, although this report is not the first to suggest such a transition. The graphic, Figure 22.13, derived from several different sources of data illustrates the evolution of energy sources over time, suggesting that reliance on renewable sources of energy will continue to grow rapidly in the coming decades.

REFERENCES

[1] Alic J. 2012. Energy and Public Opinion: Confusion Reigns. Available at http://oilprice.com/Energy/Energy-General/Energy-and-Public-Opinion-Confusion-Reigns.html. Accessed November 3, 2014.

[2] Gallup Politics. 2013. Americans Want More Emphasis on Solar, Wind, Natural Gas. http://www.gallup.com/poll/161519/americans-emphasis-solar-wind-natural-gas.aspx. Cited December 20, 2013.

[3] REN21. Renewables 2013 Global Status Report, REN21 Secretariat, ISBN 978-3-9815934, Paris, France; 2013.

[4] Channel J, Lam T, Pourreza S. Shale & Renewables: A Symbiotic Relationship, Citi Research Report; 2013.

[5] National Renewable Energy Laboratory. Renewable Electricity Futures Study (Entire Report), 4 vols. NREL/TP-6A20-52409. In: Hand MM, Baldwin S, DeMeo E, Reilly JM, Mai T, Arent D, Porro G, Meshek M, Sandor D, editors. U.S. Government Printing Office, Golden (CO); 2012.

[6] Global Wind Energy Council. 2013. Wind Energy Statistics 2012. Available at http://www.gwec.net. Accessed November 3, 2014.

[7] The World Bank website. Available at http://web.worldbank.org/WBSITE/EXTERNAL/TOPICS/EXTENRGY2/0,,contentMDK:22855502~pagePK:210058~piPK:210062~theSitePK:4114200,00.html. The World Bank Publications; Cited December 20, 2013.

[8] U.S. Energy Information Administration. Annual Energy Outlook 2013. U.S. Government Printing Office, Washington (DC); 2013.

[9] The Electric Drive Transportation Association. Available at http://www.electricdrive.org/. EDTA Publications. Referenced December 18, 2013.

[10] Yergin, D. (2011) *The Quest, Energy, Security, and the Remaking of the Modern World.* Penguin Group, New York.

[11] U.S. Energy Information Administration. International Energy Outlook 2013. U.S. Government Printing Office, Washington (DC); 2013.

[12] The German Association of Energy and Water Industries (Bundesverband der Enegie—und Wasserwirtschaft—BDEW). Available at http://www.bdew.de/. Referenced December 20, 2013.

[13] Vaisala Beijing China Office. 2014. 3Tier website. Available at http://www.3tier.com/en/. Accessed November 6, 2014.

[14] European Wind Energy Association (EWEA). 2013. Wind in Power: 2012 European Statistics. Available at http://www.ewea.org. Accessed November 3, 2014.

[15] Gipe P. Germany Passes More Aggressive Renewable Energy Law. Renewable Energy World, July 25, 2011, renewableenergyworld.com. Renewable Energy World.

[16] Jonathan W. Air pollution could become China's biggest health threat, expert warns, Friday 16 March. The Guardian. Available at http://www.guardian.co.uk/environment/2012/mar/16/air-pollution-biggest-threat-china. Accessed November 3, 2014.

[17] Wikipedia/Bobak. Available at http://en.wikipedia.org/wiki/File:Beijing_smog_comparison_August_2005.png, http://creativecommons.org/licenses/by-sa/2.5/legalcode. Accessed on January 30, 2013.

[18] World Watch Institute. Renewable Energy and Energy Efficiency in China: Current Status and Prospects for 2020, Report 182. Mastny L, editor; Island Press; 2010.

[19] Testimony of Steven M. Wiese, President, Board of Directors, Texas Renewable Energy Industries Association (TREIA) to the Texas House State Affairs Committee, Texas Renewable Energy Industries Association; February 9, 2012.

[20] Weiss J, Bishop H, Fox-Penner P, Shavel I. Partnering Natural Gas and Renewables in ERCOT, Prepared by the Brattle Group for the Texas Clean Energy Coalition; Texas Renewable Energy Industries Association; 2013.

[21] U.S. Energy Information Administration website, International Energy Statistics page. Available at http://www.eia.gov/cfapps/ipdbproject/IEDIndex3.cfm?tid=2&pid=2&aid=7. U.S. Government Printing Office; Cited December 28, 2013.

[22] Spector D. 2011. New Zealand Commits to 90% Renewable Energy By 2025, Business Insider. Available at http://www.businessinsider.com/new-zealand-renewable-energy-2011-09. Cited December 29, 2013.

[23] Shahan Z. New Zealand Environment Court Says No to Huge Wind Farm, CleanTechnia. Available at http://cleantechnica.com/2009/11/09/new-zealand-environment-court-says-no-to-huge-wind-farm/. Cited December 29, 2013.

[24] Business Day Live, Africa website. Available at http://www.bdlive.co.za. Times Media Group; Cited December 29, 2013.

[25] International Atomic Energy Agency website. 2013 Joint Nuclear Energy Management School. Available at http://www.iaea.org/nuclearenergy/nuclearknowledge/schools/NEM-school/2012/Japan/PDFs/week2/CR6_Kenya.pdf. Cited December 29, 2013.

[26] International Energy Agency website. http://www.iea.org/. US Government Printing Office; Cited May 22, 2014.

[27] Schlumberger Business Consulting website. Energy Perspectives page. http://www.sbc.slb.com/Our_Ideas/Energy_Perspectives.aspx. Cited May 22, 2014.

23

WIND ENERGY WORKFORCE, EDUCATION, AND JOBS

23.1 HISTORIC GROWTH OF THE WIND ENERGY INDUSTRY

Wind energy has been a source of jobs for centuries from the millers of the English post windmills to the water pumpers of the Great Plains. Today, the multidisciplinary wind energy workforce is estimated to employ some 300,000 people and is rapidly growing around the globe. The continuing development of wind energy in the world and United States (shown in Figs. 23.1 and 23.2) has resulted in a need for an expanded workforce.

23.2 SECTORS AND LOCATIONS OF JOBS WITHIN THE WIND ENERGY INDUSTRY

Figure 23.3 shows the wind industry supply chain and the jobs associated with it. Workforce needs range from procurement of raw materials through component manufacturing, project development, and operation and maintenance. This list is not all inclusive. For example, under components, the gearbox is not listed and under operation and maintenance, wind energy forecasting for the utilities is not listed. Yet the diagram gives an adequate overview of the wind development process.

Wind Energy Essentials: Societal, Economic, and Environmental Impacts, First Edition.
Richard P. Walker and Andrew Swift.

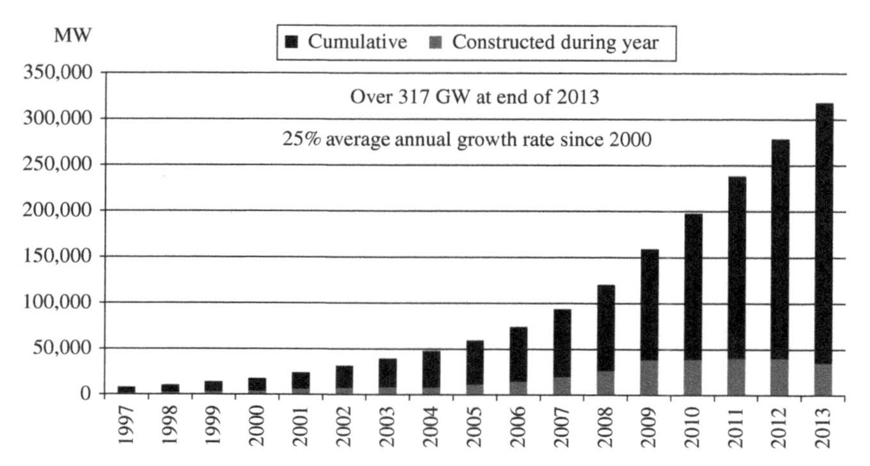

FIGURE 23.1 Worldwide energy development (Source: R. Walker, based on data from Global Wind Energy Council press releases [1]).

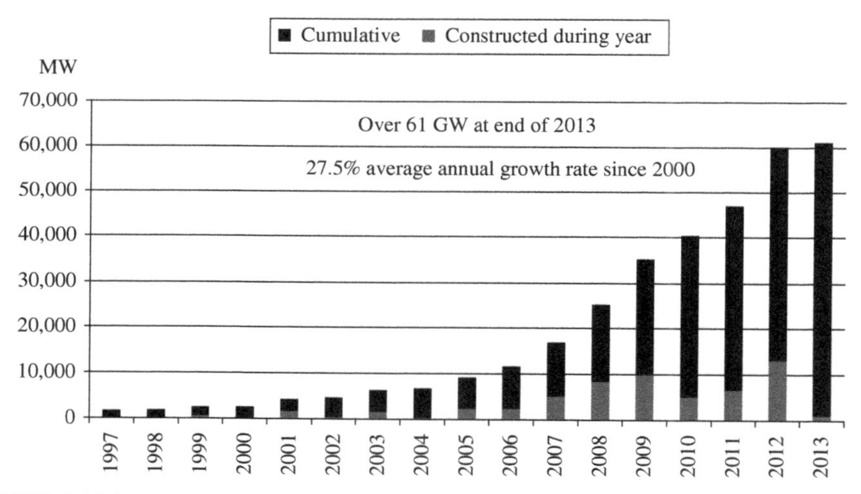

FIGURE 23.2 Installed wind power capacity in the United States (Source: R. Walker, based on data from AWEA Annual Market Reports [2]).

Figure 23.4 shows the approximate number of jobs in the United States in wind energy as of the end of 2013 and sectors of the industry these jobs are in.

Figure 23.5 shows how these jobs are distributed across the United States. While Texas leads the nation in terms of both installed wind capacity and wind energy jobs, all states have benefited from the growth of the wind energy industry.

This is a multidisciplinary and complex industry requiring a number of specialized skills, training and education.

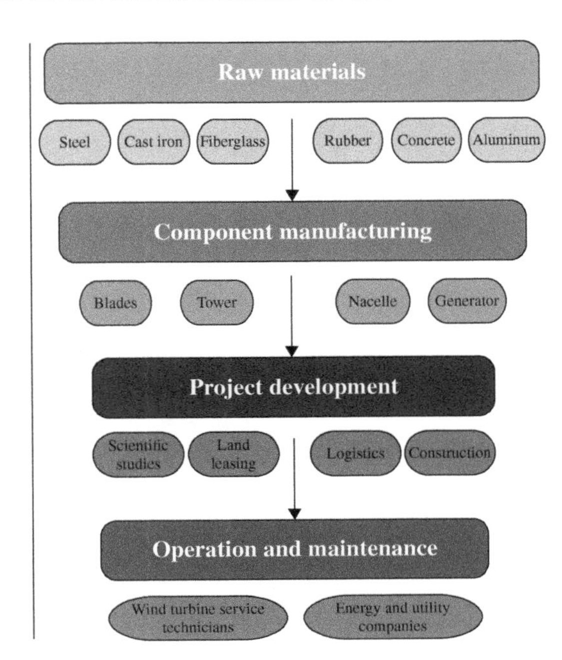

FIGURE 23.3 Wind industry supply chain (Credit: Hamilton and Liming [3]).

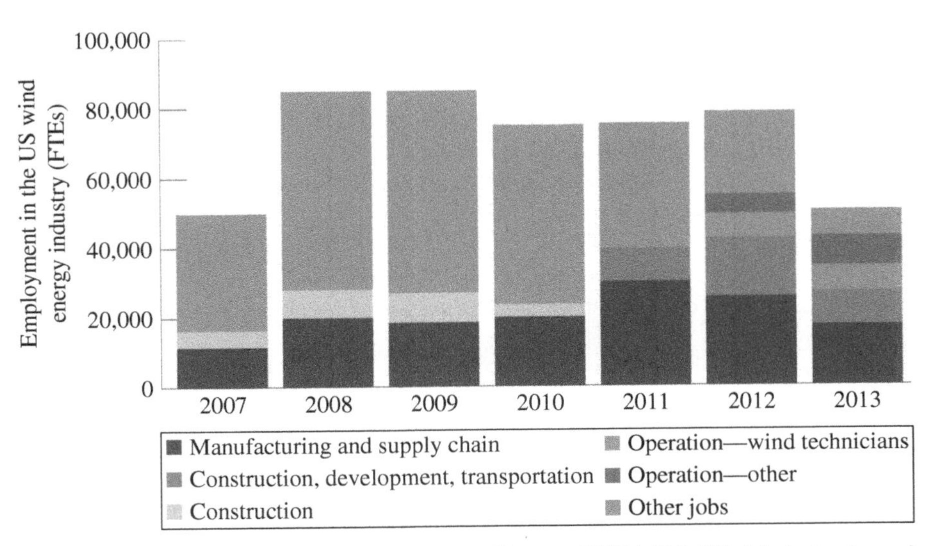

FIGURE 23.4 Total wind energy employment (Source: AWEA U.S. Wind Industry Annual Market Report 2013 [4]).

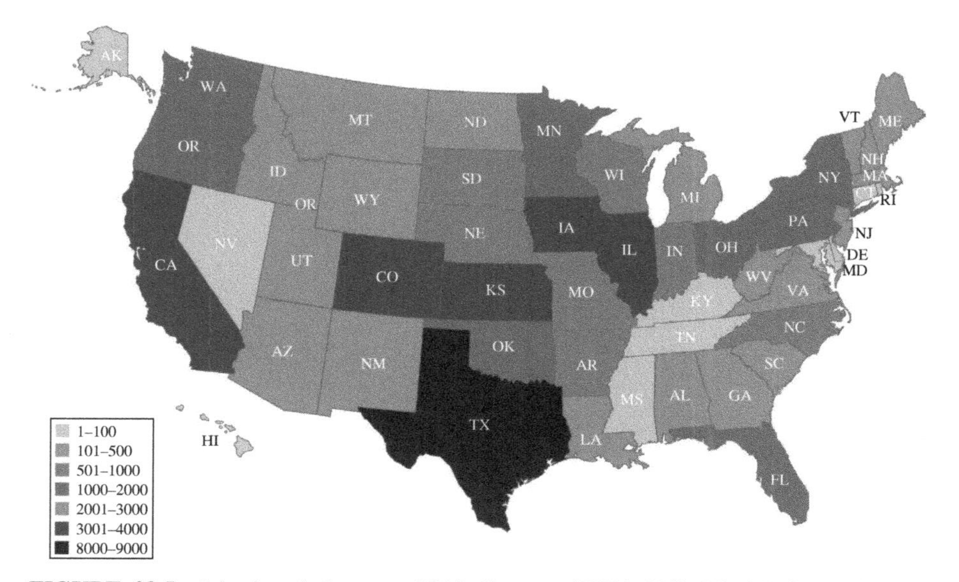

FIGURE 23.5 Jobs in wind power, 2013 (Source: AWEA U.S. Wind Industry Annual Market Report 2013 [4]).

Companies continue to install wind-related equipment manufacturing and service facilities in the United States. Figure 23.6 shows new wind manufacturing or service facilities opened in 2012 (red), planned wind manufacturing or service facilities announced in 2012 (blue), and existing facilities (green).

Other studies indicate that the number of wind energy jobs could increase substantially. One such study, released in February 2010, is the *Jobs Impact of a National Renewable Electricity Standard (RES)* [6], which would require the country as a whole to have 25% of its electricity to be produced from renewable energy by 2025. The study indicates that such a standard would lead to an additional 116,000 jobs in the wind energy industry as compared with the number of jobs without the standard.

Figure 23.7 shows the job intensity of the wind energy industry per unit of wind energy produced. The unit of measurement is job years per gigawatt-hour (GWh), and it shows that wind is a very job-intensive electric generation industry. Although, wind does not have any fuel processing, as do coal, natural gas, and nuclear, it is still very job-intensive.

Within Texas, wind energy development was especially rapid between 2005 and 2010 (see Fig. 23.8), when Texas led the nation in the annual amount of new wind generation installed each year. However, due to this rapid growth, wind developers and project owners began to experience growing transmission constraints between the windy western areas of the state to the load centers in the eastern part of the state, resulting in curtailment of many of the West Texas wind projects.

To address this issue, the state legislature in 2005 implemented a process to permit and construct transmission lines to Competitive Renewable Energy Zones (CREZ),

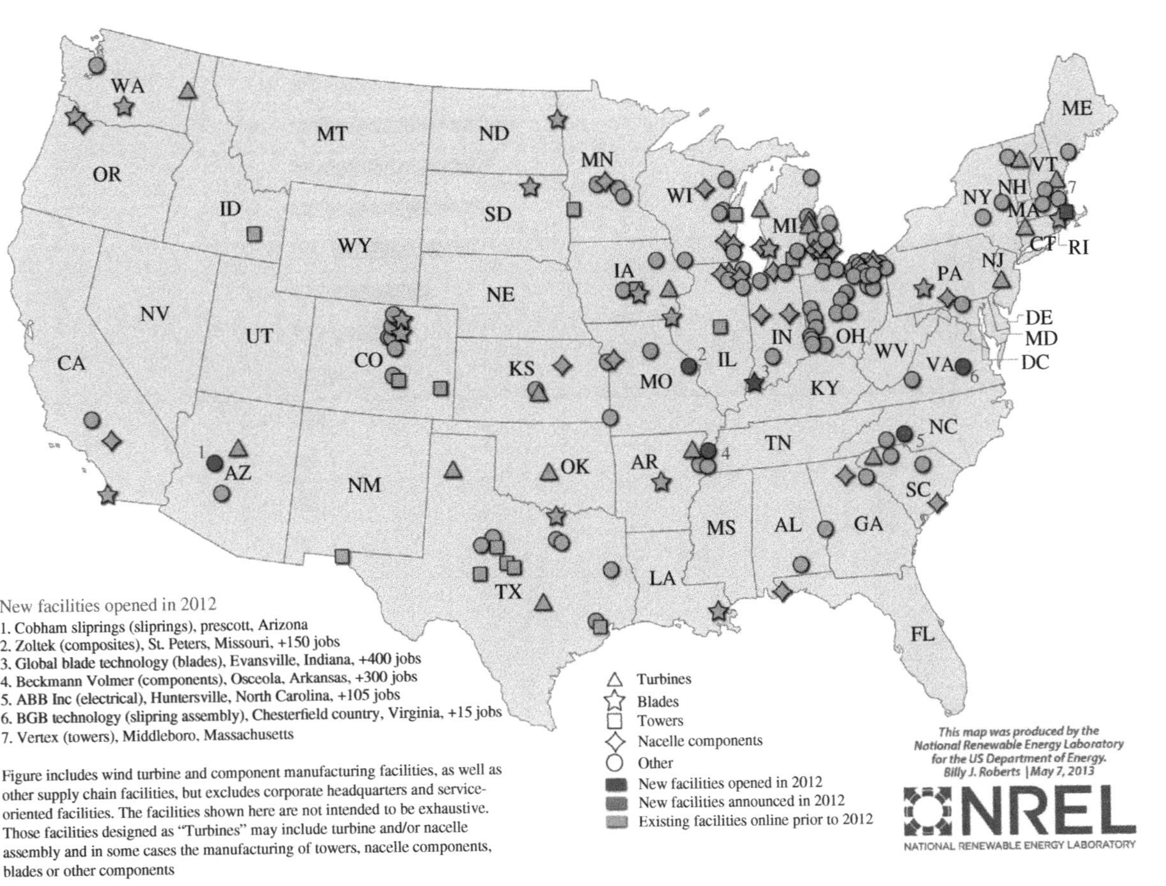

FIGURE 23.6 Wind energy facilities 2012 (Source: *2012 Wind Technologies Market Report* [5]).

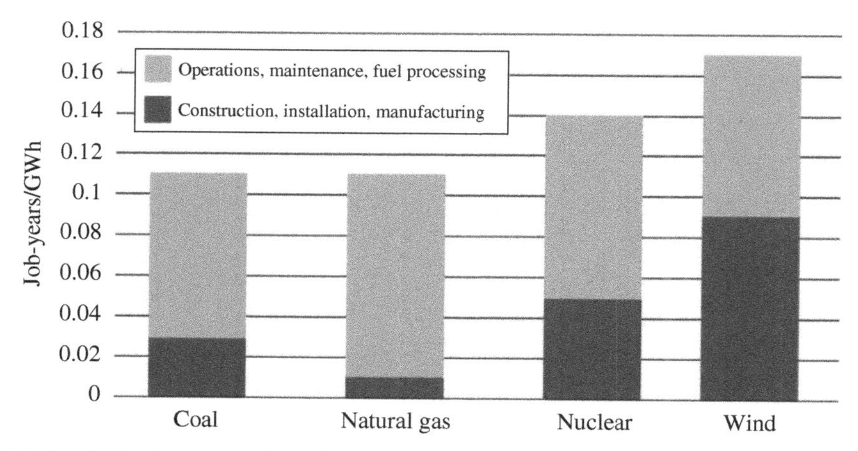

FIGURE 23.7 Average total employment for different energy technologies (Source: World Resources Institute [7]).

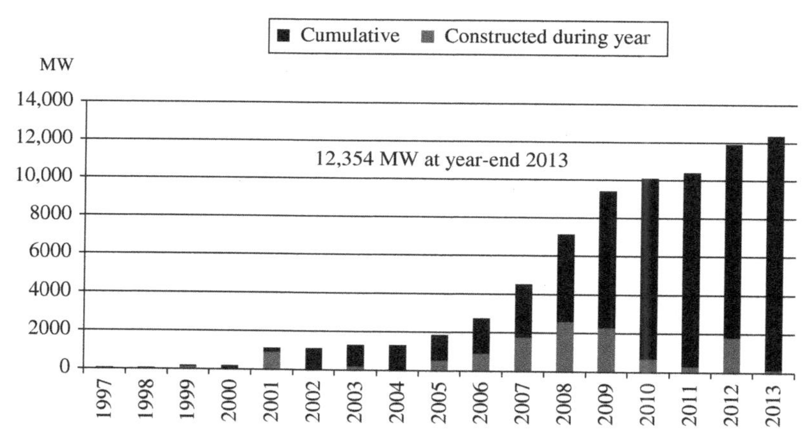

FIGURE 23.8 Texas wind energy development (Source: R. Walker, based on data from AWEA Annual Market Reports [2]).

selected based on wind resource and transmission access potential, to facilitate movement of electrical energy from renewable resource zones in the western part of the state to load centers elsewhere within the state.

The CREZ-line projects are intended to facilitate up to 18,500 MW of wind energy in Electric Reliability Council of Texas (ERCOT) through the addition of almost 3600 miles of 345-kV transmission lines, many of which are double-circuit lines (i.e., two separate transmission circuits sharing the same structures and right-of-

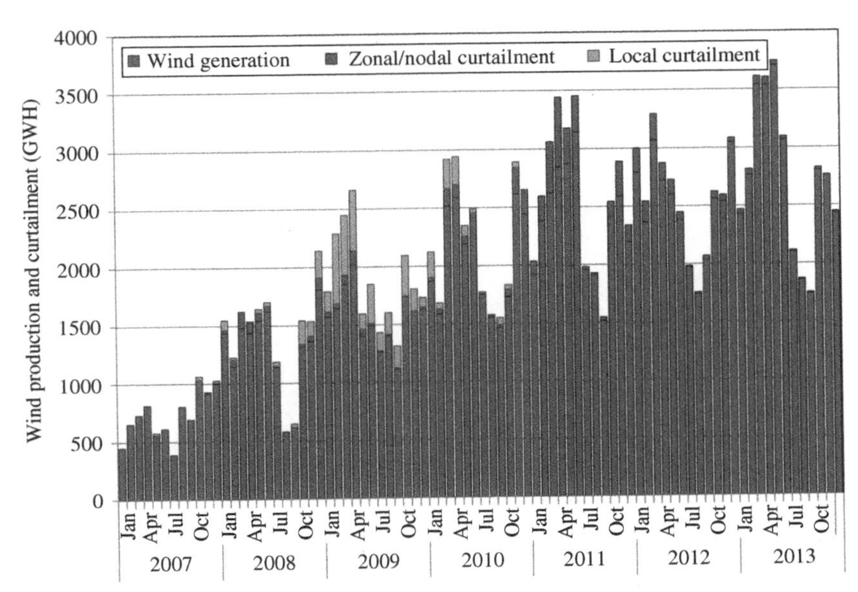

FIGURE 23.9 Curtailment of wind generation in ERCOT 2007–2013 (Source: Developed by Potomac Economics, the ERCOT Independent Market Monitor [8]).

way). As previously discussed in Chapter 12 and as shown in Figure 23.9, the completion of the CREZ lines has significantly reduced the amount of wind project curtailments due to transmission congestion.

A study undertaken by the Perryman Group, a highly respected economic and financial analysis firm headquartered in Waco, Texas, forecasted the future economic impact of the estimated $6 billion CREZ lines. The cost of these new lines will be paid by electric utility ratepayers within ERCOT as a portion of their monthly billing. However, the economic impact for the state of Texas is estimated to be a minimum of $12.5 billion in total spending, $5.3 billion in output, and 61,682 jobs or the equivalent to the Texas air transportation industry, which includes large airlines such as Southwest and American Airlines. At a maximum, this impact would be $23.6 billion in annual total spending, $10.5 billion in output each year, and 125,915 permanent jobs - the equivalent of the computer and electronics industry, which includes companies such as Texas Instruments and Dell Computers (Fig. 23.10).

The Perryman study [9] concluded "The CREZ transmission investment will help solidify Texas' position at the forefront of the wind power, renewables, and associated industries." Construction of the CREZ electric power transmission lines is presently moving forward, with completion expected in the 2013–2014 time frame (Figs. 23.11 and 23.12).

Figure 23.12 shows a map of the CREZ zones and the transmission lines that will be built to move renewable energy to the load centers in the eastern part of the state.

FIGURE 23.10 Wind development such as this construction of a multi-MW wind turbine in Texas, was constrained beginning in 2008 due to lack of transmission (Source: R. Walker).

FIGURE 23.11 CREZ electric power transmission line under construction near Snyder, Texas, July 2011, with a wind farm shown in the background (Photo Credit: Charles Norland, Norland Photographic Art).

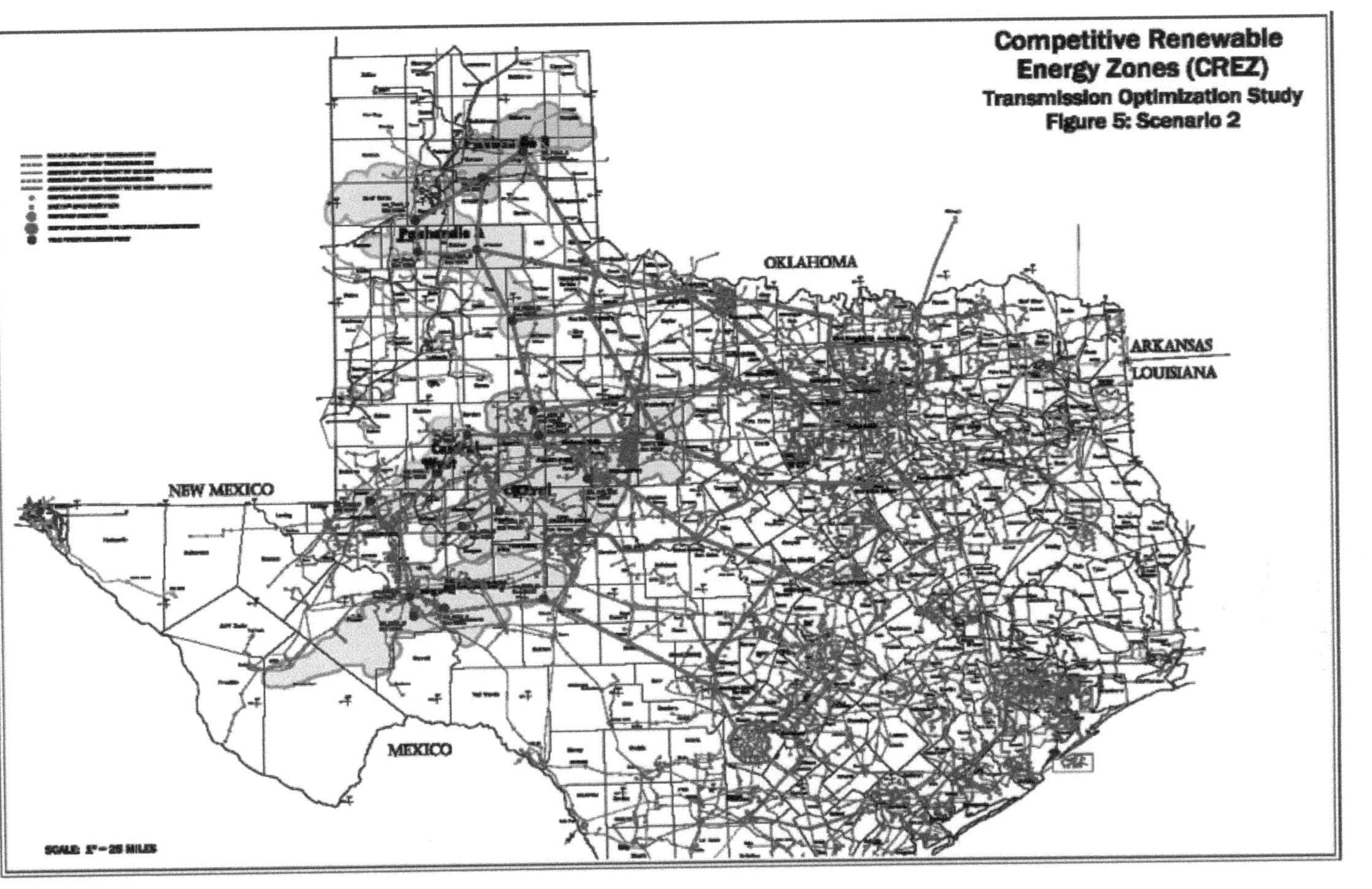

FIGURE 23.12 CREZ zone transmission map (Source: ERCOT [10]).

23.3 PROJECTED GROWTH OF THE WIND ENERGY INDUSTRY

Next, we examine the future of this industry in the United States as envisioned in the DOE's *20% Wind Energy by 2030* report [11]. The report, published in 2008, outlines the growth scenario for wind that would result in a 20% wind electricity fraction in the United States, or approximately 300 GW of installed capacity, as shown in Figure 23.13, by the year 2030. As shown in the figure, by the year 2015, offshore installations would begin, but the bulk of installations would still be land-based.

Figure 23.14 shows how turbines would be distributed throughout the states. The darker colors indicate a higher wind capacity installed. The small turbine icon located offshore indicates offshore site development. Notice that Texas wind development includes both onshore and offshore development.

Figure 23.15 shows the Indirect and Direct jobs expected to be created within manufacturing, construction, and operations. If we refer back to Figure 23.8, of the 85,000 jobs, 30,000 represent manufacturing, construction, and operations, while the remaining 55,000 jobs are considered *other indirect jobs,* which can include the following:

- Manufacturing (small components, electrical parts, and raw component suppliers)
- Developers and development services (land acquirement, permitting, and wind resource assessors)
- Financial and consultant services (financiers, accountants, and consultants)
- Contracting and engineering services (electrical, mechanical, and civil engineers)

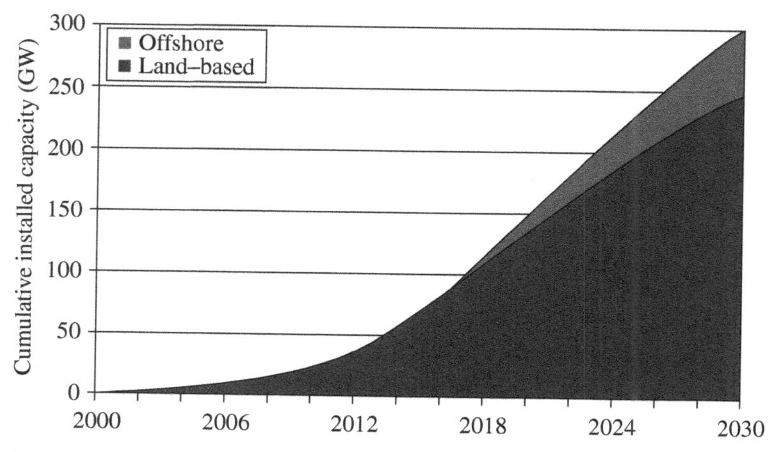

FIGURE 23.13 Estimated breakdown of offshore versus land-based wind capacity in DOE's *20% Wind Energy by 2030* scenario (Source: DOE [11]).

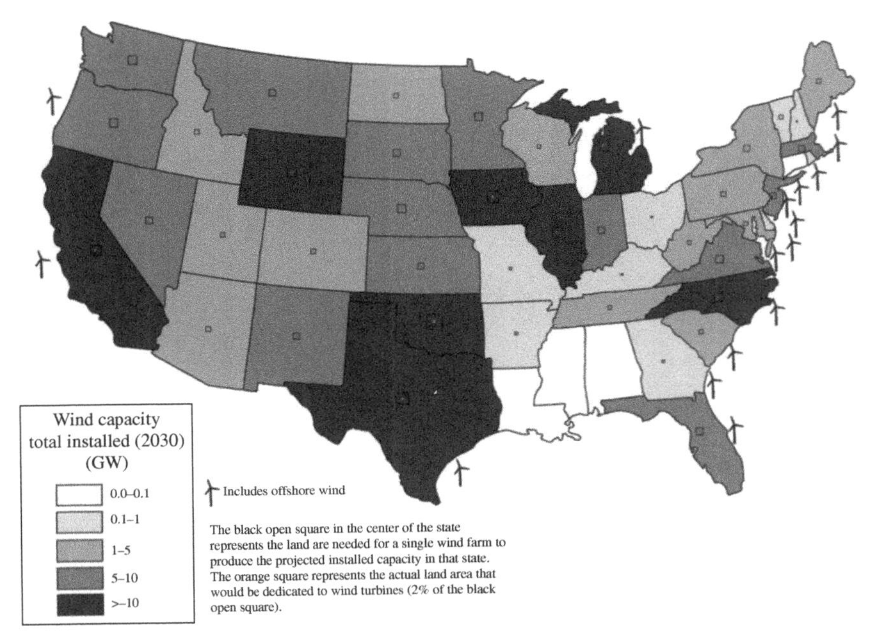

FIGURE 23.14 Projected installed wind nameplate capacity by state in 2030 (Source: DOE [11]).

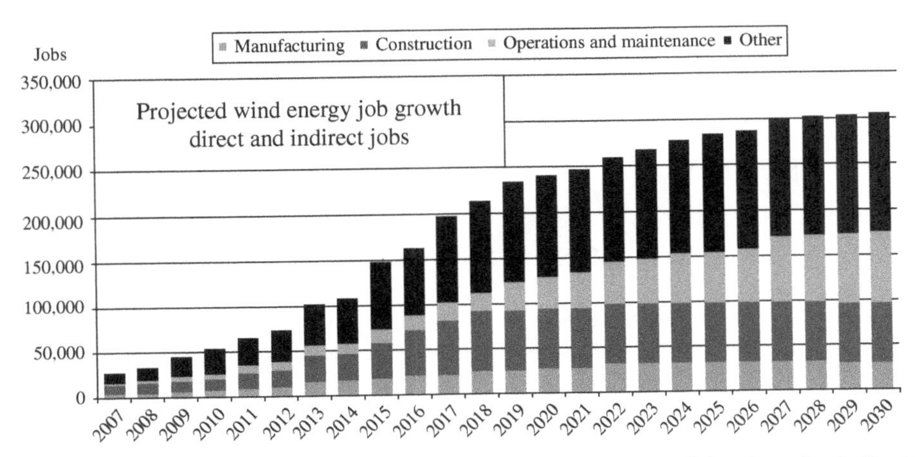

FIGURE 23.15 Direct manufacturing, construction, and operations jobs plus other indirect jobs supported by the 20% wind scenario (Source: DOE [11]).

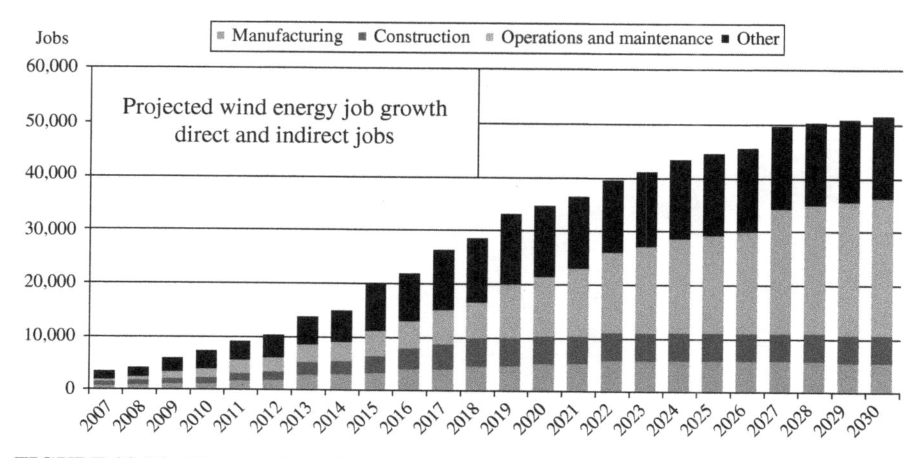

FIGURE 23.16 Estimated number of professional wind energy jobs that could benefit from university education in wind energy (Source: R. Walker et al. [12]).

- Parts-related services (repair shops and equipment manufacturers and suppliers)
- Transportation and logistics

Most of the jobs indicated in Figure 23.15 will be jobs filled by wind technicians or people with 2-year college and technical training or industry cross-training. However, some fraction of these jobs will require more training and education in order to be adequately filled. We estimate that 20,000–30,000 of the direct jobs will require and benefit from university-level training in wind energy. That would necessitate approximately 1000–1500 wind energy–educated university graduates per year as shown in Figure 23.16. Many of the indirect jobs could also benefit from wind energy education at the university level.

23.4 TYPES OF JOBS WITHIN THE WIND ENERGY INDUSTRY

What types of jobs will be needed in order to support this industry at a 20% wind penetration level in the United States? Some will be engineering jobs (as shown in Table 23.1), which lists 18 engineering-related jobs within the wind energy industry.

For people interested in direct engineering applications in wind energy such as wind turbine design or wind turbine tower and foundation design, it is recommended that those students obtain an engineering degree and supplement it with a minor or undergraduate certificate in wind energy or following up with a graduate certificate or master's degree within a wind energy specialty.

This industry is very multidisciplinary, and it will require much more than engineering skills and training in order to meet the 20% wind energy goal for the nation.

TABLE 23.1 Engineering Jobs in the Wind Industry (Engineering Discipline Shown in Parenthesis)

1. Tower and foundation design (CE)
2. Road design (CE)
3. Structural and blade design and testing (CE/ME)
4. Surveyors (CE)
5. Geotechnical engineer (CE)
6. Geotechnical testing (GEOL/CE)
7. Environmental management (CE/Environmental Engineer)
8. Construction-project management (CE)
9. Project development (CE/ME/IE/EE)
10. Safety and environmental health
11. Wind turbine design (ME)
12. Power system integration and substation design (EE/ME)
13. Interconnection design/collection system design (EE)
14. Lean manufacturing for components and assembly (IE/ME)
15. Site operations managers (ME/EE)
16. Predictive maintenance specialists (ME/IE)
17. Supervisory Control and Data Acquisition (SCADA) Project engineers (EE/CS)
18. Safety engineer (IE)

CE, civil engineering; CS, computer science; EE, electrical engineering; GEOL, geology; IE, industrial engineering; ME, mechanical engineering.

TABLE 23.2 Non-engineering Wind industry Jobs That Can Benefit From Education in Wind Energy

1. Resource assessment specialist
2. Wind power production data analyst
3. Wind power forecaster
4. Technical sales and marketing
5. Project development
6. Utility interconnection consultant
7. Community liaison
8. Regulatory/government liaison
9. Operations and maintenance manager
10. Risk management and assessment
11. Supply chain management
12. Manufacturing
13. Energy analyst
14. Geographical information system specialist
15. Wind project finance analyst
16. Marketing wind power projects
17. Wind energy tax specialist
18. Land manager
19. Property tax manager
20. Utility liaison/interconnection experts
21. Permitting specialist
22. Construction manager
23. Environmental/habitat specialist
24. Visual impact consultant

We have examined other types of jobs needed for the industry to move forward and complete the supply chain shown in Figure 23.3 from the Bureau of Labor Statistics. In Table 23.2, there is a list of 24 jobs that would require education and training in wind energy but not necessarily an engineering degree. These jobs include business, finance, environmental science, atmospheric science, and a number of related disciplines.

TABLE 23.3 Wind Energy Career Jobs, Salaries, and Job Descriptions[a]

Engineering	
Type of engineers	**Median annual wages**
Aerospace engineers	$94,780
Civil engineers	$76,590
Electrical engineers	$83,110
Electronics engineers, except computer	$89,310
Environmental engineers	$77,040
Health and safety engineers, except mining safety	$74,080
Industrial engineers	$75,110
Materials engineers	$83,190
Mechanical engineers	$77,020
Engineers, all other	$89,560
Engineering technicians, except drafters	$50,130

Manufacturing	
Occupation	**Median annual wages**
Machinists	$41,480
Computer-controlled machine tool operators, metal and plastic	$34,790
Team assemblers	$29,320
Welders, cutters, solderers, and brazers	$35,920
Inspectors, testers, sorters, samplers, and weighers	$37,500
Industrial production managers	$87,120

[a] Source: Bureau of Labor Statistics [3].

A recent 2011 report by the Bureau of Labor Statistics also examined the types of jobs and salaries related to wind energy. As shown in Tables 23.3 and 23.4, the list includes everything from engineering to manufacturing to project development occupations. There is a discussion of the project development occupations and their responsibilities in the related tables for land-acquisition specialists, asset managers, and logisticians.

To date, most of the wind energy industry education and training is "on-the-job" and much of the training may be characterized as inadequate. At a minimum, this approach will hamper the development and trajectory needed to reach the DOE target of 20% wind energy for the nation. At worst, this strategy could set the stage for a catastrophic event—for instance, a major blackout caused by avoidable errors in a particular wind-power system integration design.

Proper education and training of the wind energy workforce is imperative as this industry moves forward. Texas Tech University (TTU) has addressed this issue and is one of the leading universities in the nation in wind-related education. Other universities in the United States have long-standing wind programs, such as the University of Massachusetts at Amherst, University of California at Davis, the University of Colorado, and Colorado State University; however, these programs are typically housed within the engineering school, as opposed to being truly multidisciplinary.

TABLE 23.4 Wind Energy Career Jobs, Salaries, and Job Descriptions[a]

Project Development Occupations

Land-acquisition specialists are responsible for designing and implementing land-acquisition plans for new wind development sites. Land-acquisition specialists work closely with landowners, local governments, and community organizations to gain support for proposed wind projects. They also work with lawyers, permitting specialists, engineers, and scientists to determine whether sites are suitable for wind farm development and to lead the process of purchasing or leasing the land.

Asset managers are responsible for representing owner interests, especially by maximizing profits, in wind-farm projects. They ensure that the land is used in the most efficient way possible and oversee the project's finances, budget, and contractual requirements.

Logisticians are responsible for keeping transportation as efficient as possible. Because wind farm projects are expensive and run on tight schedules, any time spent waiting for delayed turbine components costs money. Logisticians have to work extensively with both the manufacturer and construction team to develop an optimized schedule for delivering turbine components.

There are no earnings data available for land-acquisition specialists and asset managers. However, similar occupations in commercial real estate and property management pay median salary $74,010.

Logisticians working in the management, scientific, and technical consulting services industry group, which includes many firms that work primarily in logistics, had median annual salary of $65,950, in May 2009. This wage is not specific to the wind energy industry.

Occupation	Median annual wages
Atmospheric and space scientists	$84,710
Zoologist and wildlife biologists	$56,500
Geoscientists, except hydrologists and geographers	$81,220
Environmental scientists and specialists, including health	$61,010

Construction

Occupation	Median annual wages
Construction laborers	$29,110
Operating engineers and other construction equipment operators	$39,530
Crane and tower operators	$47,170
Electricians	$49,800

[a] Source: Bureau of Labor Statistics [3].

The wind program at TTU begin in the College of Engineering and has evolved to be fully multidisciplinary, and is based on a 40-year experience in wind science and engineering—beginning with a 1970 tornado that severely affected the local community with millions of dollars in damage and a number of deaths.

The Wind Science and Engineering Ph.D. program at TTU was established in 2007 and is the first degree of its kind in the United States. The degree includes

TABLE 23.5 Wind Energy Program Enrollments and Timeline At Texas Tech University

First wind energy graduate class offered in 2005
Wind Science and Engineering Ph.D. approved in 2007
First undergraduate Wind Energy class offered in Spring 2009
Wind Energy Student Association established in Fall 2010
Bachelor of Science in Wind Energy approved in Fall 2011
2014 Enrollments
Over 140 Bachelor of Science in Wind Energy majors
Over 1000 course enrollments in Wind Energy per year
40% of enrollments are by Distance Education
20 Ph.D. Students Enrolled in Wind Science and Engineering

required courses in economics, atmospheric science, and engineering. In 2009, graduate certificates in wind energy were initiated at the university with both a technical and a managerial track offered. In 2011, a Bachelor of Science in Wind Energy was approved, in addition to an Undergraduate Minor in Wind Energy as well as an Undergraduate Certificate in Wind Energy. Professional development courses in wind energy are also available for Continuing Education Units. Table 23.5 shows the timeline of how these wind energy programs have been developed since 2005.

In summary, as the wind industry grows in the nation and the world, there are and will continue to be numerous job and career opportunities in the wind energy sector as well as the renewable energy sector in general. As pointed out in the beginning of this chapter, wind technology has a long history of generating employment for many people over several centuries and as a modern wind energy industry grows to be a substantial portion of the electric supply in the United States and other nations, job opportunities will continue. To be successful, this growing industry will require well-trained, energetic and innovative personnel to meet its potential as a major contributor to the energy mix of this nation and the world.

REFERENCES

[1] Global Wind Energy Council website. 2014. Global Cumulative Installed Wind Capacity 1996–2013. Available at http://www.gwec.net. Accessed December 29, 2013.

[2] American Wind Energy Association website. 2011. Windpower Outlook 2010. Available at http://www.awea.org. Accessed December 29, 2013.

[3] Hamilton J, Liming D. *Careers in Wind Energy*. Washington (DC): U.S. Bureau of Labor Statistics; 2011.

[4] American Wind Energy Association. AWEA U.S. wind industry annual market report 2013. American Wind Energy Association, Washington (DC); 2014.

[5] Wiser R, Bolinger M. *2012 Wind Technologies Market Report*. Berkeley (CA): U.S. Department of Energy, Lawrence Berkeley National Laboratory; 2013.

[6] Navigant Consulting, for RES Alliance for Jobs (2010). *Jobs Impact of a National Renewable Electricity Standard: Final Report*. Washington (DC): Navigant Consulting, Inc..

[7] World Resources Institute. Policy design for maximizing U.S. wind energy jobs, WRI Fact Sheet, Washington (DC); 2010, and Wei, Patadia, and Kammen (2009), University of Berkeley, Berkeley (CA).

[8] Potomac Economics, the ERCOT Independent Market Monitor. 2014. 2013 State of the Market Report for the ERCOT Wholesale Electricity Markets. Potomac Economics, Austin (TX).

[9] The Perryman Group. *Winds of Prosperity*. Waco (TX): Perryman Group; 2010.

[10] Electric Reliability Council of Texas website. 2008. Competitive Renewable Energy Zone (CREZ) Transmission Optimization Study. Available at http://www.ercot.com. Accessed November 4, 2014.

[11] U.S. Department of Energy, Energy Efficiency and Renewable Energy. *20% Wind Energy by 2030: Increasing Wind Energy's Contribution to U.S. Electricity Supply*. Washington (DC): U.S. Department of Energy, Energy Efficiency and Renewable Energy; 2008. DORE report DOE/GO-102008-2567.

[12] Walker R, Swift A, Mehta K, Seger K. Development of workforce for wind energy. Proceedings of ASME 2010 4th International Conference on Energy Sustainability; May 17–22, 2010 Phoenix; Arizona, USA; 2010. p ES2010–90348.

24

THE FUTURE OF ELECTRIC ENERGY

24.1 SUPPLY AND USE PROJECTIONS OF FOSSIL FUELS

Most people understand that given the dominance of fossil fuels for electricity production, these resources will be part of the fuel mix for years into the future. Given the concerns about greenhouse gas emissions, many observers feel that that fossil fuels will be replaced by nuclear and/or renewable sources, and some people are so concerned about climate change that they want to see a complete transition to clean energy sources within just a few years. Other observers realize that fossil fuels are still the dominant source of energy and believe that there are adequate supplies of coal and natural gas worldwide to meet much of our nation's energy demand for years into the future. This is reflected in the projection of energy sources used to generate electricity in the United States through the year 2040 found in the US-EIA's 2014 Annual Energy Outlook (Early Release version) [1], as shown in Figure 24.1. These projections indicated that the use of natural gas and renewables for electric generation will increase in the future, while the reliance on coal and nuclear energy will decrease. Many advocates for the increased use of renewable energy, believe that the use of these sustainable energy sources must increase at a much more rapid pace in order to mitigate the impacts of global climate change and other air emissions related to the use of fossil fuels.

Estimates indicate that the United States has 100–200 years of coal reserves based on current rates of consumption. If one does not believe that humans can cause global

Wind Energy Essentials: Societal, Economic, and Environmental Impacts, First Edition.
Richard P. Walker and Andrew Swift.

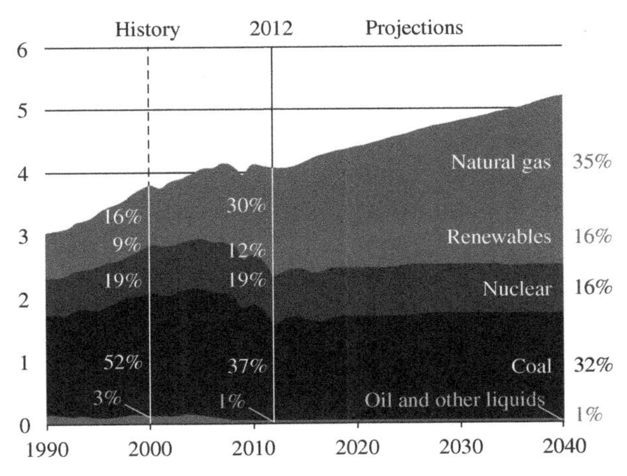

FIGURE 24.1 Electricity Generation by Fuel, 1990–2040, in trillion kilowatt-hours (Source: US-EIA 2014 Annual Energy Outlook (Early Release version) [1]).

warming, that person may see little reason for a rapid transition to sustainable energy. However, two factors will probably continue to lead to a continued push for a transition to clean energy: first, the climate impact of burning fossil fuels and adding carbon dioxide to the atmosphere may in fact be conclusively proven to be a significant problem in the future; and secondly, the growing population of the world and the rapidly growing demand for energy resources needed to provide a reasonable standard of living are a concern if based solely on fossil fuel resources.

24.1.1 Natural Gas

Natural gas, chemically known as methane, is considered a fossil fuel, but in fact it is a product of decomposition of organic material and is thus produced on a continuing basis. Its chemical composition is CH_4—showing that it is mostly a carrier of hydrogen—and thus a clean-burning hydrocarbon with low-carbon emissions. Compared with coal, natural gas power plants produce about one-half the amount of carbon dioxide per unit of electric energy produced. Over the past several decades, natural gas has been considered a premium fuel not for only power plants but also for delivery by pipeline to homes and industries. In the 1970s, US natural gas supply was limited, and there were government regulations imposed to ban its use as a boiler fuel for electricity production.

Recently, the supply situation for natural gas has been greatly increased by improved resource characterization and drilling technologies, such as directional drilling and hydraulic fracturing of natural gas–bearing shale resources. The increase in supply has been so dramatic that some media outlets have declared the US energy supply problem is solved. Most analysts believe this to be an overstatement. However, because of these new technological advances, there is general agreement that natural gas supplies and reserves have increased dramatically both in the United States and worldwide.

As has been pointed out numerous times in this book, renewable resources like wind and solar are intermittent resources and electric power requires reliable sources of generation to meet demand. Natural gas–fueled power stations typically have rapid ramp rates, meaning they can be turned on and reach maximum capacity levels rather quickly and can also quickly ramp-down to adjust for lower demand for electricity or increases in wind energy production. This makes natural gas generation a natural complement to solar and wind electric systems. Some energy analysts see this as a transition situation with natural gas serving as a transitional fuel to an all-renewable future. They may be correct. However, a more pragmatic approach may consider natural gas generation a long-time partner in the transition to a renewable energy future. Energy transitions of the past (wood to coal to oil and so forth) have typically taken place over centuries, and given the complexities of national energy policies and the inertia of energy infrastructure development, there is no reason to believe a transition to a mix of resources that is more dominated by renewable resources will be more rapid.

By taking this "partnering" approach with natural gas power systems, the adoption of renewable sources of power generation may be accelerated. The intermittency issue of wind and solar energy is addressed, there will be minimal carbon emissions, and supplies of natural gas will be extended at lower prices, as renewable sources provide the bulk of the energy resources. One example of this approach is already operational in West Texas, an area of substantial wind energy development. The Antelope power station in Abernathy, Texas, is a natural gas–fueled station using rapid-start internal combustion engines so that the facility can deliver power in unison with the output of local wind energy plants. The facility is owned by Golden Spread Electric Cooperative.

24.1.2 Six Attributes of a Future Vision for Electric Supply

Characteristics or attributes of an electric system that would lead to increased use of clean energy sources and more efficient utilization of our fossil fuel supplies might include the following:

- An electric utility supply that
  - is based on a diverse set of electric power technologies
  - maximizes renewable resources
  - uses natural gas–fueled generation to insure adequate and timely electric supply
  - evolves over time with technological innovation
- Wind energy, both onshore and offshore, would provide significant bulk electrical power requirements.
- Solar energy would provide on-peak energy, matching the peak load requirements of most utilities.
- Modern transmission infrastructure would be put in place to move resources in an optimal manner, improving reliability at lower costs.

- Smart Grid technologies would help to integrate these resources with customer demands, and improvements in energy efficiency would lower costs.
- Overall costs would remain reasonable as wind and solar energy costs continue to decline, and these fixed-price resources stabilize electric rates that presently rely on variable fuel costs, ever-changing environmental controls, and overall regulatory uncertainty.

24.2 EIGHTY PERCENT RENEWABLES BY 2050

In 2008, researchers at the Department of Energy (DOE) and the National Renewable Energy Laboratory (NREL), supported by many other energy and environmental experts, produced a report titled "20% Wind Energy by 2030: Increasing Wind Energy's Contribution to U.S. Electricity Supply" [2]. Recently, DOE and NREL have completed an analysis of scenarios for an 80% renewable electric energy mix by the year 2050 in the United States. According to NREL's website, the report titled "The Renewable Electricity Futures Study" [3] "is an initial investigation of the extent to which renewable energy supply can meet the electricity demands of the continental United States over the next several decades." NREL further states that "this study explores the implications and challenges of very high renewable electricity generation levels—from 30% up to 90%, focusing on 80%, of all US electricity generation from renewable technologies—in 2050" and that "at such high levels of renewable electricity generation, the unique characteristics of some renewable resources, specifically geographical distribution and variability and uncertainty in output, pose challenges to the operability of the nation's electric system."

Key findings of *The Renewable Electricity Futures Study* include the following:

- Renewable electricity generation from technologies that are commercially available today, in combination with a more flexible electric system, is more than adequate to supply 80% of total US electricity generation in 2050 while meeting electricity demand on an hourly basis in every region of the country.
- Increased electric system flexibility, needed to enable electricity supply–demand balance with high levels of renewable generation, can come from a portfolio of supply- and demand-side options, including flexible conventional generation, grid storage, new transmission, more responsive loads, and changes in power system operations.
- The abundance and diversity of US renewable energy resources can support multiple combinations of renewable technologies that result in deep reductions in electric sector greenhouse gas emissions and water use.
- The direct incremental cost associated with high renewable generation is comparable with published cost estimates of other clean energy scenarios. Improvement in the cost and performance of renewable technologies is the most impactful lever for reducing this incremental cost.

As pointed out in this book, the availability of wind and solar resources alone is more than sufficient to provide all of the electric needs for the country; however, issues of cost, reliability, grid integration, grid stability, transmission, and so forth need to be resolved. In discussions with some of the contributors to the 80% renewables by 2050 scenario outlined in *The Renewable Electricity Futures Study*, it is interesting to note that wind will contribute between one-fourth and one-third of the renewable resources in the 80% scenario as it is currently envisioned. The idea of generating 80% of the electric energy in the country from renewable sources by 2050 is ambitious but provides a useful tool for scenario planning, as did the *20% Wind Energy by 2030* report. As noted in Chapter 3 of this book, currently only about 10% of US electric energy needs are met with sources of renewable energy, so an 80% renewable scenario will require many significant policy changes.

24.3 WIND ENERGY RESEARCH AND DEVELOPMENT

The major goal of research and development for wind energy is to improve the ability of wind to be an integrated part of the generation mix for the electric utility system in the United States. The goal of that research and development is to lower costs and to improve performance and the reliability of wind energy systems as they are integrated with the electric utility grid. This means investing in new and innovative technological solutions, lowering operating and maintenance costs, improving performance, and decreasing the resulting cost of energy from these systems. Current research topics include wake studies to improve performance of downwind turbines operating in the wake of upwind turbines, better understanding of the interaction between the atmospheric boundary layer and the rotor of wind turbines with ever-increasing size, improved gearbox and generator technologies, improved power electronic systems, improved blade performance and structural efficiency, and integrating other renewable resources such as solar energy. For example, recently published research using hourly data shows that capacity factors of combined wind/solar plants could approach 60% as compared with the 35–45% capacity factor of wind-only or 20–25% capacity factor of solar-only renewable electric power systems [4]. The overarching objective is to make wind more reliable on the utility system while decreasing its cost.

24.4 SMART GRID/INFRASTRUCTURE UPDATES

As mentioned in Chapter 15 while discussing energy efficiency, there is research and development underway to integrate new meters and computer resources on the electric grid system allowing consumers to better monitor and manage their electricity use. This concept is called the Smart Grid and could result in significant improvements in reliability and the efficient use of electricity, as well as significant peak-shaving opportunities for the utilities. Imagine, for example, a customer being able to see in real time the cost of electricity being consumed in one's home and having the ability to program the shutting down of appliances to reduce monthly electric bills while optimizing use of services and minimizing energy and capacity requirements from the utility grid—all controlled through the Internet or with the use of a smartphone or similar device locally or from a remote source. The Pecan Street Demonstration project in Austin Texas is an innovative demonstration of smart grid technology. One can go to http://www.pecanstreet.org/ for more information on this topic.

The Smart Grid represents the intersection of electric energy, information technology, and telecommunications. The concept, illustrated in Figure 24.2, includes modernization of the electric delivery system so that it monitors, protects, and automatically optimizes the operation of its interconnected elements—from the central and distributed generator through the high-voltage network and distribution system; to industrial users and building automation systems; to energy storage installations; and to end-use consumers and their thermostats, electric vehicles, appliances, and other household devices. A key element in this concept is the Smart Meter, a digital meter that records individual real-time customer electric energy consumption. These meters allow for an automated transfer of information between the home and the electric energy provider and will help revolutionize the electric grid, by allowing two-way communication between the utility and the consumer.

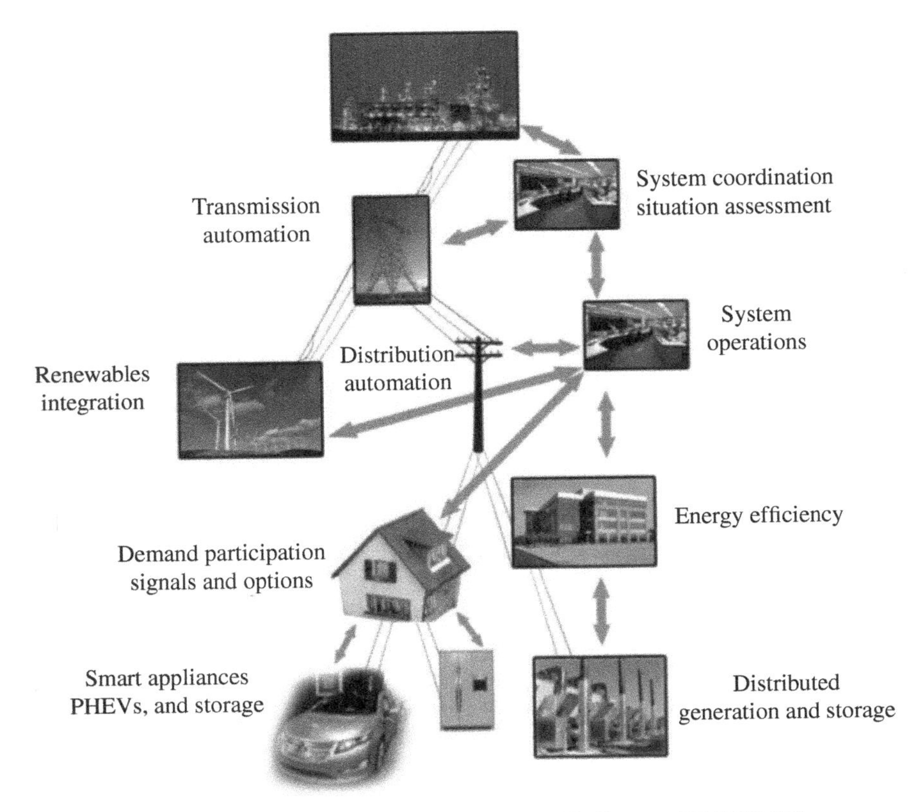

FIGURE 24.2 Schematic of the Smart Grid® (Source: US-DOE [5]).

24.5 THE CONSUMER OF THE FUTURE/CONSUMER APPLICATIONS

The electric consumer of the future will demand smart devices that can be interactive and controlled in real time, both within the home or remotely through the Internet or other communication devices. These advances and sophisticated applications provide new opportunities for technological innovation, new companies, new jobs, and new opportunities to improve the delivery of electric services while reducing costs.

A summary of the potential impacts of Smart Grid and Smart Meter technology on the consumer are outlined in the following bullet list:

- Smart Meters will enable customers to make choices about when to use or not use electric energy based on real-time cost.
- Smart Meters can help cut the cost of electric bills by showing the current rate of energy usage. During peak times, the price of electricity is generally higher, and during nonpeak times, the prices are typically lower. By knowing the peak rates, customers can curtail energy consumption and reduce their electric bills.

24.6 ENERGY STORAGE

Energy storage is considered by many to be "the Holy Grail" of renewable energy, meaning that if one could provide reasonably cost-effective energy storage solutions, wind and solar energy could be stored at times of overproduction for use at times of underproduction as needed by the consumer—finally solving the intermittency issues of renewable resources. To date, these systems are typically too expensive; however, there are already a number of possible storage systems under development, as discussed in the following sections.

24.6.1 Vehicle to Grid Solutions

The advent of the electric vehicle or hybrid-electric vehicle provides an opportunity for electric storage in a vehicle such that if that vehicle could be connected to an electric outlet at the owner's residence, the battery would become part of the grid system and could therefore store energy. These systems have been called *Vehicle to Grid* (V2G) energy-storage systems. Computer technology and power electronic controllers could provide the integration of these technologies and could even allow the customer to use the battery pack in the vehicle to supply backup power in the event of grid power outage in an emergency. This could revolutionize emergency power systems. Additionally, inexpensive off-peak power could be used to charge the battery providing inexpensive power for transportation, while periods of high electric demand could engage the batteries providing opportunities for peak-shaving.

24.6.2 Battery Technology

Battery technology continues to improve. Technologies such as the lead acid battery are being integrated with small home-based wind and solar systems, while larger liquid and lithium-ion batteries are being experimentally connected to the utility grid to smooth power fluctuations and to deliver conditioned power on demand.

24.6.3 Compressed Air Energy Storage

The concept of compressing air using excess energy from wind, or cheap off-peak power, and storing it in underground salt caverns or specially made tanks, has been investigated for several decades. The stored air is then expanded converting the compressed air to mechanical and electrical power. These systems are large-scale and have been tried in several places on an experimental basis, but typically the siting requirements and the initial cost have kept these from having more widespread applications. The first such plant was constructed in Germany in 1978, while the only U.S. Compressed Air Energy Storage plant (as of the writing of this book) is a 110-MW project located in McIntosh, Alabama, although several projects are currently being evaluated.

24.6.4 Hydrogen

Many researchers have considered using excess electrical energy to power an electrolysis process to produce gaseous hydrogen, which could then either be burned or combined in a fuel cell to produce electricity on demand. These hydrogen systems have had significant research attention but costs have typically been too high and there is a significant loss of energy in the conversion process from electricity to hydrogen and back to electricity, which together have limited large-scale implementation.

24.7 CLEAN ENERGY TRANSPORTATION: THE ROLE FOR WIND ENERGY AND OTHER RENEWABLES IN REDUCING NATIONAL DEPENDENCE ON PETROLEUM IMPORTS

Presently, the transportation system of the world relies heavily on petroleum-based products, mostly oil converted to gasoline. As we look toward a clean energy future, there is significant research going on in a number of areas related to clean energy transportation. Since winds in the Great Plains of the United States are typically strongest in the nighttime hours, using wind-generated electricity to charge electric vehicles at night has been suggested by many futurists.

24.7.1 Electric and Hybrid-Electric Vehicles

Electric and hybrid-electric vehicles are being produced. General Motors has recently produced the Chevy Volt, a hybrid vehicle but mostly electric and capable of being plugged in. For many years, Toyota has produced the Prius, a hybrid-electric vehicle that uses a gasoline engine to charge batteries when the charge is low and to extend the vehicle range. One of the advantages of electric vehicles, besides limiting gasoline consumption, is that they can recover the energy of acceleration during stopping by reversing the action of the electric motors attached to the wheels making them short-term generators and recovering some of the energy. This is called "regenerative braking" and adds to the efficiency of these vehicles. Electric and hybrid-electric vehicles can also significantly reduce the amount of air emissions resulting from transportation in urban areas plagued by smog. Some cities, such as Los Angeles, allow the use of electric or hybrid-electric vehicles occupied by a single rider in High Occupancy Vehicle (HOV) lanes normally reserved only for vehicles with two or more riders. To date, the purchase price of such vehicles is still high but there is considerable continuing research and development in this area, and the operating cost of such vehicles are typically lower than that of gasoline-fueled vehicles.

24.7.2 Hydrogen

Hydrogen production was discussed earlier; however, a number of researchers have investigated using hydrogen directly as a gaseous fuel to drive internal combustion engines or as a fuel to drive a fuel cell for an all-electric hydrogen-powered vehicle. As with the other technologies discussed, although technically feasible, costs have not allowed the hydrogen car to go into mass production. One of the advantages of burning hydrogen, either in a fuel cell or with oxygen, is that most of the combustion products are water. Direct combustion leads to some oxides of nitrogen, but the carbon-based emissions are avoided entirely.

24.7.3 Biofuels

Ethanol plants are abundant throughout the country and provide between 10 and 20% of the fuel mix in certain areas depending on the time of the year. There is much discussion on the costs and subsidies for ethanol, which can be quite high. Additionally, the "fuel-for-food" (substitution of food production from corn for ethanol production for fuel) discussion about biofuels continues to be controversial. However, other potential sources of biofuel, such as algae, hold significant promise. One of the most interesting concepts is to channel emissions from industrial processes through algae ponds, which can then use CO_2 emissions as "food" for the algae rather than adding more CO_2 directly to the atmosphere.

24.8 ENERGY AND CLEAN WATER: USING WIND TO DESALINATE AND PURIFY WATER

Every community needs the availability of energy and clean water. In many of the arid parts of the United States, water resources are available but are often contaminated with minerals and salts making them unusable. Reverse osmosis technologies to desalinate the water are available; however, there is a high electricity cost associated with the reverse osmosis process. Projecting electric energy costs into the future is often difficult due to the variability of fuel costs and has hampered development of reverse osmosis water purification in communities that have saline water sources available but lack freshwater supplies. Wind and other renewable sources may provide part of a solution in certain areas by using wind power to drive reverse osmosis systems in the evening and store the water. This would decouple the time of use of wind from its generation to its consumption. The produced and purified water could be stored in large tanks for use as needed. Several West Texas communities have investigated this concept, to include the town of Seminole, Texas, and researchers at Texas Tech University have published several papers on this topic [6, 7].

24.9 ELECTRIFICATION IN THIRD-WORLD COUNTRIES

As pointed out earlier, up to two billion people, or about one-third of the Earth's population, do not have adequate electrical supplies. The development of renewable resources could provide a partial solution to many of the Third-World countries lacking adequate electrical supplies if costs can be reduced for these communities. One advantage of a renewable solution delivered at the point of use is that the complex electric grid that was part of the expansion of electric power in the United States and other industrialized countries would not be necessary to deliver basic electric service to many of these remote areas. The combination of a wind generator, photovoltaic panel, and a battery can provide significant improvement to the quality of life for these communities in remote locations—providing lighting, communication, and refrigeration services without grid-delivered electric power.

It is unclear at present how these technologies will impact the future of renewable energy and electric power delivery, but with growing world populations and an ever-growing need for power and energy resources, renewable energy will certainly be a substantial part of the future electric energy mix.

REFERENCES

[1] U.S. Energy Information Administration. AEO2014 Early Release Overview. Washington (DC): U.S. Government Printing Office; 2013.

[2] U.S. Department of Energy, Energy Efficiency and Renewable Energy. 20% Wind Energy by 2030: Increasing Wind Energy's Contribution to U.S. Electricity Supply, DOE/GO-102008-2567. Washington (DC): U.S. Government Printing Office; 2008.

[3] National Renewable Energy Laboratory.*Renewable Electricity Futures Study*. In: Hand M, Baldwin S, DeMeo E, Reilly JM, Mai T, Arent D, Porro G, Meshek M, Sandor D, editors, *Renewable Electricity Generation and Storage Technologies*. 4 vols, NREL report NREL/TP-6A20-52409. Golden (CO): U.S. Government Printing Office; 2012.

[4] Pattison C. Firming Wind Energy with Solar Photovoltaics, Wind Science and Engineering [PhD dissertation]. Lubbock: Texas Tech University; 2011.

[5] U.S. Department of Energy. Smart Grid System Report. Washington (DC): U.S. Government Printing Office; 2009.

[6] Swift A, editor. *Wind and Water, Redeveloping Regional Resources for the New Economy.* 2nd ed. Lubbock: Texas Tech University, Wind Science and Engineering Research Center; 2005.

[7] Swift A, Rainwater K, Chapman J., et al. Wind Power and Water Desalination Technology Integration, DWPR Report No. 146. Denver: U.S. Bureau of Reclamation, Technical Service Center; 2009.

APPENDIX A

WIND ENERGY REFERENCE TABLES FOR UNITS, CONVERSIONS, SYMBOLS, AND ENERGY EQUIVALENTS

A.1 SYSTEMS OF UNITS

A.1.1 The US Customary System of Units

The United States typically uses the US Customary System of Units (sometimes called the British system) in which the base units are foot (ft) for length, pound (lb) for force, and second (s) for time.

A.1.2 The International System of Units

In 1960, the *Eleventh General Conference on Weights and Measures* formally adopted the International System of Units, for which the abbreviation is SI in all languages as the international standard. *The Wind Energy Industry typically uses SI units.*

A.2 BASE UNITS AND THEIR SYMBOLS

Quantity	Name of unit	Symbol
Length	Meter	m
Mass	Kilogram	kg
Time	Second	s
Electric current	Ampere	A

(*Continued*)

Wind Energy Essentials: Societal, Economic, and Environmental Impacts, First Edition.
Richard P. Walker and Andrew Swift.

(*Continued*)

Quantity	Name of unit	Symbol
Thermodynamic temperature	Kelvin	K
Amount of substance	Mole	mol
Luminous intensity	Candela	cd

A.3 DERIVED UNITS

Quantity	Derived SI unit	Symbol	Special name
Area	Square meter	m^2	—
Volume	Cubic meter	m^3	—
Linear velocity	Meter per second	m/s	—
Angular velocity also rotational velocity	Radian per second	rad/s	
Linear acceleration	Meter per second squared	m/s^2	—
Angular acceleration	Radian per second squared	rad/s^2	
Frequency	(Cycle) per second	Hz	Hertz
Density	Kilogram per cubic meter	kg/m^3	—
Force	Kilogram·meter per second squared	N	Newton
Moment of force also torque	Newton·meter	N·m	—
Pressure	Newton per meter squared	Pa	Pascal
Work	Newton·meter	J	Joule
Energy	Newton·meter	J	Joule
Power	Joule per second	W	Watt

A.4 DIMENSIONS OF PHYSICAL QUANTITIES

		Common units	
Physical quantity	Dimension	SI system	US customary system
Length	L	m, mm	in., ft
Area	L^2	m^2, mm^2	in.^{2}, ft^2
Volume	L^3	m^3, mm^3	in.^{3}, ft^3
Angle	1 (L/L)	rad	rad
Time	T	s	s
Linear velocity	L/T	m/s	ft/s
Linear acceleration	$1/T^2$	m/s^2	ft/s^2

A.5 DIMENSIONS OF PHYSICAL QUANTITIES (CONTINUED)

		Common units	
Physical quantity	Dimension	SI system	US customary system
Angular velocity also rotational velocity	1/T	rad/s	rad/s
Angular acceleration	$1/T^2$	rad/s^2	rad/s^2
Mass	M	kg	slug
Force	ML/T^2	N	lb
Moment of force	ML^2/T^2	N·m	ft·lb
Pressure	M/LT^2	Pa, kPa	psi, ksi
Stress	M/LT^2	Pa, MPa	psa, ksi
Energy	ML^2/T^2	J	ft·lb
Work	ML^2/T^2	J	ft·lb
Power	ML^2/T^3	W	hp
Linear impulse	ML/T	N·s	lb·s
Momentum	ML/T	N·s	lb·s
Specific weight	M/L^2T^2	N/m^3	lb/ft^3
Density	M/L^3	kg/m^3	$slug/ft^3$
Second moment of area	L^4	m^4, mm^4	In.^{4}, ft^4
Moment of inertia	ML^2	$kg{\cdot}m^2$	$slug{\cdot}ft^2$

Gravitational constant "*g*" = acceleration of gravity at 45° latitude; $g = 32.2\ ft/s^2 = 9.81\ m/s^2$.

1. Angular measure: "Radian"
   a. A unit of angular measure derived by determining the circular angle subtended by an arc of length equal to the radius of the circle.
   b. By convention; π radians = [circumference/diameter] = [circumference/2·radius].
   c. "Degrees" are also used for angular measure, but degrees are NOT dimensionless, thus radians are preferred in calculations.
   d. A circle has $360° = 2\pi$ rad.
2. Also used are RPM (revolutions per minute) and degree per second for rotational velocity.
   a. Both should be converted to radians/second for calculations.
3. 1 kg = 2.2 lbm [lbm is the symbol for "pound mass"—on Earth; 1 lbm weighs 1 pound (lb)].
4. 1 slug = 32.2 lbm [slug is the unit of mass in English system].
5. Other English abbreviations are assumed known or available in an English dictionary.

A.6 GREEK LETTERS

Greek symbols are often used in technical writing. The list provided is a reference guide.

Greek letter	Greek name
Α, α	Alpha
Β, β	Beta
Γ, γ	Gamma
Δ, δ	Delta
Ε, ε	Epsilon
Ζ, ζ	Zeta
Η, η	Eta
Θ, θ	Theta
Ι, ι	Iota
Κ, κ	Kappa
Λ, λ	Lambda
Μ, μ	Mu
Ν, ν	Nu
Ξ, ξ	Xi
Ο, ο	Omicron
Π, π	Pi
Ρ, ρ	Rho
Σ, σ	Sigma
Τ, τ	Tau
Υ, υ	Upsilon
Φ, φ	Phi
Χ, χ	Chi
Ψ, ψ	Psi
Ω, ω	Omega

A.7 SELECTED RULES FOR WRITING SI QUANTITIES

1. **All physical quantities and measurements consist of two parts: "A Number" and "Unit."**
   a. For example, 5 feet; 5 inches; 5 miles.
   b. Use prefixes (kilo..., mega..., etc., to keep numerical values generally between 0.1 and 1000).
   c. Use of prefixes *hector-*, *deca-*, *deci-*, and *centi-* should generally be avoided except for certain areas or volumes where the numbers would be otherwise awkward.
2. **Unit designations:**
   a. Avoid ambiguous double solidus (e.g., write N/m^2 not N/m/m).
   b. Exponents refer to entire unit (e.g., mm^2 means $(mm)^2$).

A.8 MULTIPLE UNITS

			Prefix	
Numerical	Factor by which unit is multiplied	Common usage	Name	Symbol
—	10^{18}		exa	E
—	10^{15}	Quadrillion	peta	P
—	10^{12}	Trillion	tera	T
Etc.	10^{9}	Billion	giga	G
1,000,000	10^{6}	Million	mega	M
1,000	10^{3}	Thousand	kilo	k
100	10^{2}	Hundred	hecto[a]	—
10	10	Ten	deca[a]	—
0.1	10^{-1}	—	deci[a]	—
0.01	10^{-2}	—	centi[a]	—
0.001	10^{-3}	—	milli	m
0.0000001	10^{-6}	—	micro	μ
Etc.	10^{-9}	—	nano	n
—	10^{-12}	—	pico	p
—	10^{-15}	—	femto	f
—	10^{-18}	—	atto	a

[a]Generally, avoid these terms when possible. See aforementioned rule for more information.

A.9 CONVERSION FACTORS BETWEEN THE SI AND US CUSTOMARY SYSTEMS

Quantity	US Customary to SI	SI to US Customary
Length	1 in = 25.40 mm	1 m = 39.37 in.
	1 ft = 0.3048 m	1 m = 3.281 ft
	1 mi = 1.609 km	1 km = 0.6214 mi
Area	1 in.^{2} = 645.2 mm^2	1 m^2 = 1550 in.^{2}
	1 ft^2 = 0.0929 m^2	1 m^2 = 10.76 ft^2
Volume	1 in.^{3} = 16.39(10^3)mm^3	1 mm^3 = 61.02(10^{-6})in.^{3}
	1 ft^3 = 0.02832 m^3	1 m^3 = 35.31 ft^3
	1 gal = 3.785 L (liter)	1 L = 0.2642 gal
Velocity	1 in./s = 0.0254 m/s	1 m/s = 39.37 in./s
	1 ft/s = 0.3048 m/s	1 m/s = 3.281 ft/s
	1 mi/h = 1.609 km/h	1 km/h = 0.6214 mi/h
Acceleration	1 in./s^2 = 0.0254 m/s^2	1 m/s^2 = 39.37 in./s^2

(*Continued*)

(*Continued*)

Quantity	US Customary to SI	SI to US Customary
	1 ft/s² = 0.3048 m/s²	1 m/s² = 3.281 ft/s²
Mass	1 slug = 14.59 kg	1 kg = 0.06854 slug
Force	1 lb = 4.448 N	1 N = 0.2248 lb
Distributed load	1 lb/ft = 14.59 N/m	1 kN/m = 68.54 lb/ft
Pressure or stress	1 psi = 6.895 kPa	1 kPa = 0.1450 psi
	1 ksi = 6.895 MPa	1 MPa = 145.0 psi
Bending moment or torque	1 ft·lb = 1.356 N·m	1 N·m = 0.7376 ft·lb
Work or energy	1 ft·lb = 1.356 J	1 J = 0.7376 ft·lb
Power	1 ft·lb/s = 1.356 W	1 W = 0.7376 ft·lb/s
	1 hp = 745.7 W	1 kW = 1.341 hp

A.10 WORK AND ENERGY DEFINITIONS

- Work and energy are equivalent.
- Work = Force × Distance
- Power is the rate at which work is done or energy is produced or consumed; Power = Work/Time.
- Efficiency = $E_{out}/E_{in} = P_{out}/P_{in}$
- **Multiply** *efficiency ratios to calculate compound efficiencies.*

A.11 ENERGY CONVERSIONS

- 1 BTU = 1.05 kJ = heat to raise 1 lb of water, 1°F (British Thermal Unit)
- A kitchen match ~1 BTU[1]
- 1 kWh = 3412 BTU = 3600 kJ
- 1 J = 0.24 cal; 1 cal = heat required to raise 1 g of water, 1°C
- 1 SCF. NTL. GAS (standard cubic foot of natural gas) = 1000 BTU[1]
- 1 metric ton coal = 2200 lbs = 28 million BTU[1]
- 1 gallon gasoline = 1.25×10^5 BTU[1]
- 1 barrel of oil = 42 gal = 5.8×10^6 BTU[1]
- 1 quad = 10^{15} BTU
- 1 "sun" = 1000 W/m² (average)

[1] Approximate value of heat released when burned in air.

A.12 OTHER ENERGY UNITS

	English	SI
Heat	BTU	Joule
Mechanical energy	ft·lb	N·m = joule, J
Mechanical power	Horsepower	joule/sec., J/s; or watt, W
Electric energy	—	Kilowatt·hr, kWh
Electric power	—	1 watt = 1 volt·1 amp

BIBLIOGRAPHY

1. Penner SS, Icerman L. Energy: Non-Nuclear Technologies. Volume 2, New York: Addison-Wesley Publishing Company; 1975.
2. Riley WF, Sturches LD. Engineering Mechanics: Statistics. New York: John Wiley & Sons; 1993.
3. Mechtly EA. The International System of Units: Physical Constants and Conversion Factors (NASA SP-7012). National Aeronautics and Space Administration; 1964.

APPENDIX B

LIST OF ACRONYMS

AC or ac	alternating current
ACHP	Advisory Council on Historic Preservation
AEP	annual energy production
AES	Alternative Energy Standard
APLIC	Avian Power Line Interaction Committee
ARPA	Archaeological Resources Protection Act
ARRA	American Recovery and Reinvestment Act of 2009
AWEA	American Wind Energy Association
AWWI	American Wind Wildlife Institute
BGEPA	Bald and Golden Eagle Protection Act
BLM	U.S. Bureau of Land Management
Btu or BTU	British thermal unit
BWEC	Bats and Wind Energy Cooperative
CAES	compressed air energy storage
CanWEA	Canadian Wind Energy Association
CCCT	combined cycle combustion turbine
CCN	certificate of convenience and necessity
CCS	carbon capture and storage
CE	Civil Engineering
CFLs	compact florescent lights
CHP	combined heat and power
CO_2	carbon dioxide
COE	U.S. Army Corps of Engineers

Wind Energy Essentials: Societal, Economic, and Environmental Impacts, First Edition.
Richard P. Walker and Andrew Swift.

CREZ	Competitive Renewable Energy Zone
CSAPR	Cross-State Air Pollution Rule
CT	combustion turbine
CWA	Clean Water Act
CZMA	Coastal Zone Management Act
dB	decibel
DC or dc	direct current
DHS	Department of Homeland Security
DoD	Department of Defense
DOE or USDOE	United States Department of Energy
DSM	demand-side management
EE	Electrical Engineering
EIA or USEIA	United States Energy Information Administration
EIS	environmental impact statement
EMF	electric and magnetic field (or electromagnetic field)
EPA or USEPA	United States Environmental Protection Agency
EPC	engineering procurement and construction
ERCOT	Electric Reliability Council of Texas
ERDA	Energy Research and Development Administration
ESA	Endangered Species Act
EWEA	European Wind Energy Association
FAA	Federal Aviation Administration
FGD	flue-gas desulfurization
FIT	feed-in tariff
GDP	gross domestic product
GE	General Electric
GEOL	geology
GHGs	greenhouse gases
GW	gigawatt or 1,000 MW
GWh	gigawatt-hour or 1,000 MWh
GWP	global warming potential
HVDC	high-voltage direct current
IA	interconnection agreement (or transmission interconnection agreement)
IC	internal combustion
IE	Industrial Engineering
IEC	International Electrotechnical Commission
IEEE	Institute of Electrical and Electronics Engineers
IGCC	integrated gasification combined cycle
IPP	independent power producer
ISO	independent system operator
ITC	investment tax credit
kW	kilowatt
kWh	kilowatt-hour
LED	light-emitting diode
LMP	locational marginal price
LVRT	low-voltage ride through
MACRS	Modified Accelerated Cost-Recovery System
MBTA	Migratory Bird Treaty Act

MCPE	market clearing price of energy
MMBtu	one million British thermal units
MPH or mph	miles per hour
MPS or mps	meters per second
MRO	Midwest Reliability Organization
MSW	municipal solid waste
MW	megawatt or 1000 kW
MWh	megawatt-hour or 1000 kWh
NAAQS	National Ambient Air Quality Standards
NCF	net capacity factor
NEPA	National Environmental Policy Act
NERC	North American Electric Reliability Corporation
NESC	National Electric Safety Code
NETL	National Energy Technology Laboratory
NEXRAD	next generation radar (weather radar)
NHPA	National Historic Preservation Act
NOAA	National Oceanic and Atmospheric Administration
NOx	nitrous oxide
NPDES	National Pollutant Discharge Elimination System
NPV	net present value
NREL	National Renewable Energy Laboratory
NUG	nonutility generator
NWCC	National Wind Coordinating Collaborative
OCC	Oklahoma Corporation Commission
O&M	operations and maintenance
OPEC	Organization of the Petroleum Exporting Countries
OTEC	ocean thermal energy conversion
PLJV	Playa Lake Joint Venture
PPA	power purchase agreement
PTC	production tax credit
PUC	Public Utility Commission
PURPA	Public Utility Regulatory Policies Act
PV	photovoltaic
QOL	quality of life
RCRA	Resource Conservation and Recovery Act
REA	Rural Electrification Administration
REC	renewable energy credit
REPI	Renewable Energy Production Incentive
RES	renewable electricity standard
RFP	request for pricing or request for proposal
RPM	revolutions per minute
RPS	renewable portfolio standard
RTO	Regional Transmission Organization
SCADA	supervisory control and data acquisition
SCCT	simple cycle combustion turbine
SGSP	salinity-gradient solar pond
SHPO	State Historic Preservation Office
SO_2	sulfur dioxide
SPP	Southwest Power Pool

SWPPP	Stormwater Pollution Prevention Plan
TRE	Texas Reliability Entity
TSA	turbine supply agreement
TSP	transmission service provider
TSR	tip speed ratio
USFS	U.S. Forest Service
USFWS	U.S. Fish and Wildlife Service
USGS	United States Geological Service
UWIG	Utility Wind Integration Group (now, UVIG)
VOC	volatile organic compound
WECC	Western Electricity Coordinating Council
WTE	waste-to-energy

APPENDIX C

GLOSSARY

Accommodation doctrine: A legal principle that generally states that the owner of mineral rights can use the portions of the surface reasonably necessary for the exploration and production of minerals as long as it is done in an nonnegligent manner complying with statutory limitations and conducts these activities with due regard to the owner of the surface rights and his or her activities.

Ad valorem tax: A tax based on the assessed value of real estate or personal property. Ad valorem taxes can be property tax or even duty on imported items. Property ad valorem taxes are the major source of revenue for state and municipal governments.

Alternating current: Electricity that changes direction periodically; the period is measured in cycles per second, or Hertz (Hz).

Ambient noise level: The composite of noise from all sources near and far; the normal or existing level of environmental noise at a given location.

Ampacity: The current-carrying capacity of conductors or equipment, expressed in amperes.

Ampere or Amp: The unit of electric current flow that indicates how much electricity flows through a conductor.

Amperage: A unit of electrical current, equal to Coulombs per second; the flow rate of electrons moving through a circuit.

Wind Energy Essentials: Societal, Economic, and Environmental Impacts, First Edition.
Richard P. Walker and Andrew Swift.

Anemometer: An instrument used to measure the speed of the wind; the most common type used in the wind energy industry is the three-cup anemometer, a series of cups mounted at the end of arms that rotate in the wind; other types of anemometers include the pressure-tube anemometer, which uses the pressure generated by the wind to measure its speed, and the hot-wire anemometer, which uses the rate at which heat from a hot wire is transferred to the surrounding air to measure wind speed.

Angle of attack: The angle of relative air flow to the wind turbine blade chord.

Audible noise: Generally considered to have frequencies between 20 and 20,000 Hz, although some show the range extending down to 16 Hz.

Avoided cost of energy: A value of electricity produced approximately equivalent to the value of fuel not burned as a result of purchasing renewable energy.

Balancing authority: The responsible entity that integrates resource plans ahead of time, maintains load-interchange-generation balance within a balancing authority area and supports interconnection frequency in real time.

Barotrauma: Trauma caused by rapid or extreme changes in air pressure.

Bathymetry: The topographical map of the seafloor from water depth measurements.

Behind-the-meter: Locating wind or solar facilities at the site of an electric load, displacing purchased electric power, and metering any electricity returned to the grid.

Betz limit or Betz coefficient: The theoretical maximum efficiency at which a three-bladed horizontal-axis wind generator can operate.

Biomass: Plant materials and animal waste used especially as a source of fuel; can include fuel used to generate electricity, fuel used to produce heat, or fuel used for transportation vehicles.

Biomass cofiring: The mixing of biomass or biomass-derived fuel with traditional fossil fuel as a fuel source for a traditional power plant; this often involves mixing agricultural or forestry waste with coal for use in coal-fired power plants.

Book depreciation: Asset depreciation based on generally accepted accounting principles; is a function of the assumed life of the asset.

British thermal unit (or Btu): The amount of heat energy needed to raise the temperature of 1 pound of water by 1 degree F.

Capacity factor: The ratio of the actual quantity of energy produced over a specific time interval to the maximum possible energy produced by a power plant over the same time interval.

Carbon neutral: The state at which the net amount of carbon dioxide emitted into the atmosphere by a building, business, or technology is reduced to zero because it is balanced by actions to reduce or offset such emissions.

Chord: The width of a wind turbine blade at a given location along the length.

Coriolis effect: The effect of the rotation of the Earth on wind direction caused by deflection of air currents; this effect causes clockwise rotation around a high-pressure area and counterclockwise rotation around a low-pressure area.

Current: The movement or flow of electrons through a conductor, as measured in amperes; represented by the symbol I. An ampere is a specific number of electrons that moves past a given point in one second.

Cut-in speed: The minimum wind speed required for a wind turbine to begin generating electricity.

Cut-out speed: That wind speed at which a wind turbine controller shuts the turbine down to prevent damage to the turbine in high wind speed conditions.

Cycling: Changing the output of a power plant either up or down over a short time period.

Drag: With regard to wind energy, the force exerted on an object by moving air; can also refer to a type of wind generator or anemometer design that uses cups instead of blades with airfoils.

Decibel: A unit used to measure the intensity of a sound, or the degree of loudness; decibel measurements use a logarithmic scale.

Dispatchable: The ability to turn on, turn off, ramp-up, or ramp-down an electric generator as needed to match electricity production with electricity usage.

Distribution lines: Electric lines with operating voltages of 34.5 kV or less; the distribution system carries energy from the local substation to individual households, using both overhead and underground lines.

Efficiency: The extent to which time, effort, or energy is well used for an intended task or purpose; often used with the specific purpose of relaying the capability of a specific application of effort to produce a specific outcome effectively with a minimum amount or quantity of waste, expense, or unnecessary effort.

Electrical energy: Electrical energy is the generation or use of electric power over a period of time expressed in kilowatt-hours (kWh) or megawatt-hours (MWh).

Energy conservation: Using less energy by doing without the desired energy service.

Energy efficiency: Using less energy to provide equal or more of a desired energy service.

Environmental externality: An unintended and uncompensated environmental effect of production and consumption that do not accrue to the parties involved in the activity, such as particulate matter emissions from coal-fired power plants that may result in breathing problems or increased medical costs for individuals living near the power plant.

First Law of Thermodynamics: (or the Law of Conservation of Energy) States that the energy in a process is neither created nor destroyed and the total energy remains constant.

Fracking: Hydraulic fracturing is the propagation of fractures in a rock layer, as a result of the action of a pressurized fluid. Induced hydraulic fracturing, or "fracking," is a technique used to release petroleum, natural gas (including shale gas, tight gas, and coal seam gas), or other substances for extraction.

Freewheeling: The occurrence of a wind generator that has lost connection to electric load, which puts it in danger of self-destruction from overspeeding.

Furling: The act of a wind turbine yawing out of the wind to prevent overloading in high wind speed conditions.

Gantt chart: A type of bar chart that is a useful tool for project planning; it is typically used to identify all steps or actions that need to occur to develop and construct a project, graphically illustrating how long each step is expected to take and which steps are dependent upon the completion of another step in the process.

Generator: A general name given to a machine for transforming mechanical energy into electrical energy.

Generation company: Again in states that have deregulated sectors of the electric utility business, the generation sector of the electric utility industry may have been deregulated. In such cases, power plants have to compete on price and other factors such as the reliability of their power or the environmental attributes of their power.

Global warming potential: A measure used to assess the threat posed by various greenhouse gases; how much heat one molecule of a gas will trap relative to a molecule of carbon dioxide.

Greenhouse effect: The trapping of the Sun's warmth in a planet's lower atmosphere due to the greater transparency of the atmosphere to visible radiation from the Sun than to infrared radiation emitted from the planet's surface.

Heat rate: A measure of a power plant's overall thermal performance or energy efficiency for some period of time, usually expressed as the number of Btu required to produce one kilowatt-hour of electricity.

Hertz (Hz): Measurement unit of frequency in cycles per second. The US electric grid operates at a frequency of 60 Hz.

Horizontal-axis wind turbine (or HAWT): The most common design of utility-scale wind turbines in which the shaft is parallel to the ground (i.e., horizontal) and the blades are perpendicular to the ground.

Hub: The center of a wind generator rotor that holds the blades in place and attaches to the shaft.

Ice shedding: A situation in which ice that has accumulated on wind turbines begins to fall off or be thrown off of turbine blades either to increasing temperatures or to centrifugal forces.

Independent power projects (IPPs): IPPs are power plants not affiliated with an electric utility within the electric market that they are selling power into.

Many wind energy projects are IPPs. IPPs can operate in states where vertically integrated utilities are still commonplace or in deregulated markets.

Infrasound: Sound with a frequency too low to be detected by the human ear and is generally considered to have frequencies below 20 Hz, although there is not always a clear delineation between infrasound and low-frequency sound.

Interruptible load: Refers to electricity normally being consumed by an industrial customer of the utility who has agreed to reduce some or all of its electricity consumption during emergency situations in exchange for a lower electricity price during periods of normal operation.

Kilovolt or kV: A unit of electrical pressure equal to 1000 V.

Kilowatt or kW: A unit of electrical power equal to 1000 W. Electric power is often expressed in kilowatts.

Kilowatt-hour or kWh: A unit of electrical energy equal to 1000 watt-hours or a power demand of 1000 watts for 1 h. Sales of electricity from the local utility company to one's home or business are expressed in kWh.

Landfill gas: Gas generated during anaerobic decomposition of organic waste in landfills; gases produced during decomposition include methane, carbon dioxide, oxygen, nitrogen, and various non-methane organic compounds.

Leading edge: The edge of a wind turbine blade that faces toward the direction of rotation.

Lift: The force exerted by moving air on asymmetrically shaped wind turbine blades at right angles to the direction of relative movement.

Load-following: The constant adjustment of generation levels as demand for electricity fluctuates throughout the day.

Low-frequency: sound is generally considered to have frequencies in the range of 10–200 Hz.

L_{max}**:** The maximum sound level measured.

L_{eq}**:** Equivalent continuous sound or an average sound energy over a given time period.

L_{10}**:** Sound level exceeded 10% of the time, which is generally considered to be the sound level that will annoy people.

L_{90}**:** Sound level exceeded 90% of the time, which is generally considered to be a measure of ambient background noise.

L_{dn}**:** Day–night average sound level or the average sound level for a 24-h period.

Marginal cost: The increase or decrease in costs as a result of one more or one less unit of output. In terms of electricity, the marginal cost is usually directly related to the cost of fuel and the efficiency of the power plant used to produce the next kilowatt-hour.

Megawatt or MW: A unit of electrical power, equal to 1000 kW or one million watts. The maximum production capability of large wind turbines is usually expressed in MW.

Megawatt-hour or MWh: A unit of electrical energy equal to 1000 kWh. Sales of electricity from large wind energy projects to utility companies are usually expressed in MWh.

Mesonet: A network of weather stations that provide mesoscale meteorological data.

Merchant plant: An electric power–generating station that is not included in the rate base of a traditional utility and that sells power in a competitive wholesale market; such projects may also sell power at current market prices rather than through a long-term purchased power agreement.

Methane hydrate: (or methane clathrate) Crystalline solids consisting of gas molecules, usually methane, each surrounded by a cage of water molecules, representing a huge source of fossil fuel; found in deep-sea sediments several hundred meters thick directly below the seafloor and in association with permafrost in the Arctic.

Multiplier effect: (or ripple effect) Expenditures in one sector of the region's economy that lead to additional positive impacts throughout many other sectors.

Nacelle: The box or protective covering at the top of a wind turbine that covers its inner workings (generator, gearbox, yaw motors, drive shaft, etc.).

Net capacity factor: The ratio of actual energy production during a given period of time to potential or hypothetical production if all turbines ran at their rated capacity for the entire period of time, without occurring any losses due to items such as electrical losses, turbulence, wake effect, planned or forced maintenance, and blade degradation.

Net metering: Situations in which electric utilities allow homes or businesses to install "behind-the-meter" renewable energy systems such as solar photovoltaic or small wind generators and using electric meters that keep track of any excess energy produced by the renewable energy system that is fed back into the electric grid; such practices may be voluntary on the part of the utility or required by state or local regulators.

NEXRAD: A network of 159 high-resolution Doppler weather radars operated by the National Weather Service, an agency of the National Oceanic and Atmospheric Administration. NEXRAD detects precipitation and atmospheric movement or wind.

Noise: Implies the presence of sound but also implies a response to sound

noise is often defined as unwanted sound.

Nonspinning reserve: Refers to off-line generation capacity that can be ramped-up and synchronized to the grid within some specified time period but may also include interruptible load that can be removed from the system in a specified time.

Ohm: The unit of electrical resistance to current flow in a circuit. Resistance is one ohm when a voltage of one volt will send a current of one ampere through, that is, one ohm equals one volt per ampere.

Ohm's Law: Defines the relationship between current, voltage, and resistance. Three ways that Ohm's Law can be expressed are as follows

Current (in amperes) = Voltage (in volts)/resistance (in ohms) or $I = V/R$
Resistance (in ohms) = Voltage (in volts)/current (in amperes) or $R = V/I$
Voltage (in volts) = Current (in amperes) × resistance (in ohms) or $V = I \times R$.

Off-peak power: Power generated or consumed off the peak of the daily electric generation /demand curve.

Operating reserve: The amount of generating capability above firm system demand that is required to provide for regulation, load forecasting error, forced outages, scheduled outages, and local area protection; reserves are analogous to an insurance policy to help when the unexpected occurs; operating reserves can consist of spinning reserve and nonspinning reserve.

Power: The rate at which work is done, or for electricity, the rate at which electrical energy is generated or consumed (energy per unit of time), usually measured in watts, kilowatts, or megawatts.

Power curve: The relationship between the power output and wind speed, depicted either graphically or as a table.

Power purchase agreement (or purchased power agreement): An agreement between an electric power generator and an electric power purchaser setting out pricing and terms for purchase of power from the generating facility over some set period of time.

Power quality: The concept of powering and grounding sensitive equipment in a manner that is suitable to the operation of that equipment.

Production tax credit: An incentive offered in the United States to encourage the production of renewable energy; originally created by the Energy Policy Act of 1992; as the name implies, the amount of the tax credit is tied to the amount of energy produced as opposed to the amount of money invested in the project.

Pro forma model: "Pro forma" is a Latin term meaning "for the sake of form." In the investing world, it describes a method of calculating financial results in order to emphasize either current or projected figures.

Quad: One quadrillion British thermal units (Btu) of energy.

Ramping: Changing the output of an electric generator, either up or down, to accommodate fluctuations in load or in wind generation output.

Ramp rate: A measure of how quickly the generator can change output and is normally expressed in megawatts per minute. Some generators, such as the state-of-the-art natural gas–fired generators, have high ramp rates, whereas coal-fired plants and nuclear power plants may have low ramp rates. Therefore, regions with higher percentages of natural gas–fired generation may be able to better accommodate higher levels of wind penetration than other regions with greater reliance on coal or nuclear power.

Rated power: The maximum power that a generating device is intended to produce.

Rayleigh distribution: A continuous probability distribution that can be useful in estimating the frequency or distribution of various wind speeds when only the average annual wind speed is known for a given location.

Regenerative braking: Recovering the energy of acceleration during stopping by reversing the action of the electric motors attached to the wheels making them short-term generators and recovering some of the energy.

Regulation or regulation service: An ancillary transmission service one balancing authority may provide to another balancing authority to correct deviations in the frequency of the electric grid and to balance out transactions between neighboring systems.

Renewable energy: Any energy resource that is naturally regenerated over a short time scale and derived directly from the Sun (such as thermal, photochemical, and photoelectric), indirectly from the Sun (such as wind, hydropower, and photosynthetic energy stored in biomass), or from other natural movements and mechanisms of the environment (such as geothermal and tidal energy). Renewable energy does not include energy resources derived from fossil fuels, waste products from fossil sources, or waste products from inorganic sources.

Renewable energy credit (REC) or Green Tag: A tradable commodity used to track renewable energy purchases of retail electric providers and sales of renewable energy under the electricity provider's green energy tariff. An REC basically represents the environmental attributes of 1000 kWh or 1 MWh of electricity produced by a qualifying source of renewable energy, such as wind turbines.

Renewable portfolio standard or renewable electricity standard: A governmental mandate requiring utilities under the jurisdiction of the governmental entity to obtain some portion of their energy from qualifying sources of renewable energy.

Resistance: A measure of how hard it is for electric current to move through a material; the opposition to the flow of electrical charge. The unit of measurement is ohm, represented as Ω.

Retail electricity provider (REP): In some states that have deregulated sectors of the electric utility business, customers may choose between many REPs that have to compete based on price, marketing, and attributes of the energy they sell, such as companies that sell a "green energy" product.

Ripple effect: (or multiplier effect) Expenditures in one sector of the region's economy that lead to additional positive impacts throughout many other sectors.

Root: The area of a wind turbine blade located nearest to the hub; this is generally the thickest and widest part of the blade.

Scrubber: Flue-gas desulfurization (FGD) equipment used at a power plant to remove SO_2.

Setback: The distance between a wind turbine and some object, such as an occupied structure or a public road.

Shadow flicker: Alternating changes in light intensity caused by a moving wind turbine blade casting shadows on the ground or on stationary objects, such as a window at a dwelling.

Site suitability certificate: Certification that all major components of the wind turbines being used for a wind energy plant project are expected to have a useful life of at least 20 years given the site's wind regime and the developer's proposed spacing and location of wind turbines.

Smart grid: Integrates new meters and computer resources on the electric grid system allowing consumers to better monitor and manage their electricity use, as well as allowing the utility company to better monitor and control the utility network.

Smart meter: A digital meter that records individual real-time customer electric energy consumption. These meters allow for an automated transfer of information between the home and the electric energy provider.

Sound: Describes wave-like variations in air pressure that occur at frequencies that can stimulate receptors in the inner ear and, if sufficiently powerful, be appreciated at a conscious level.

Sound frequency: The frequency of a sound is measured in vibrations per second, or Hertz (Hz), and determines how the sound may be categorized.

Spinning reserve: Refers to online reserve capacity that is synchronized to the electric grid and that can rapidly meet electric demand within seconds or minutes.

Stray voltage: A voltage resulting from the normal delivery and/or use of electricity (usually smaller than 10 volts) that may be present between two conductive surfaces that can be simultaneously contacted by people or animals.

Substation: An electrical facility containing several types of equipment, usually including one or more transformers and two or more electric lines entering the facility. Substations are used for purposes such as the following: (1) the connection of generators, transmission or distribution lines, and loads to each other, (2) the transformation of power from one voltage level to another, (3) controlling system voltage and power flow, or (4) isolation of sections of an electric line that may be experiencing overloads or faults.

Sustainable development: Development that meets the needs of the present without compromising the ability of future generations to meet their own needs.

Tax abatement: An agreement between a business owner and a local taxing authority to reduce ad valorem taxes for some period of time as an enticement for the business owner to locate a new facility within the taxing authority's area.

Tenant farmer: A person farming a tract of land who is not the owner of the property; a tenant farmer usually would not share in royalties derived from wind turbines and thus may oppose placement of wind turbines on the property.

Transformer: An electrical device that raises or lowers voltage to facilitate the connection of electric lines of differing voltages or the interconnection of a generator producing power at one voltage to an electric line of a differing voltage. In a wind farm, padmount transformers are used to increase the voltage coming from the generator to the higher voltage used for the collection system (typically 34.5 kV). Transformers may also be located in the project substation to further increase the voltage to that of the transmission grid.

Transmission and distribution company (T&D company): Similar to TSP, but company would also own distribution lines and associated facilities; companies owning the distribution system are normally regulated and are often charged with the responsibility for maintaining highly reliable service to retail customers.

Transmission lines: The high-voltage electric lines used for transport of generator-produced electric energy to loads, usually through multiple paths. Operating voltages of 69, 115, or 138 kV are normally used to distribute energy within regions or areas and consist of overhead lines for the most part. Higher voltages such as 230, 345, 500, or 765 kV provide the "backbone" of most transmission grids in the United States and interconnect the various reliability regions in the United States or interconnect generating stations to large substations located close to load centers. The vast majority of transmission-voltage electric lines are located aboveground due to the significantly higher cost of placing the lines underground, with exceptions typically being in the heart of very large cities where no space is available for aboveground lines.

Transmission service provider (TSP): The company or business unit that owns transmission lines and associated equipment such as transmission-voltage substations and thus would provide transmission services to wind energy projects, to other sources of electric generation, and to retail electric providers. TSPs' rates are usually regulated by a state regulatory authority or the Federal Energy Regulatory Commission (FERC) and are required to provide nondiscriminatory or "open access" transmission service to all generators.

Unit commitment: A process used by electric system operators while planning to meet the upcoming day's projected electricity requirements.

Unit conversion process: The process of converting from one set of units to another. For common conversion units, one can use an Internet search engine and simply type in the conversion needed. More complex conversions may need to be done by hand. One method often used is demonstrated here. For example, to convert a wind speed of 10 miles per hour to meters per second, set up sequential multiplication factors as shown in the following equation

$$\frac{10\,\text{miles}}{\text{h}} \times \frac{1\,\text{h}}{3600\,\text{s}} \times \frac{5280\,\text{feet}}{1\,\text{mile}} \times \frac{0.305\,\text{m}}{1\,\text{foot}} = \frac{4.47\,\text{m}}{\text{second}}\ [\text{answer}]$$

Vehicle to Grid (V2G) energy storage: Allows the battery in an electric vehicle when plugged into the electrical socket to provide energy back into the electric system to reduce peak loads or to provide backup power.

Vertical-axis wind turbine: A wind turbine design in which the rotating shaft is perpendicular to the ground and the turbine blades rotate parallel to the ground.

Vertically integrated utility: A utility company that provides generation, transmission, distribution, and retail electric sales for its customers; states typically maintain regulatory authority over the rates and quality of service for vertically integrated utilities.

Volt: The unit of electric force or pressure; the pressure that will cause a current of one ampere to flow through a resistance of one ohm.

Voltage: The electrical force or pressure that causes current to flow in a circuit, as measured in volts.

Watt: The unit used for measuring electrical power usage or production. It is the amount of energy expended per second by a current of one ampere under the pressure of one volt or the power needed for one ampere to flow through one ohm.

Wind generator or wind turbine: A machine that transforms mechanical energy from the wind into electrical energy.

Windmill: A device driven by the power in the wind and typically used for milling grain or pumping water.

Wind penetration: The percentage of total power or energy being produced that comes from wind generation.

Wind rose: A graphical representation or diagram depicting the statistical distribution of wind direction and speed at a given site; useful for determining the preferred direction of wind turbine rows and spacing between wind turbines.

Wind shear (vertical): The change in horizontal wind speed with a change in height.

Wind shear (directional): The change in wind direction with a change in height.

Wind turbulence: The rapid disturbances or irregularities in the wind speed, direction, and vertical component.

INDEX

Wind Energy Essentials: Societal, Economic, and Environmental Impacts, First Edition.
Richard P. Walker and Andrew Swift.